# PROJECT MANAGEMENT with CPM, PERT and Precedence Diagramming

# PROJECT MANAGEMENT with CPM, PERT and Precedence Diagramming

THIRD EDITION

**Joseph J. Moder**
*Professor, Department of Management Science and Computer Information Systems, University of Miami*

**Cecil R. Phillips**
*Vice President, Kurt Salmon Associates, Inc.*

**Edward W. Davis**
*Professor, Colgate Darden Graduate School of Business Administration, University of Virginia*

VAN NOSTRAND REINHOLD COMPANY
New York

Library of Congress Catalog Card Number: 82-16035
ISBN: 0-442-25415-6



Manufactured in the United States of America

Published by Van Nostrand Reinhold Company Inc.
135 West 50th Street
New York, New York 10020

Van Nostrand Reinhold Company Limited
Molly Millars Lane
Wokingham, Berkshire RG11 2PY, England

Van Nostrand Reinhold
480 Latrobe Street
Melbourne, Victoria 3000, Australia

Macmillan of Canada
Division of Gage Publishing Limited
164 Commander Boulevard
Agincourt, Ontario M1S 3C7, Canada

15 14 13 12 11 10 9 8 7 6 5 4

**Library of Congress Cataloging in Publication Data**

Moder, Joseph J.
Project management with CPM, PERT, and precedence diagramming.

Rev. ed. of: Project management with CPM and PERT. 2nd ed. 1970.
Includes bioliographical references and index.
1. Industrial project management. 2. Critical path analysis. 3. PERT (Network analysis) I. Phillips, Cecil R. II. Davis, Edward Willmore, 1941–
III. Title. IV. Title: Project management with C.P.M., P.E.R.T., and precedence diagramming.
T56.8.M63 1983 658.4'04 82-16035
ISBN 0-442-25415-6

# PREFACE

The year 1983 marks well over a decade since the publication of the 2nd edition of this text, and a quarter century since the initial development of PERT and CPM. During this period, the methodology has matured considerably and the emphasis on the early "frills" of these techniques has all but disappeared. While the concepts of PERT probabilities and the elegant time-cost tradeoff of CPM were useful and helped to create credibility and interest in these techniques during their development and early application, emphasis has shifted towards problems deemed to be of more importance to practitioners. We have tended to follow this movement in this third edition of the text. While we have not eliminated the so-called "frills," we have reduced the space devoted to them somewhat, and have added new material that may improve their applicability.

Since the major benefit of network based methods of project planning and control is the development of the project network itself, the most important change in this third edition is the emphasis on node and precedence diagrams as alternatives to the arrow diagram representation of a project plan. While the arrow diagram was the mainstay of PERT/CPM and their immediate offsprings in the 1960s, the 1970s saw a gradual shift towards node diagrams, and more recently to precedence diagrams. As the reader will learn, this is a mixed blessing. The enrichment of network logic afforded by precedence diagramming comes only at the expense of more complex results in the scheduling time analysis computations. We have included in this edition a comprehensive treatment of this problem so that the reader can intelligently assess its potential in specific application areas.

Another topic of increasing importance in the application of network methods is the scheduling of activities under constraints on resource availablities. This chapter of the text has been considerably expanded, and updated to reflect the current methods being used in this area. Finally, the use of computers in this field has tended to follow the continued improvement and cost reduction in computer hardware. The chapter on computer processing has also been expanded and updated.

The basic organization of the chapters remains essentially the same as previous editions, with Part I basic and Part II advanced subjects. The 6 chapters of Part I comprise a complete course in the fundamentals of the planning and scheduling features of critical path methods, including cost control methods and practical applications. One consistent set of terms and symbols is used throughout Part I. Thus, industrial and commercial users of the methods may study only Part I in preparation for most practical applications. A good two- or three-day training course for industrial personnel can be based on Part I, with selected portions from Part II as appropriate to the needs of the group.

The material in Part II is suitable for more advanced users and college level courses in departments of management science and business administration, industrial and systems engineering, and civil engineering. While certain portions of the text require some statistics and linear programming background, the prerequisites are generally satisfied by upper level undergraduate students of business or engineering.

Of interest to college instructors will be the added exercises at the end of most chapters, and the inclusion of many problem solutions at the end of the book.

Joseph J. Moder
*Coral Gables, Florida*

Cecil R. Phillips
*Atlanta, Georgia*

Edward W. Davis
*Charlottesville, Virginia*

# NOMENCLATURE

$AF$ = activity (free) float or slack.

$a$ = the "optimistic" performance time estimate used in PERT—the time which would be bettered only one time in twenty, i.e., the fifth percentile (where specifically noted, it will also be used to denote the zero percentile used in conventional PERT).

$b$ = the "pessimistic" performance time estimate used in PERT—the time which would be exceeded only one time in twenty, i.e., the ninety-fifth percentile (where specifically noted, it will also be used to denote the 100 percentile used in conventional PERT).

$C$ = cost slope for an activity used in time-cost trade-off procedures.

$C_d$ = direct costs associated with the performance of an activity in time $d$, the "crash" performance time.

$C_D$ = direct costs associated with the performance of an activity in time $D$, the "normal" performance time.

$d$ = "crash" activity performance time—the minimum time in which the activity can be performed.

$D$ = "normal" activity performance time—the one which minimizes the activity direct costs; also used to denote the mean activity performance time based on a single time estimate.

$E$ = earliest (expected) event occurrence time.

$ES$ = earliest (expected) activity start time.

$EF$ = earliest (expected) activity finish time.

$F$ = total path float or slack.

$FF$ = the lead/lag time associated with a finish-to-finish constraint used in precedence diagramming.

$FS$ = the lead/lag time associated with a finish-to-start constraint used in precedence diagramming.

$L$ = latest allowable event occurrence time.

$LS$ = latest (allowable) activity start time.

$LF$ = latest (allowable) activity finish time.

$m$ = the "most likely" performance time estimate used in PERT—the modal value of the performance time distribution.

$SF$ = the lead/lag time associated with a start-to-finish constraint used in precedence diagramming.

$SS$ = the lead/lag time associated with a start-to-start constraint used in precedence diagramming.

$t$ = actual activity performance time, determined after the activity has actually been completed.

$t_e$ = mean activity performance time based on the three (PERT) time estimates, $a$, $m$, and $b$.

$T$ = actual occurrence time of a specific network event, determined after the event has actually occurred.

$T_d$ = total (expected) project duration time achieved by using "crash" activity performance times on all critical path activities.

$T_D$ = total (expected) project duration time achieved by using all "normal" activity performance times.

$T_S$ = scheduled event occurrence time.

$V_t$ = the estimated variance of the actual performance time, $t$, based on the PERT formula $[(b - a)/3.2]^2$.

$(V_t)^{1/2}$ = square root of $V_t$, called the standard deviation of the actual activity performance time, $t$.

$V_T$ = the estimated variance of the actual occurrence time, $T$, of a specific network event.

$(V_T)^{1/2}$ = square root of $V_T$, called the standard deviation of the actual event occurrence time, $T$.

$Z$ = standard normal deviate, equal to the difference between a random variable, such as $T$, and its expected or scheduled time, such as $T_S$, divided by the standard deviation of the random variable, such as $(V_T)^{1/2}$.

# CONTENTS

# PROJECT MANAGEMENT with CPM, PERT and Precedence Diagramming

# I

# BASIC TOPICS

# 1

# INTRODUCTION

*Management* is a process concerned with the achievement of goals or objectives. *Project management* involves the coordination of group activity wherein the manager plans, organizes, staffs, directs, and controls to achieve an objective with constraints on time, cost, and performance of the end product. This text will deal primarily with these planning and control functions, and only peripherally with organization matters. *Planning* is the process of preparing for the commitment of resources in the most effective fashion. *Controlling* is the process of making events conform to schedules by coordinating the action of all parts of the organization according to the plan established for attaining the objective. It can also be said that project management is a blend of art and science: the art of getting things done through and with people in formally organized groups; and the science of handling large amounts of data to plan and control so that project duration and cost are balanced, and excessive and disruptive demands on scarce resources are avoided. This text will deal primarily with the science of project management, with occasional excursions into the art when it has a direct relationship with the science.

It is appropriate at this point to elaborate on the term *project.* A project is a set of tasks or activities related to the achievement of some planned objective, normally where the objective is unique or non-repetitive. Thus, a project is usually distinguished from repetitive or continuous production processes by the characteristic of uniqueness, or the "one-shot" nature of the objective. However, the term *project* will be interpreted quite broadly in this text in order to encompass the many possible applications of critical path methods, or network planning techniques, that have come about since the development of PERT and CPM in the late fifties. Projects may involve routine procedures, such as the monthly closing of accounting books. In this case, network planning techniques are useful for detailed analysis and optimization of the operating plan. Usually, however, these techniques are applied to one-time efforts. Although similar work may have been done previously, it is not being repeated in the identical manner on a production basis. Consequently, in order to accomplish the project goal or objective, the manager must plan and schedule largely on the basis of his experience with similar projects, applying his judgment to the particular conditions of the project at hand. During the course of the project, he must continually replan and reschedule because of unexpected progress, delays, or technical conditions.

Until the advent of critical path methods, there was no generally accepted formal procedure to aid in the management of projects. Each manager had his own scheme, which often involved limited use of bar charts—a useful tool in production management but inadequate for the complex interrelationships associated with contemporary project management. The development of network based planning methods in the late fifties provided the basis for a more formal and general approach toward a discipline of project management. Critical path methods involve both a graphical portrayal of the interrelationships among the elements of a project, and an arithmetic procedure which identifies the relative importance of each element in the over-all schedule. Since their development, critical path methods have been applied with notable success to research and development programs, all types of construction work, equipment and plant maintenance and installation, introduction of new products or services or changeovers to new models, development of major transportation and energy related systems, strategic long-term planning, management information systems developments, production planning, emergency planning, and even the production of motion pictures, conduct of political campaigns, and complex surgery. According to our definition, all of these activities are classed as projects.

In all of these projects, management is concerned with developing a "good" (or at least a workable) plan of the activities that make up the project, including a specification of their interrelationships. Also, management is interested in scheduling these activities in an acceptable time span, considering the manpower and other resources required to carry out the program as it progresses in time. Management is also concerned with monitoring the expenditure of time and money in carrying out the scheduled program, as well as the resulting "product" quality or performance, for cases where achievement of the project objective(s) can be measured on a continuous or ordinal scale(s). For the most part, critical path methods have concentrated on the time parameter and to a somewhat lesser extent on the cost parameter. The performance parameter is a much more difficult and varied problem. It will not be treated explicitly in this text.

## DEVELOPMENT OF THE NETWORK PLAN CONCEPT

The network diagram is essentially an outgrowth of the bar chart which was developed by Gantt in the context of a World War I military requirement. The bar chart, which is primarily designed to control the time element of a program, is depicted in Figure 1-1a. Here, the bar chart lists the major activities comprising a hypothetical project, their scheduled start and finish times, and their current status. The steps followed in preparing a bar chart are as follows:

1. Analyze the project and specify the basic approach to be used.
2. Break the project down into a reasonable number of activities to be scheduled.
3. Estimate the time required to perform each activity.
4. Place the activities in sequence of time, taking into account the requirements that certain activities must be performed sequentially while others can be performed simultaneously.
5. If a completion date is specified, the diagram is adjusted until this constraint is satisfied.

The primary advantage of the bar chart is that the plan, schedule, and progress of the project can all be portrayed graphically together. Figure 1-1a shows the five-activity plan and 15-week schedule, and current status (end of third week) indicates, for example, that activity B is slightly behind schedule. In spite of this important advantage, bar charts have not been too successful on one-time-through projects with a high engineering content, or projects of large scope. The reasons for this include the following:

1. Planning and scheduling are considered simultaneously.
2. The simplicity of the bar-chart precludes showing sufficient detail to enable timely detection of schedule slippages on activities with relatively long duration times.
3. The bar-chart does not show explicitly the dependency relationships among the activities. Hence, it is very difficult to impute the effects on project completion of progress delays in individual activities.
4. The bar-chart is essentially a manual-graphical procedure. It is awkward to set up and maintain for large projects, and it has a tendency to quickly become outdated and lose its usefulness.

With the above disadvantages in mind, along with certain events of the mid-fifties, the stage was set for the development of a network-based project management methodology. Some of the notable events were the emergence of general systems theory and the second generation of large digital computers, and the initiation of very large and technically demanding programs in weapons systems, power generation systems, etc. Because of the enormous size of many of these

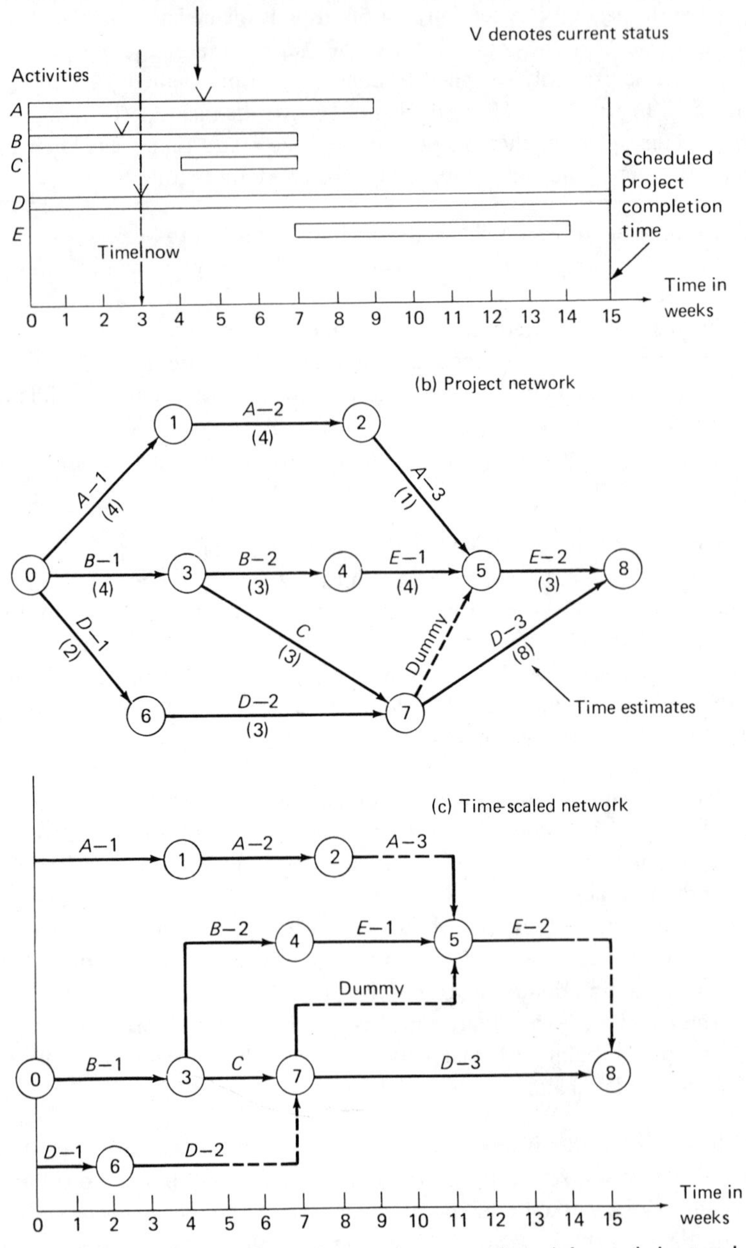


**Figure 1-1** Comparison of Gantt chart, project network, and time-scaled network.

programs that contain thousands of significant activities, taking place in widely dispersed locations, something like network-based methodology literally had to emerge.

The heart of network-based planning methods is a graphical portrayal of the plan for carrying out the program. Such a graph, called a network, shows the dependency relationship among the project activities using the simple logic that *all activities preceding a given activity must be completed before the given activity may begin.* For example, in Figure 1-1b, activities *A*-3 and *E*-1 *must* both be completed before *E*-2 *may* begin. Activities without predecessors (*A*-1, *B*-1, and *D*-1) may begin anytime after the project starts, and when *all* of the activities without successors (*D*-3 and *E*-2) are completed, the project is complete.

In the project network shown in Figure 1-1b, the lines denote activities which usually require time, manpower, and facilities to complete. Each activity originates and terminates in a unique pair of nodes called *events;* time flows from the tail to the head of each arrow. Events denote a point in time; their occurrence signifies the completion of all activities terminating in the event in question. For example, the occurrence of event 7 signals the completion of activities 3-7 (3 → 7) *and* 6-7. Dashed line arrows, called *dummies*, show precedence relationships only; they usually require no time, manpower, etc., to perform. Such a relationship is shown by activity 7-5. This networking scheme is widely used today and is called activity-on-arrows, or merely an arrow diagram. Another important way of networking is to reverse the role of the arrow and the node; the result is called activity-on-nodes, or a node diagram. Detailed treatment of both of these schemes of networking will be given in Part I of this text.

Figure 1-1b shows the network plan for carrying out this hypothetical project. It should be noted that planning is separate from scheduling. The latter is initiated by adding estimates of activity duration times to Figure 1-1b, and determining the impact of these times on overall project duration. One method of showing this scheduling step is given in Figure 1-1c, where the project plan of Figure 1-1b is drawn to scale on a time base. It is not too unlike the original Gantt bar chart. It clearly illustrates, however, several major differences in the traditional bar chart and time-scaled network diagrams. First, the network shows greater detail. For example, activity *D* is broken down into activities *D*-1, *D*-2, and *D*-3. The second and more important difference is that the interdependency of the activities is clearly shown. For example, activity *E* can start as soon as activity *B* is finished; however, the last portion of activity *E*, denoted by *E*-2, cannot begin until activities *E*-1, *A*-3, *C*, and *D*-2 are all completed.

In Figure 1-1c, activities *A*-3, *D*-2, and *E*-2 have the last portion of their arrows dashed. Based on the estimated activity duration times, these dashed lines denote that these activities could be completed prior to the occurrence time of their succeeding events. For this reason, the paths along which these activities lie are referred to as *slack paths*, or paths with *float time;* that is, these paths require less time to perform than the time allowed for them. For example, the estimated time to perform activities *D*-1 and *D*-2 is 2 + 3 = 5 weeks, as indicated on the time scale. However, the time interval between the occurrence of the initial and terminal events of this path, i.e., events 0 and 7, is 7 weeks. Thus, we say that this path has two weeks of "slack" or "float." Now activities *B*-1, *C*, and *D*-3 have no dashed portions; the sum of their expected performance times, 15, is the same as the time interval between the occurrence of events 0 and 8. For this reason, this path is referred to as the *critical path;* it is the longest path through the network.

From Figure 1-1c it is easy to see how the third of the above listed disadvantages of the bar chart can be overcome. The impact of delays in completion of any activity can be easily ascertained. For example, any delays in activities *B*-1, *C*, and *D*-3 will result in a corresponding delay in project completion, since they have no slack. However, delays in activities *D*-1 plus *D*-2 can be up to 2 weeks before any delay in project completion would result. These concepts are treated in detail in Chapter 4.

Although the network may be drawn to a time scale, as shown in Figure 1-1c, the nature of the network concept precludes this luxury in most applications, at least in the initial planning stage. Thus, the length of the arrow is unimportant. You can "slide" activities back and forth on a bar chart with ease, because the dependency relations are not shown explicitly. However, if a network is drawn on a time scale, a change in the schedule of one activity will usually displace a large number of activities following it, and, hence, may require a considerable amount of redrawing each time the network is revised or updated. There are practical applications for time-scaled networks especially where the network is closely related to production planning and where scheduling to avoid overloading of labor or facilities is a prime consideration. In these cases the time-scaling effort is worthwhile, for it clearly illustrates conflicting requirements.

## THE SYSTEMS APPROACH TO PROJECT MANAGEMENT

Cleland and King[1]* point out the strong influence of general systems theory, as it had evolved in the mid-fifties, on the development of network-based project management methodology. The systems concept instills the desire to achieve *overall effectiveness of the organization*, in an environment which invariably involves *conflicting organizational objectives.* A systems-oriented manager realizes that he can achieve the overall goals of the organization only by viewing the entire system. He must seek to understand and measure departmental interrelationships, and to integrate them in a fashion which enables the organization to efficiently pursue its goals. To accomplish this, increasing use is required of objective scientific analysis in solving decision problems. These methods rely on

*These numbers refer to References given at the end of each chapter.

models–formal abstractions of real world systems–to predict the outcomes of various available alternatives in complex decision problems. In effect, the systems approach may be viewed here as a logically consistent method of reducing (by network based methods) a large part of a complex problem to a simple output. This output can be used by a decision maker, in conjunction with other considerations, to arrive at a best decision. It permits him to put aside those things which are best handled by systems analysis, and to focus on those aspects of the problem which are most deserving of his individual attention. In this way, the manager is able to get the "big picture" in its proper perspective, rather than requiring him to devote attention to a myriad of minor, seemingly unrelated aspects of the total system.

Systems concepts have also brought about changes in the way plans are executed. This change is primarily in organizational structures and the emergence of the *project manager*, who must deal with traditional organizational philosophy in a very delicate way. Traditional philosophy is based on a vertical flow of authority and responsibility relationships, which emphasizes only parts and segments of the organization. It does not place sufficient importance on the interrelationships and integration of activities involved in the total array. In dealing with these problems, the project manager, whose scope of interest and influence cuts across these traditional vertical lines of authority, has been placed in a (formal or informal) matrix form of organization. These concepts will be treated in Chapter 6 of this text.

The strong influence of general systems theory on the evolution of the overall network-based methodology of project management can also be seen by an analogy to systems control theory. The latter is illustrated in Figure 1-2a, where it is applied to controlling the temperature of a room to a prescribed time-temperature profile, noted as INPUT. The control problem is created by random OUTSIDE heat load DISTURBANCES impinging on the room to be controlled, causing deviations from the desired INPUT. These deviations are detected by a thermometer which monitors the state of the system, and transmits this information to the thermostat along the negative feedback loop. The thermostat sends deviations between the INPUT and the OUTPUT to the controller, which in turn drives the furnace in a manner to eliminate the undesired temperature deviations.

The analogy for network-based project management methodology is shown in Figure 1-2b. The control system INPUT is the network-based project plan, including a time and cumulative cost schedule of the project activities. The control problem here is created by random OUTSIDE DISTURBANCES to material procurements, administrative/technical approvals, personnel and technical problems, etc. The effects of these disturbances on the project are detected by a periodic monitoring of activity time and cost status in the form of field reports. The latter are used to update the project network to determine the current time and cost status of the project. Deviations from the INPUT plan are the basis for

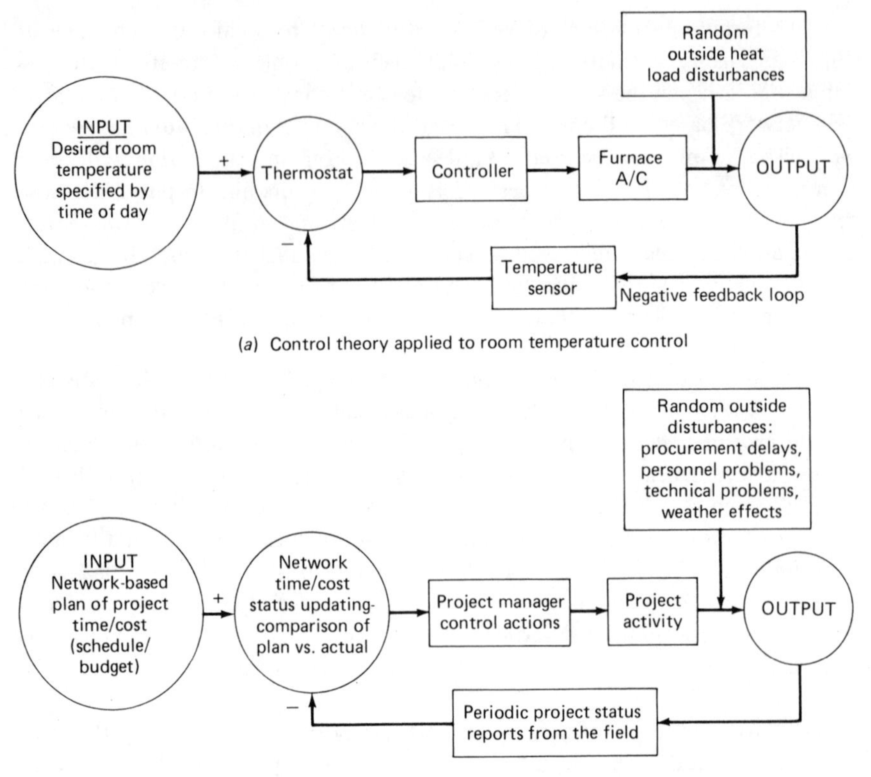


(*a*) Control theory applied to room temperature control



(*b*) Network-based project management control methodology

**Figure 1-2 Control theory analogy to network-based project management methodology.**

action by the project manager who initiates control decision actions, as needed, to bring the project back into line. This is the basic methodology to be developed in this text.

## HISTORICAL DEVELOPMENT OF NETWORK-BASED PROJECT MANAGEMENT METHODOLOGY

As mentioned above, the initiation of large development programs in the 1950s, along with the simultaneous development of the digital computer and general systems theory, set the stage for the development of network-based project management methodology. Actually, it was a rediscovery of procedures that can at least be traced back a half century to a Polish scientist named Karol Adamiecki (1931).[2] He developed and published a methodology in a form that he called the Harmonygraph. This is a 90-degree rotated bar chart type graph with a vertical time scale, and a column (strip) for each activity in the project. It is illustrated in Figure 1-3, using the same network adopted in Figure 1-1. Note that the activity strips are ordered so that the predecessors of any given activity will always be found to its left. The activity strips each contain a movable tab whose length is proportional to the estimated duration of the activity, and whose location along the time scale denotes the scheduled start and finish time for each activity. Thus, starting at the first column on the left in Figure 1-3 and working to the right, it is easy to schedule each activity so that all time precedence constraints are satisfied. This is carried out by noting that constraining completion times of all predecessor activities will have already been determined, and will appear nearby to the left. For example, activity $E$-2, in Figure 1-3, has an earliest allowable start time of 11, since the earliest finish times of its predecessors $A$-3, $E$-1, and Dummy, are 9, 11 and 7, respectively. The largest, or latest, of these times is 11, which becomes the earliest allowable start time for activity $E$-2.

The Harmonygraph is equivalent to an arrow diagram of the project plan, and hence it could alleviate all of the shortcomings of the bar chart listed above. By incorporating sliding tabs on each activity strip, it even overcomes the difficulty of keeping a bar chart up to date. These tabs facilitate the simulation of alternative schedules and network updating to reflect current status deviations from original schedules.

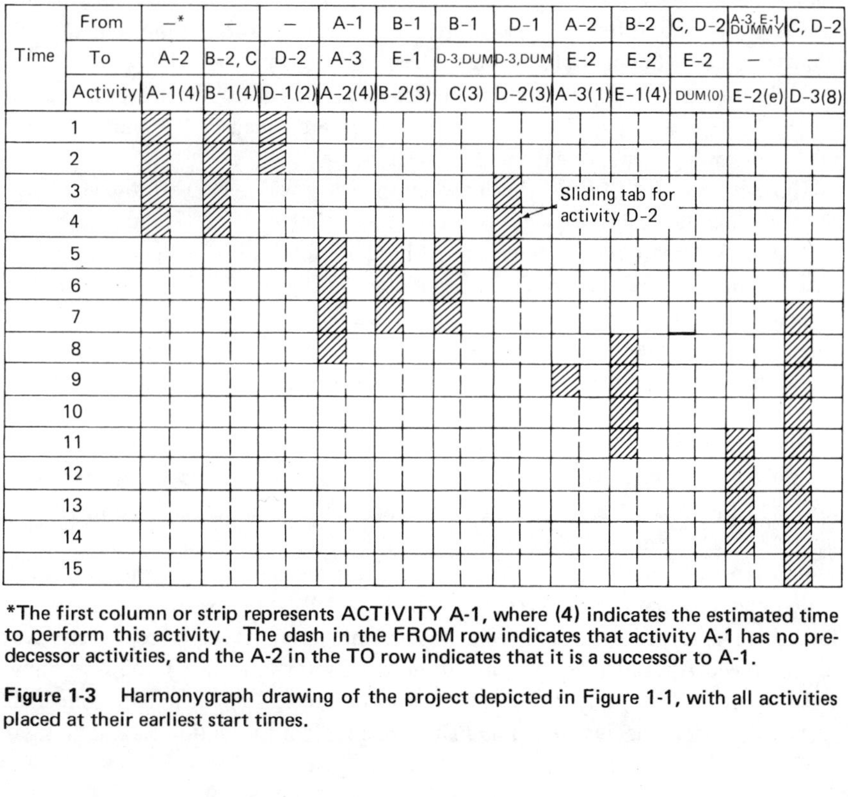

| Time | From | —* | — | — | A-1 | B-1 | B-1 | D-1 | A-2 | B-2 | C, D-2 | A-3, E-1, DUMMY | C, D-2 |
|---|---|---|---|---|---|---|---|---|---|---|---|---|---|
| | To | A-2 | B-2, C | D-2 | A-3 | E-1 | D-3,DUM | D-3,DUM | E-2 | E-2 | E-2 | — | — |
| | Activity | A-1(4) | B-1(4) | D-1(2) | A-2(4) | B-2(3) | C(3) | D-2(3) | A-3(1) | E-1(4) | DUM(0) | E-2(e) | D-3(8) |
| | 1 | | | | | | | | | | | | |
| | 2 | | | | | | | | | | | | |
| | 3 | | | | | | | | | | | | |
| | 4 | | | | | | | | | | | | |
| | 5 | | | | | | | | | | | | |
| | 6 | | | | | | | | | | | | |
| | 7 | | | | | | | | | | | | |
| | 8 | | | | | | | | | | | | |
| | 9 | | | | | | | | | | | | |
| | 10 | | | | | | | | | | | | |
| | 11 | | | | | | | | | | | | |
| | 12 | | | | | | | | | | | | |
| | 13 | | | | | | | | | | | | |
| | 14 | | | | | | | | | | | | |
| | 15 | | | | | | | | | | | | |

Sliding tab for activity D-2

*The first column or strip represents ACTIVITY A-1, where (4) indicates the estimated time to perform this activity. The dash in the FROM row indicates that activity A-1 has no predecessor activities, and the A-2 in the TO row indicates that it is a successor to A-1.

**Figure 1-3** Harmonygraph drawing of the project depicted in Figure 1-1, with all activities placed at their earliest start times.

One can only conjecture why this elegant solution to the problem was ignored, and had to be reinvented some years later. First, the author, while developing a fine methodology, did very little to sell its merits with impressive applications. But the most important reason was undoubtedly the timing of the development; in 1930 it was ahead of its time by a quarter of a century.

After these 25 years passed, an explosion occurred in interest in the problem of scheduling large projects. Three developments in the U.S. and Great Britain will be described below. Although they were independent, concurrent studies, they all based their respective methodologies on a network representation of the project plan, such as that shown in Figure 1-1b. They all essentially rediscovered the Harmonygraph in the systems approach age of the computer, and added one essential feature: They broke away from a strict graphical procedure by adding a tabular and arithmetic approach to the scheduling process.

The unpublished British development is discussed by Lockyer.[3] He describes how the Operational Research Section of the Central Electricity Generating Board investigated the problem involved with the overhaul of generating plants. By 1957, they had devised a technique which consisted essentially of identifying the "longest irreducible sequence of events." This term was later shortened to "major sequence," and corresponds to what we have referred to above as the critical path. Successive experimental applications of this methodology in the period 1958-60 resulted in very impressive project duration reductions of 42% and then 32% of the previous average times.

While this study was being conducted in Great Britain, two corresponding developments were taking place in the U.S. They are called PERT and CPM, and the historical highlights of each will be given below.

The development of PERT began when the Navy was faced with the challenge of producing the Polaris missile system in record time in 1958. Several studies[4, 5] indicated that there was a great deal to be desired with regard to the time and cost performance of such projects conducted during the 1950's. These studies of major military development contracts indicated that actual costs were, on the average, two to three times the earliest estimated costs, and the project durations averaged 40 to 50 percent greater than the earliest estimates. Similar studies of commercial projects indicated average cost and time overruns were 70 and 40 percent, respectively. While many people feel that original estimates must be optimistic in order to obtain contracts, a more important reason for these failures was the lack of adequate project management planning and control techniques for large complex projects.

It was recognized that something better was needed in the form of an integrated planning and control system for the Polaris Weapons System program. To face this challenge, a research team was assembled consisting of representatives of Lockheed Aircraft Corporation (prime contractor of Polaris), the Navy Special Projects Office, and the consulting firm of Booz, Allen and Hamilton. This research project was designated as PERT, or Program Evaluation Research Task.

By the time of the first internal project report, PERT had become Project Evaluation and Review Technique.[6] This research team evolved the PERT system from a consideration of techniques such as Line-of-Balance,[7] Gantt charts, and milestone reporting systems.

Time was of the essence in the Polaris program, so the research team concentrated on planning and controlling this element of the program. A major accomplishment of the PERT statistical procedure is the utilization of probability theory for managerial decision making. Scheduling systems have, traditionally, been based upon the idea of a fixed time for each task. In the PERT system three time estimates are obtained for each activity—an optimistic time, a most likely time, and a pessimistic time. This range of times provides a measure of the uncertainty associated with the actual time required to perform the activity sometime in the future. With the PERT procedure, it is possible, on the basis of these estimates, to derive the probabilities that a project will be completed on or before a specified schedule date. The misleading notion of a definite time for the completion of a project, or subproject, can be replaced by statements of the possible range of times and the probabilities associated with each. The result is a meaningful and potentially useful management tool. By adding to this information an appraisal of the consequences of not meeting a scheduled date and the cost of expediting a project in various ways, management can better plan at the outset of a project.

PERT also emphasizes the control phase of project management by various forms of periodic project status reports. The work of the original PERT research team has been extended into the areas of planning and controlling costs,[8,9] and to a lesser degree, into the areas of the performance or quality of the product.[10]

CPM (Critical Path Method) grew out of a joint effort conducted in the period Dec. 1956–Feb. 1959 by the duPont Company and Remington Rand Univac.[11,12] The objective of the CPM research team was to determine how best to reduce the time required to perform routine plant overhaul, maintenance, and construction work. In essence, they were interested in determining the optimum tradeoff of time (project duration) and total project cost. This objective amounts to the determination of the duration of a project which minimizes the sum of the direct and indirect costs, where, for example, direct costs include labor and materials, while indirect costs include the usual items, such as supervision, as well as "cost" of production time lost due to plant downtime.

The activities comprising this type of project are characteristically subject to a relatively small amount of variation compared to the activities of the Polaris program. Hence, unlike PERT, CPM treats activity performance times in a deterministic manner and has as its main feature the ability to arrive at a project schedule which minimizes total project costs.

The pioneering PERT and CPM groups did not know of each other's existence until early 1959, when the momentum of each effort was too great to influence the other. What has finally emerged from them is essentially a methodology similar to the Harmonygraph, with the addition of tabular computer outputs that give the start and finish times and slack (float) for each activity, and the need to sort this output in a variety of ways for different people to use. The sophisticated time-cost trade-off optimization algorithm of CPM, and the probability of meeting a schedule feature of PERT really play a very minor role in the applications today. However, it is felt by Kelley,[13] one of the pioneers of CPM, that we might not have CPM or PERT today if it were not for these sophisticated frills. They added that "something extra" to heighten interest and motivation and, more importantly, to enhance credibility. Although it may only be a spurious correlation, it is interesting to note that the Harmonygraph lacked such a frill, and it never made it in the world of applications.

This brief historical treatment of network based project management methodology is by no means complete. For example, the work of John Fondahl,[14] commencing in 1958 is certainly noteworthy. It dealt with a node-diagram approach to the basic CPM problem. Also, about this time, the development of the "method-of-potentials" took place in France. It was based on a network logic which constrained the *start* of one activity to lag a specified amount of time after the *start* of its predecessor activity. In a sense, this was a forerunner to the development of "precedence diagramming" which took place in the U.S. in the 1960s. Other studies of note will be cited in subsequent chapters.

## SUMMARY OF NETWORK-BASED PROJECT MANAGEMENT METHODOLOGY

Network-based project management methodology is a dynamic planning and control procedure, as was illustrated in Figure 1-2 above. This concept is shown in embellished form in Figure 1-4, which embodies the following steps, each of which will be described below:

STEP 1. Project Planning
STEP 2. Time and Resource Estimation
STEP 3. Basic Scheduling
STEP 4. Time-cost Trade-offs
STEP 5. Resource Allocation
STEP 6. Project Control

### Step 1. Project Planning

The activities making up the project are defined, and their technological dependencies upon one another are shown explicitly in the form of a network diagram. This step is shown in box (1) of Figure 1-4, and is the subject of Chapter 2, Developing the Network. Three alternative methods of networking will be

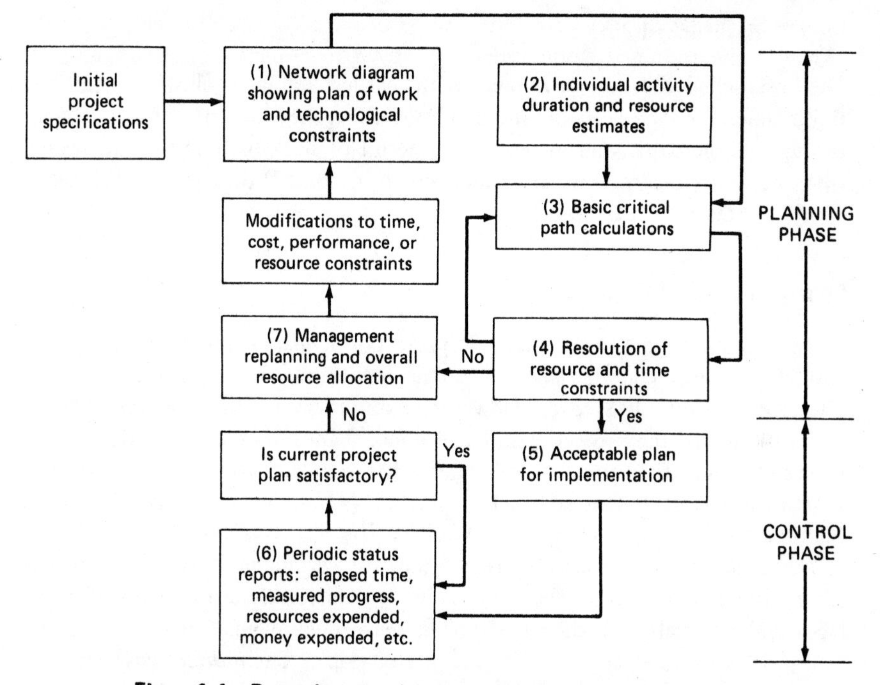


**Figure 1-4** Dynamic network-based planning and control procedure.

taken up, called *arrow*, *node*, and *precedence diagramming*. This is the most important step in the entire PERT/CPM procedure. The disciplined approach of expressing a plan for carrying out a project in the form of a network accounts for a large portion of the benefits to be derived from PERT/CPM. The development of the network is, in a sense, the simulation of alternative ways of carrying out the project. Experience has shown that it is preferable to make planning errors on paper, rather than in "bricks and mortar." It should also be added that if useful results are not obtained from these methods, it is usually because of inadequately prepared networks.

### Step 2. Time and Resource Estimation

Estimates of the time required to perform each of the network activities are made; these estimates are based upon assumed manpower and equipment requirements and availability, and other assumptions that may have been made in planning the project in Step 1. This step is shown in box (2) of Figure 1-4. Single-time estimation is taken up in Chapter 3. The three-time estimation method associated with PERT is treated in Chapter 9.

### Step 3. Basic Scheduling

The basic scheduling computations give the earliest and latest allowable start and finish times for each activity, and as a byproduct, they identify the critical path through the network, and indicate the amount of slack or float time associated with the noncritical paths. This step, shown in box (3) of Figure 1-4, is taken up in Chapter 4.

### Step 4. Time-Cost Trade-offs

If the scheduled time to complete the project as determined in Step 3 is satisfactory, the project planning and scheduling moves on to a consideration of resource constraints in Step 5. However, if one is interested in determining the cost of reducing the project completion time, then time-cost trade-offs of activity performance times must be considered for those activities on the critical and near critical paths. This step, shown in box (4) of Figure 1-4, is taken up in Chapter 8.

The costs associated with a project can, for certain purposes, be classed as either direct or indirect. The *direct costs* typically include the items of direct labor and materials, or if the work is being performed by an "outside" company, the direct costs are taken as the subcontract price. The *indirect costs* may include, in addition to supervision and other customary overhead costs, items such as the interest charges on the cumulative project investment, and penalty (or bonus) costs for completing the project after (or before) a specified date. The time-cost trade-off problem is directed to the task of determining a schedule of project activities which considers explicitly the indirect as well as the direct costs, and attempts to minimize their sum.

### Step 5. Resource Allocation

The feasibility of each schedule must be checked with respect to manpower and equipment requirements, which may not have been explicitly considered in Step 3. This step, shown in box (4) of Figure 1-4, is taken up in Chapter 7. The routine scheduling computations of Step 3 determine the slack along each network path. This indicates where certain activity schedules can be moved forward or backwards in time without affecting the completion time of the project. This movement can then be used to arrive at schedules which satisfy outside constraints placed on the quantity of resources available as a function of time.

Establishing complete feasibility of a specific schedule requires frequent repetition of the basic scheduling computations, as shown by the recycle path from box (4) to box (3). It may also require replanning and overall adjustment of resources, as shown by the path from box (4) to box (7). Hence, establishing an acceptable project plan for implementation may require the performance of a number of cycles of Steps 3 and 4, and possibly Steps 1 and 2 as well.

### Step 6. Project Control (Time and Cost)

When the network plan and schedule have been developed to a satisfactory extent, they are prepared in final form for use in the field. The project is controlled by checking off progress against the schedule, as indicated in box (6), and by assigning and scheduling manpower and equipment, and analyzing the effects of delays. Whenever major changes are required in the schedule, as shown in box (7), the network is revised accordingly and a new schedule is computed. The subject of time control is taken up in Chapter 4, and cost control in Chapter 5.

In addition to its value as a means of planning a project to optimize the time-cost relationship, the critical path network provides a powerful vehicle for the control of costs throughout the course of the project. Most cost accounting systems in industry are functionally-oriented, providing cost data by cost centers within the company organization rather than by project. By the utilization of the project network as a basis for project accounting, expenditures may be coded to apply to the activities, or groups of activities within a project, thus enabling management to monitor the costs as well as the schedule progress of the work.

Although the theory of network cost control is relatively simple, it is just beginning to be employed as a practical supplement to basic critical path technology, primarily because of the necessary involvement in established cost accounting procedures, and the fairly recent widespread availability of general computer programs. Thus, each organization interested in network cost control has been faced with the inconvenience and expense of developing new accounting procedures and the adoption of appropriate computer programs. In the mid-60s several large agencies of the U.S. Government[8] required the use of cost control supplements to basic CPM and PERT requirements, and generalized computer programs were developed. The U.S. Army Corp of Engineers still has this requirement on complex projects. They have not, however, succeeded in making cost control supplements as widely used as the basic PERT/CPM procedures, but they did provide some impetus for continued development of cost control systems. Also, as the installation of computers has expanded further in industry, changes in accounting procedures have become more frequent. Indeed, this expansion is becoming a necessity in many functionally organized companies that are experiencing problems of coordinating project activities because of rapid expansion of the volume of their work. Much of this has been brought about by the frequent occurrence of multibillion dollar projects contracted on a cost basis. Such firms are currently expending a great deal of effort to develop network-based management information systems to alleviate this problem.

An introduction to the concepts and practical problems of network cost control is presented in Chapter 5. A discussion of the available computer programs for all types of network analysis is contained in Chapter 11.

The basic procedures incorporated in Steps 1 through 6 can be performed, at least to some extent, by hand. Such methods will be presented in this text because they are useful in their own right, and also because they are an excellent means of introducing the more complex procedures that require the use of computers. It is particularly important that one be able to perform, by hand, the basic critical path calculations indicated in box (3) of Figure 1-4, since this is the first step in the evaluation of a proposed network plan for carrying out a project. A very simple method of hand calculation will be presented in Chapter 4. Hand methods for the resolution of relatively simple time and resource constraints will also be presented; they will then lead into more complex procedures for which computers are a necessity. Hand and computer methods of preparing periodic status reports will also be presented. A description of available computer programs is given in Chapter 11.

## USES OF CRITICAL PATH METHODS

Since the successful application of PERT in the Polaris program, and the initial success of CPM in the chemical and construction industries, the use and further development of critical path methods has grown at a rapid rate. The applications of these techniques now cover a wide spectrum of project types.[15,16]

Research and development programs range from pure research, applied research, and development to design and production engineering. While PERT is most useful in the middle of this spectrum, variations of it have been used in the production end of this spectrum. However, overall usage is not as widespread now as in the 1960s. PERT is not particularly useful in pure research, and in fact some say it should be avoided here because it may stifle ingenuity and imagination, which are the keystones of success in pure research.

Maintenance and shutdown procedures, an area in which CPM was initially developed, continues to be a most productive area of application of critical path methods. Construction type projects continue to be the largest individual area in which these methods are applied. It is extremely useful in this field of application to be able to evaluate alternate project plans and resource assumptions on paper rather than in mortar and bricks.

More recent applications of critical path methods include the development and marketing of new products of all types, including such examples as new automobile models, food products, computer programs, Broadway plays, and complex surgical operations.

In addition to an increase in the variety of applications of critical path methods, they are being extended to answer questions of increasing sophistication. The important problem of resource constraints has been successfully expanded to include multiple resource types associated with multiple projects. Cost control, project bidding, and incentive contracting are also areas where significant developments are taking place.

## ADVANTAGES OF CRITICAL PATH METHODS

It is fitting to close this chapter with an enumeration of the advantages that one might expect from the use of critical path methods in the planning and controlling of projects.

1. *Planning* Critical path methods first require the establishment of project objectives and specifications, and then provide a realistic and disciplined basis for determining how to attain these objectives, considering pertinent time and resource constraints. It reduces the risk of overlooking tasks necessary to complete a project, and also it provides a realistic way of carrying out more long-range and detailed planning of projects, including their coordination at all levels of management.
2. *Communication* Critical path methods provide a clear, concise, and unambiguous way of documenting and communicating project plans, schedules, and time and cost performance.
3. *Psychological* Critical path methods, if properly developed and applied, can encourage a team feeling. It is also very useful in establishing interim schedule objectives that are most meaningful to operating personnel, and in the delineation of responsibilities to achieve these scheduled objectives.
4. *Control* Critical path methods facilitate the application of the principle of management by exception by identifying the most critical elements in the plan, focusing management attention on the 10 to 20 per cent of the project activities that are most constraining on the schedule. It continually defines new schedules, and illustrates the effects of technical and procedural changes on the overall schedule.
5. *Training* Critical path methods are useful in training new project managers, and in the indoctrination of other personnel that may be connected with a project from time to time.

## REFERENCES

1. Cleland, D. I. and W. R. King, *Systems Analysis and Project Management*, McGraw Hill Book Co., Inc., 1968.
2. Adamiecki, Karol, "Harmonygraph," *Przeglad Organizacji* (*Polish Journal on Organizational Review*), 1931.
3. Lockyer, K. G., *Introduction to Critical Path Analysis*, Pitman Pub. Co., 3rd Edition, 1969, Ch. 1.
4. Marshall, A. W., and W. H. Meckling, "Predictability of the Costs, Time and Success of Development," RAND Corp., Report P-1821, December, 1959.

5. Peck, M. J. and F. M. Scherer, "The Weapons Acquisition Process: An Economic Analysis," Division of Research, Graduate School of Business Administration, Harvard University, Cambridge, Mass., 1962.
6. Malcolm, D. G., J. H. Roseboom, C. E. Clark, and W. Fazar, "Applications of a Technique for R and D Program Evaluation," (PERT) *Operations Research*, Vol. 7, No. 5, 1959, pp. 646-669.
7. Turban, Efraim, "The Line of Balance–A Management by Exception Tool," *The Journal of Industrial Engineering*, Vol. 19, No. 9, September 1968, pp. 440–448.
8. The Office of the Secretary of Defense and the National Aeronautics and Space Administration, *DOD and NASA Guide*, *PERT Cost Systems Design*, U.S. Government Printing Office, Washington, D.C., June, 1962, Catalog Number D1. 6/2:P94.
9. Moder, J. J. and S. E. Elmaghraby, *Handbook of Operations Research Models and Applications*, Van Nostrand Reinhold Co., New York, NY, 1978, Chapter 10.
10. Malcolm, D. G., "Reliability Maturity Index (RMI)–An Extension of PERT into Reliability Management," *The Journal of Industrial Engineering*, Vol. 14, No. 1, January-February 1963, pp. 3-12.
11. Walker, M. R. and J. S. Sayer, "Project Planning and Scheduling," Report 6959, E. I. duPont de Nemours and Co., Wilmington, Delaware, March 1959.
12. Kelley, J., "Critical Path Planning and Scheduling: Mathematical Basis," *Operations Research*, Vol. 9, No. 3, May-June 1961, pp. 296-321.
13. Kelley, J., *CPM–In Conception* or *An Arrowing Experience*, Unpublished talk.
14. Fondahl, J. W., "A Noncomputer Approach to the Critical Path Method for the Construction Industry," Dept. of Civil Engineering, Stanford University, Stanford, Calif., 1st Edition, 1961, 2nd Edition, 1962.
15. Berkwitt, G. J., "Management Rediscovers CPM," *Dun's Review*, May 1971, Dun & Bradstreet Pub. Corp.
16. Davis, E. W., "CPM Use in Top 400 Construction Firms," *Journal of the Construction Division*, ASCE, Vol. 100, No. CO1, Proc. Paper 10395, March 1974, pp. 39-49.

## EXERCISES

1. Discuss various applications of critical path methods. For example, suppose you are in charge of the preparation of a proposal for a large and involved project, or the coming church social, or the preparation of a new college curriculum, or the development of a new product and manufacturing facility. Would critical path methods be of assistance in these undertakings? If so, in what ways?

2. Can you think of any complex projects in which critical path methods would not be of any particular value? Give examples and discuss why.

# 2

# DEVELOPMENT OF THE NETWORK

The first step in utilizing critical path methods is the identification of all the activities involved in the project and the graphical representation of these activities in a flow chart or network. This step is usually called the "planning phase," because the identification of the project activities and their interconnections requires a thorough analysis of the project, and many decisions are made regarding the resources to be used and the sequence of the various elements of the project.

In one sense the network is only a graphical representation or model of a project plan. The plan may have previously existed in some other form–in the minds of the project supervisors, in a narrative report, or in some form of bar chart. In practice, however, the preparation of a network usually influences the actual planning decisions and results in a plan that is more comprehensive, contains more detail, and is often different from the original thoughts about how the project should proceed. These changes derive from the discipline of the networking process, which requires a greater degree of analytical thinking about the project than does a narrative, a bar chart, or other types of project descriptions.

Thus, the contruction of the network often becomes an aid to and an integral part of project planning, rather than an after-the-fact graphical exercise. Indeed, the planning phase has proven to be the most beneficial part of critical path applications. In developing a detailed and comprehensive project network, users often make significant improvements over their original ideas; they do a better job of early coordination with suppliers, engineers, managers, subcontractors, and all the other groups associated with the project; and they end up with a documented plan that has strong psychological effects on the future management of the project.

One psychological effect of network preparation is that it demonstrates to supervisors and other key personnel that the management is vitally concerned about the coordination and timeliness of all project activities, and that a means of more closely monitoring these factors has been drafted. Thus, an intangible but highly significant factor—the initial motivation of the project team—can be favorably influenced by the networking effort.

The planning phase is also the most time-consuming and difficult part of most critical path method applications. This is due primarily to the inherent analytical problems in any project planning effort. One may expect some difficulty in using the network format at first, but it is soon realized that the network discipline is more of an aid to thinking than it is a set of stringent rules for drawing a chart.

Actually there are only about five rules commonly used in drawing networks, and these provide most of the flexibility needed in describing project plans. (More advanced networking schemes with even greater flexibility are described in Chapter 10.) The accuracy and usefulness of a network is dependent mainly upon intimate knowledge of the project itself, and upon the general qualities of judgment and skill of the planning personnel. Skill in using the network techniques can be quickly acquired in only one or two applications. The time required to learn to use networks drawn by someone else is even less.

This chapter is limited to those basic rules and procedures of network development which are required to prepare the first draft of a network. In Chapter 3 the addition of time estimates and the development of a final working draft are considered, including the problem of obtaining the most useful level of network detail.

There are several different graphical schemes used in drawing networks. This chapter and the rest of Part I will be based upon the most common scheme practiced among industrial users of critical path methods. This is not necessarily the best scheme, however. Two of the other schemes, the *node* scheme and *precedence* diagramming, have much to recommend them, especially under certain conditions. Readers who are not limited by contract, the computer programs available, or other reasons to use a particular scheme are urged to study all of the schemes, which are presented in this chapter and in Chapter 3. The reader should then select the scheme best suited to the circumstances. As mentioned above, in the application of critical path methods the greatest effort and expense are associated with the preparation of the network. Selection of the most useful and economical networking scheme is, therefore, worthy of special attention.

## PREPARATION FOR NETWORKING

Some experienced network users will say that all one needs to begin networking is a large piece of paper, several sharp pencils, and a large eraser. Actually, there is a bit more to it than that. Several general questions need to be raised and answered before detailed project planning should begin. Among these questions are:

1. What are the project objectives?
2. Who will be charged with the various responsibilities for accomplishing the project objectives?
3. What organization of resources is available or required?
4. What are the likely information requirements of the various levels of management to be involved in the project?

Of course, these questions are fundamental to project management and should not be passed over lightly. In some research and development projects, the development of new products, and in other cases, a discussion of the basic objectives of the project can reveal disagreements among the key persons involved. Similarly, open discussions of responsibilities and resources can bring to light erroneous assumptions or misunderstandings in these areas. Naturally, it is well to resolve these matters before proceeding with networking (and certainly before beginning the project).

An optional step in network preparation is the development of a list of work elements of the project. Such a list can be useful in discussing responsibilities and resources, as mentioned above, and it can serve as a reference for networking. Although experienced networkers usually forego the listing of activities, beginners with the technique will find that a list is helpful.

## BASIC TERMS

Several of the most common terms in networking are defined and illustrated below. Terms associated with scheduling computations are explained in later chapters.

*Definition:*

An *activity* is any portion of a project which consumes time or resources and has a definable beginning and ending. Activities may involve labor, paper work, contractual negotiations, machinery operations, etc. Commonly used terms synonymous with "activity" are "task" and "job." In the arrow scheme of net-

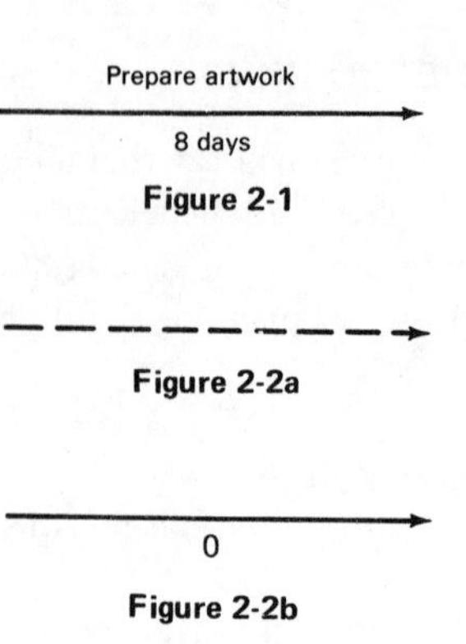


Figure 2-1

Figure 2-2a

Figure 2-2b

working, activities are graphically represented by arrows, usually with descriptions and time estimates written along the arrow (Figure 2-1).

*Definition:*

An arrow representing merely a dependency of one activity upon another is called a *dummy* activity. A dummy carries a zero time estimate. It is also called a "dependency arrow." Dummies are often represented by dashed-line arrows (Figure 2-2a) or solid arrows with zero time estimates (Figure 2-2b).

*Definition:*

The beginning and ending points of activities are called *events.* Theoretically, an event is an instantaneous point in time. Synonyms are "node" and "connector." If an event represents the joint completion of more than one activity, it is called a "merge" event. If an event represents the joint initiation of more than one activity, it is called a "burst" event. An event is often represented graphically by a numbered circle (Figure 2-3), although any geometric figure will serve the purpose.

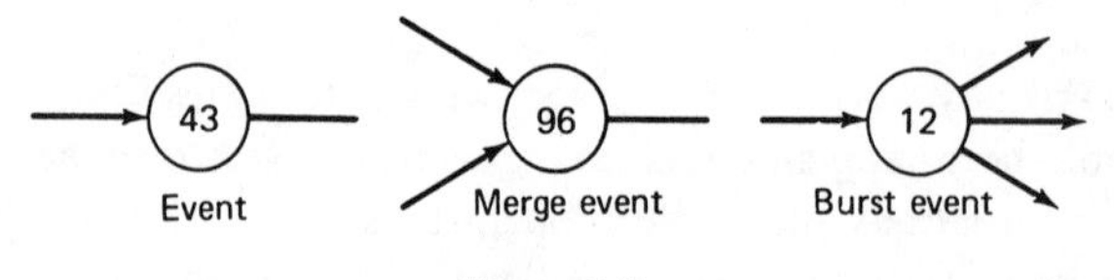


Figure 2-3

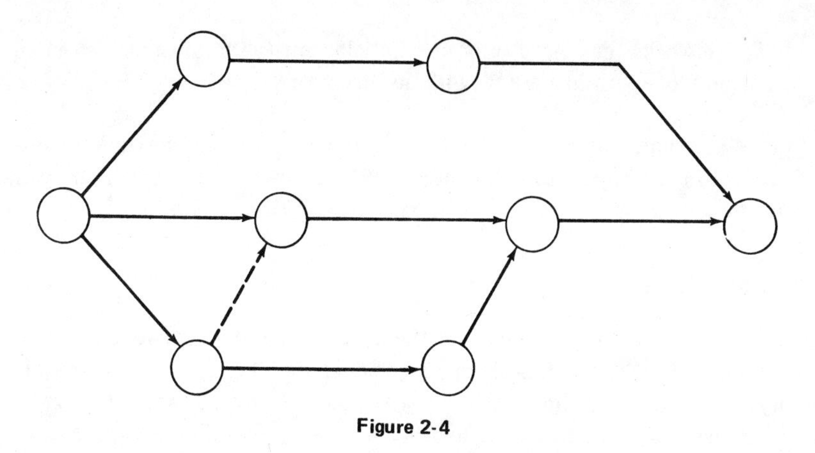

Figure 2-4

*Definition:*

A *network* is a graphical representation of a project plan, showing the interrelationships of the various activities. Networks are also called "arrow diagrams" (Figure 2-4). When the results of time estimates and computations have been added to a network, it may be used as a project schedule.

## NETWORK RULES

The few rules of networking based on the arrow scheme may be classified as those basic to all arrow networking systems, and as those imposed by the use of computers or tabular methods of critical path computation.

### Basic Rules of Network Logic

RULE 1. Before an activity may begin, all activities preceding it must be completed. (Activities with no predecessors are self-actuating when the project begins.)

RULE 2. Arrows imply logical precedence only. Neither the length of the arrow nor its "compass" direction on the drawing have any significance. (An exception to this rule is discussed under "Time-scaled Networks" below.)

### Additional Rules Imposed By Some Computers or Tabular Methods

RULE 3. Event numbers must not be duplicated in a network.

RULE 4. Any two events may be directly connected by no more than one activity.

RULE 5. Networks may have only one initial event (with no predecessor) and only one terminal event (with no successor).

Rules 4 and 5 are not required by all computer programs for network analysis, as discussed in the appendix to Chapter 4. Hand computation on the network does not require Rules 3, 4, or 5, but the network must not have loops.

## EMPHASIS ON LOGIC

At this point, it should be noted that the construction of a network should be based on the logical or technical dependencies among the activities. That is, the activity "approve shop drawings" must be preceded by the activity "prepare shop drawings," because this is the logical and technically necessary sequence.

A common error in this regard is to introduce activities into the network on the basis of a sense of time, or a "feel" for appropriate sequencing. For example, in the maintenance of a pipeline the activity "deactivate lines" might be placed after "procure pipe," because it is felt that that is the right time to deactivate the lines. (Figure 2-5a.) Rather, the deactivation activity should be placed in the network in the proper technological sequence, such as just before "remove old pipe" (Figure 2-5b). Then in the scheduling process (to be covered in Chapter 4) the best *time* to initiate the deactivation so as to minimize the down time on the pipelines can be determined.

Such emphasis on strict logic is one of the principles of networking introduced by the originators of both CPM and PERT. It is a fundamental part of the networking discipline that causes planners to think about their projects in a thorough, analytical manner. In this process old methods of performing similar projects may be questioned or disregarded, clearing the way for new and perhaps better approaches.

Application of strict logic also tends to result in a unique network for a given set of activities, that is, the network that represents the true technical dependencies of the project. If a planner introduces his or her personal "feel" for how a

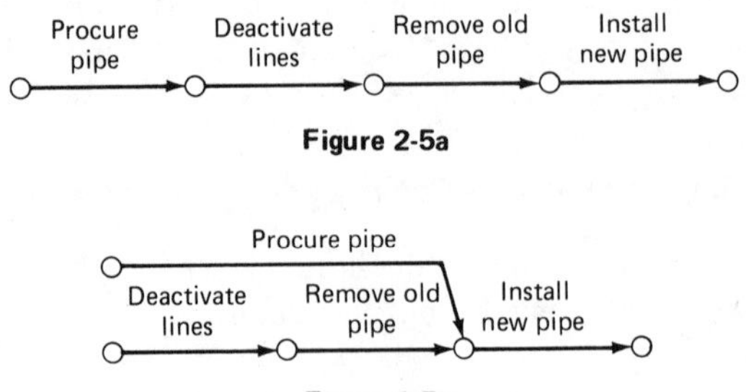


Figure 2-5a



Figure 2-5b

project should proceed, without strict attention to the logical dependencies, than a *subjective* network is produced. Subjective networks may go unnoticed for a while, but when the project gets underway and progress is reported, the subjective relationships tend to be revealed in embarrassing ways. Activities begin before the network says it is possible, others are delayed for activities that are supposed to be independent, etc. The subjective network thus loses credibility with the users and may be abandoned early in the updating process.

## INTERPRETATION OF RULES

Rules 1 and 2 may be interpreted by means of the portion of a network shown in Figure 2-6a. According to Rule 1, this diagram states that "before activity *D* can begin, activities *A*, *B*, and *C* must be completed." Note that this is not intended to imply that activities *A*, *B*, and *C* must be completed simultaneously.

Note also the definition of the events. Event 5 represents the "beginning of activity *A*." Event 6, however, means "the completion of activities *A*, *B*, and *C*, and the beginning of activity *D*." Because of the multiple meanings of events, discussion of networks in terms of activities is favored over event-oriented terms.

## COMMON PITFALLS

The most common network error involves Rule 1. As an illustration, consider the diagram of activities *A*, *B*, *C*, and *D* shown in Figure 2-6a. Suppose that activity *D* depended on the completion of *B* and *C* and on the completion of the *first half* of *A*, completing the second half of *A* being independent of *B*, *C*, and *D*. To diagram this situation correctly, we must divide activity *A* into two activities and introduce a *dummy* activity, as shown in Figure 2-6b. The dummy has been used here to correct a problem of *false dependency;* that is, activity *D* was

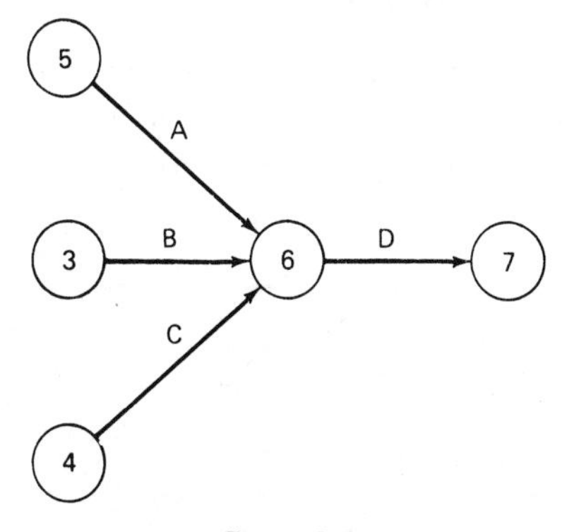


**Figure 2-6a**

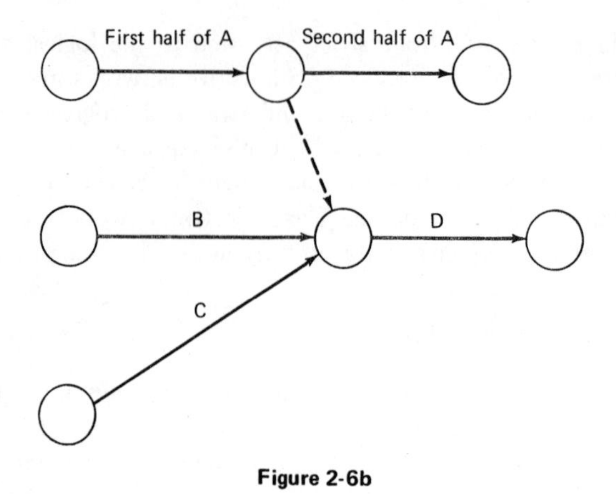


Figure 2-6b

only partially dependent on the activities preceding *D*. False dependencies represent the most subtle networking problems and must be guarded against constantly, especially at merge and burst points. (The problem of false dependencies is largely avoided in the node scheme of drawing networks, which is discussed later.)

Another network condition that must be avoided is illustrated in Figure 2-7. Activities *J*, *F*, and *K* form a *loop*, which is an indication of faulty logic. The definition of one or more of the dependency relationships is not valid. Activity *J* cannot begin until *C* and *K* are completed. But *K* depends on *F*, which depends on *J*. Thus *J* could never get started because it depends on itself. Loops, which in practice may occur in a complex network through oversight, may be remedied by redefining the dependencies to relate them correctly.

## SATISFYING COMPUTER RULES

Networking rules 3, 4, and 5 are related to the procedures for coding networks for computer analysis. Rule 3 involves another subtle problem that all computer programs have in understanding a network. Consider the diagram in Figure 2-8a.

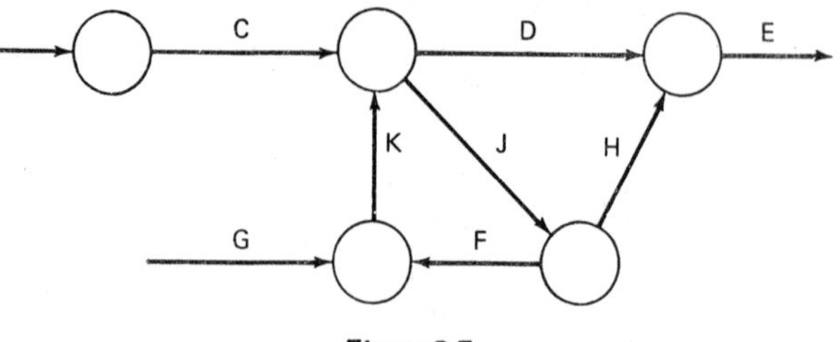


Figure 2-7

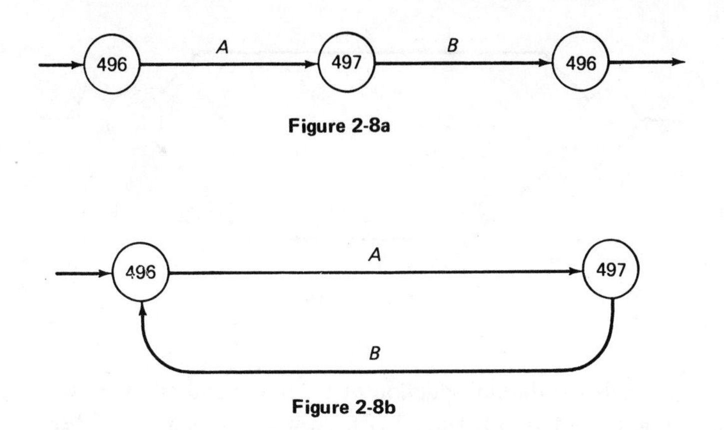


Figure 2-8a

Figure 2-8b

An attempt to process this situation on a computer would cause it to halt and print out an indication of loop, for the computer would read event 496 as a precedent to itself (Figure 2.8b). Therefore, in employing computers it is important to keep track of event numbers used and not used. (Some computer programs impose the additional restriction that each event number must be larger than any predecessor number. Such a restriction is cumbersome because it greatly inhibits network revision and updating; one tends to "run out of numbers" or lose track of the numbers used.

Rule 4 is violated when the condition shown in Figure 2-9 occurs. Activities *A* and *B* may be called *duplicate activities*, since a computer (or tabular method of computation), using only event numbers for identification, may not be able to distinguish the two activities, as indicated below:

| *Network Description* | *Computer Code* |
|---|---|
| Activity *A* | 6–7 |
| Activity *B* | 6–7 |

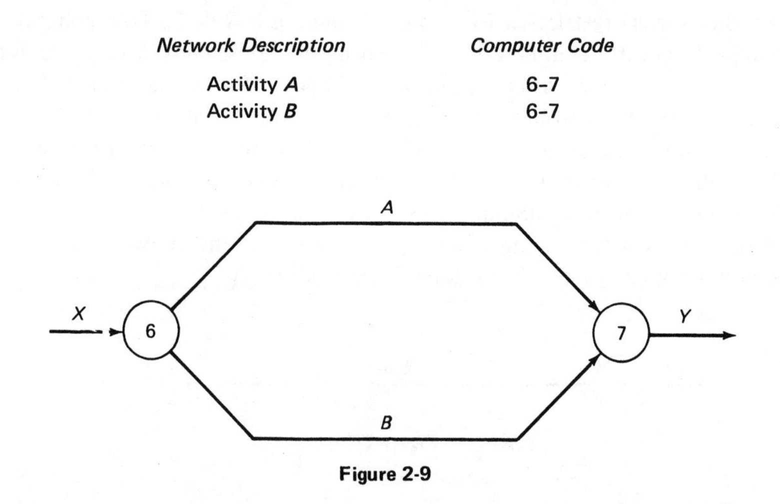


Figure 2-9

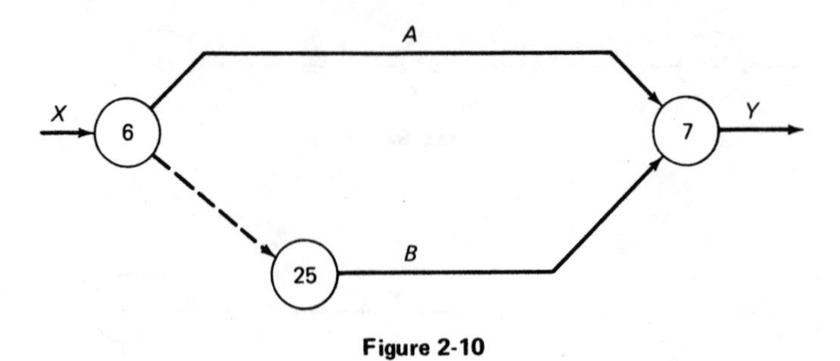


Figure 2-10

One remedy calls for the introduction of a dummy and another event in series with either activity *A* or *B* (Figure 2-10). Now the computer can distinguish between the activities by their different codes.

| *Network Description* | *Computer Code* |
|---|---|
| Activity *A* | 6–7 |
| Activity *B* | 25–7 |
| Dummy | 6–25 |

Note that the above solution *does not change the logic* of the network. Nor would the logic be changed if the dummy had been placed at the other end of *B*, or at either end of *A*. If the reader feels that a change in logic has occurred, he should review the section Interpretation of Rules.

Another way to correct duplicate activities is to combine them (Figure 2-11). This solution is simple and effective, but it may destroy some of the desired detail in the network; the question of detail is treated in Chapter 3.

Another special restriction for computer analysis is Rule 5. To accommodate this requirement it is common practice to bring all "loose ends" to a single initial and a single terminal event in each network, using dummies if necessary. For example, one may wish to network a current project that is already past the initial event. In this case the network would have a number of open ended, parallel paths at the "time now" point. These loose ends would be connected to a single initial event by means of dummies, as shown in Figure 2-12.

When methods employing hand computations on the network (or certain computer programs) are utilized, Rule 5 is not necessary.

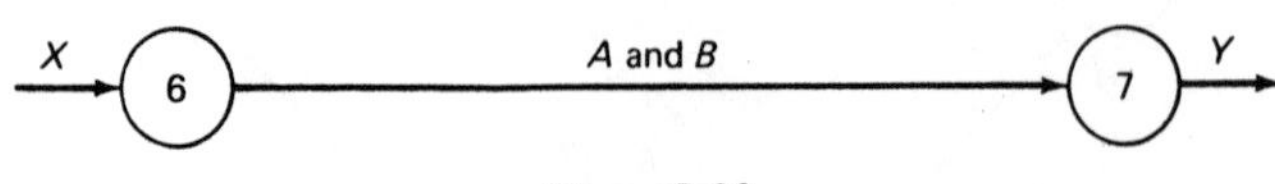


Figure 2-11

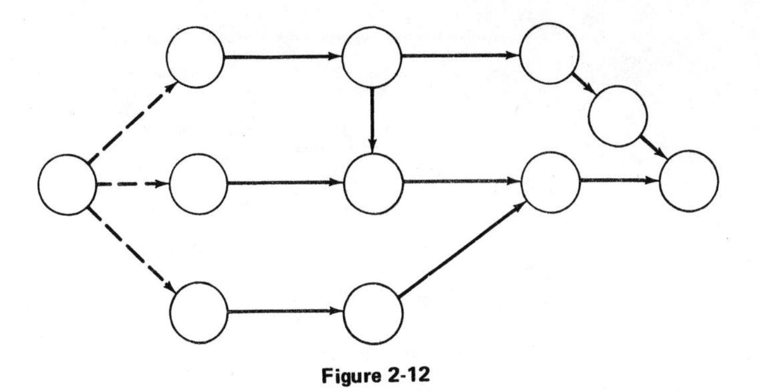

Figure 2-12

## USING DUMMIES EFFICIENTLY

While the need for dummies in certain cases has been pointed out, it is preferable to avoid unnecessary dummies. For example, consider the diagram in Figure 2-13. Evidently, activity *D* depends on *C*, *B*, and *A*. But the dependency on *A* is clear without the dummy 2–4, which is *redundant.* Such dummies should be eliminated to avoid cluttering the network and to simplify computations. (Some computer services base their charges on the number of activities, including dummies.)

In other cases it may be necessary to introduce dummies for clarity. For example, suppose a particular event is considered a *milestone* in the project, a point that represents a major measure of progress in the project. This point may be assigned a scheduled or target date, which may be noted on the network as shown in Figure 2-14. However, the ambiguity of events, especially merge or burst events, can cause confusion about the target date notation. Does the date represent the scheduled completion of activity *A*, activity *B*, activities *A* and *B*, or the start of activity *C*? In these cases it may be desirable to eliminate the possible misunderstanding by introducing dummies, as in Figure 2-15. In this example it is now clear that the date refers only to the completion of activity *A*. It is

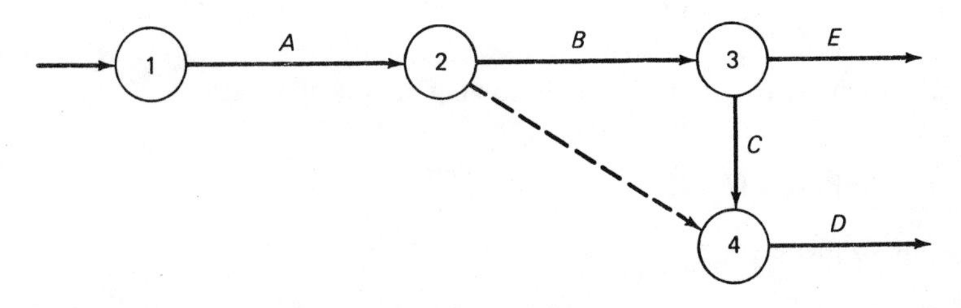


Figure 2-13

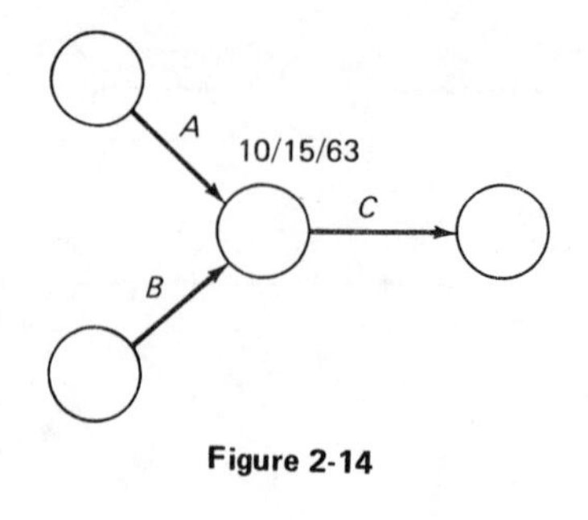


Figure 2-14

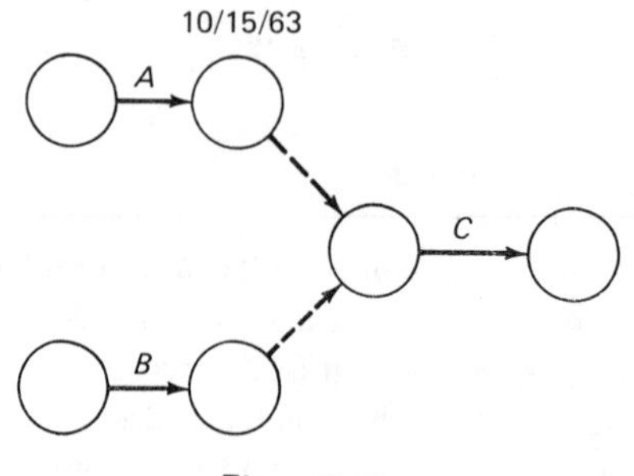


Figure 2-15

emphasized that this is only an illustration of the flexibility afforded by judicious use of dummies, and not a required practice at merge points or at milestones.

## ACTIVITY DESCRIPTIONS

Thus far in this chapter, most activities in the illustrative diagrams have been described by letter codes on the arrows. This has been done both for convenience and to emphasize the logic of the network representations. In practice it is much more common to print several descriptive words on the arrow. This avoids the need for cross-reference with a separate list of activity descriptors.

The descriptions themselves must be unambiguous; they must mean the same thing to the project manager, the field superintendent, the various subcontractors, and others expected to use the network. Descriptions should also be brief and, where possible, should make use of quantitative measures or reference points. Examples are shown in the following sample applications.

## SAMPLE APPLICATIONS

### A Machinery Installation

Consider a project involving the installation of a new machine and training the operator. Assume that the training of the operator can begin as soon as he is

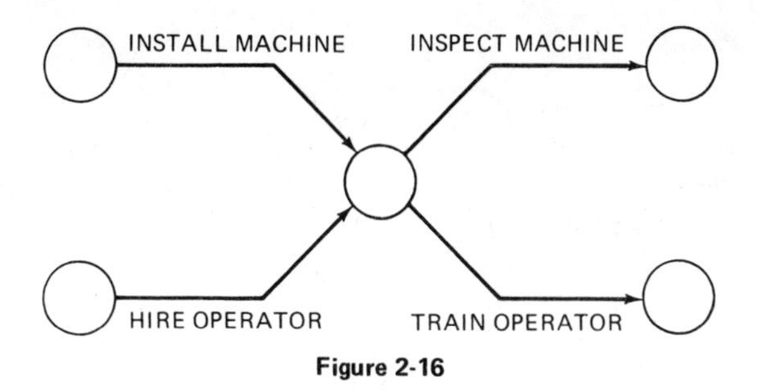


Figure 2-16

hired and the machine is installed. The training is to start immediately after installation and is not to be delayed for inspection of the machine. The inspection is to be made after the installation is complete. One might attempt to network this project as shown in Figure 2-16.

However, this network says that the inspection cannot begin until the operator is hired, which is a false dependency. To correct this representation in the network a dummy is added, as shown in Figure 2-17.

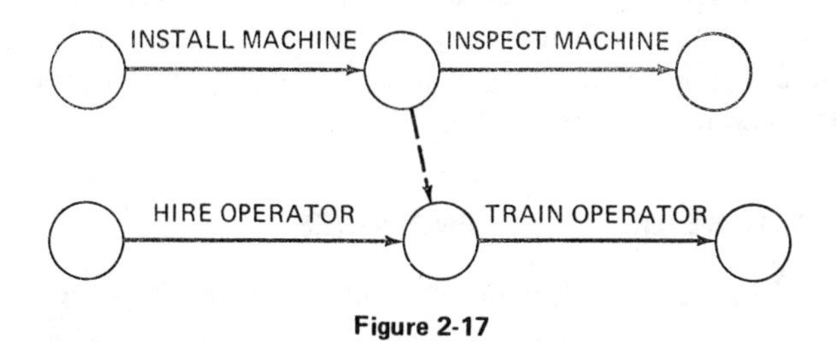


Figure 2-17

## A Market Survey

Consider now a project to prepare and conduct a market survey. Assume that the project will begin by planning the survey. After the plan is completed, data collection personnel may be hired, and the survey questionnaire may be designed. After the personnel have been hired and the questionnaire designed, the personnel may be trained in the use of the questionnaire. Once the questionnaire has been designed, the design staff can select the households to be surveyed.

Also, after the questionnaire has been designed it may be printed in volume for use in the survey. After the households have been selected, the personnel trained, and the questionnaires printed, the survey can begin. When the survey is complete the results may be analyzed. This project may be networked as shown in Figure 2-18. Note that dummy 4-3 is essential, whereas dummy 6-5 is necessary only if a computer is to be used.

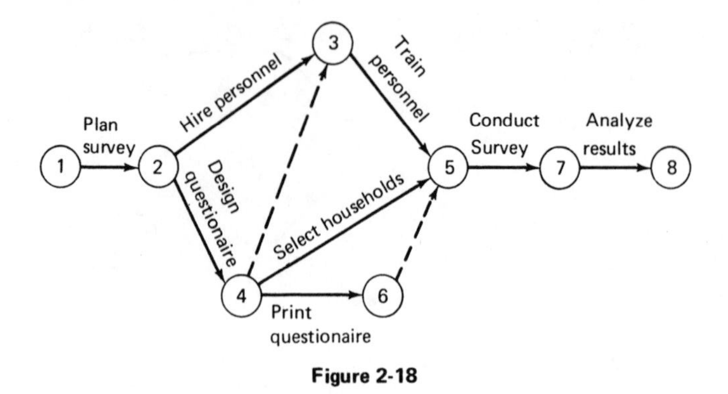


Figure 2-18

**Development of a New Product**

The network in Figure 2-19 represents a plan for the development and marketing of a new product, in this case a new computer program.[2] Note that the dummy 5-7 is used to conform to Rule 4. The dummy 2-4, however, is used to show that activities 4-5 and 4-7 depend on activities 1-2 and 3-4, while activity 2-6 depends only on activity 1-2. Another dummy is shown between events 0 and 1; this dummy is not technically necessary and could have been omitted by combining events 0 and 1. Note also that this network has two critical paths, denoted by the heavy activity lines.

## NATURAL AND RESOURCE DEPENDENCIES

The network in Figure 2-18 illustrates the fact that there are two types of activity dependency. Note that most of the activity dependencies are caused by the nature of the activities themselves; for example, personnel cannot be trained until they have been hired, and the questionnaires cannot be used in the survey until they have been printed. Such dependencies among activities may be called *natural*, and this is the most common type of dependency.

Also note, however, that the selection of households is dependent upon the design of the questionnaire, but only because one group of people is assigned to do both jobs (the "design staff" in the project description). This staffing limitation, and the implication that the design staff could not do both jobs simultaneously causes the two activities to be drawn in series (dependent) rather than in parallel. A dependency of this type is not "natural," but is caused by the resource limitation. Thus, it may be called a *resource dependency*. The resources involved may be personnel, machinery, facilities, funds, or other types of resources.

Usually it is best to include in the first network draft all resource dependencies that are known and firmly established as ground rules for the project. These

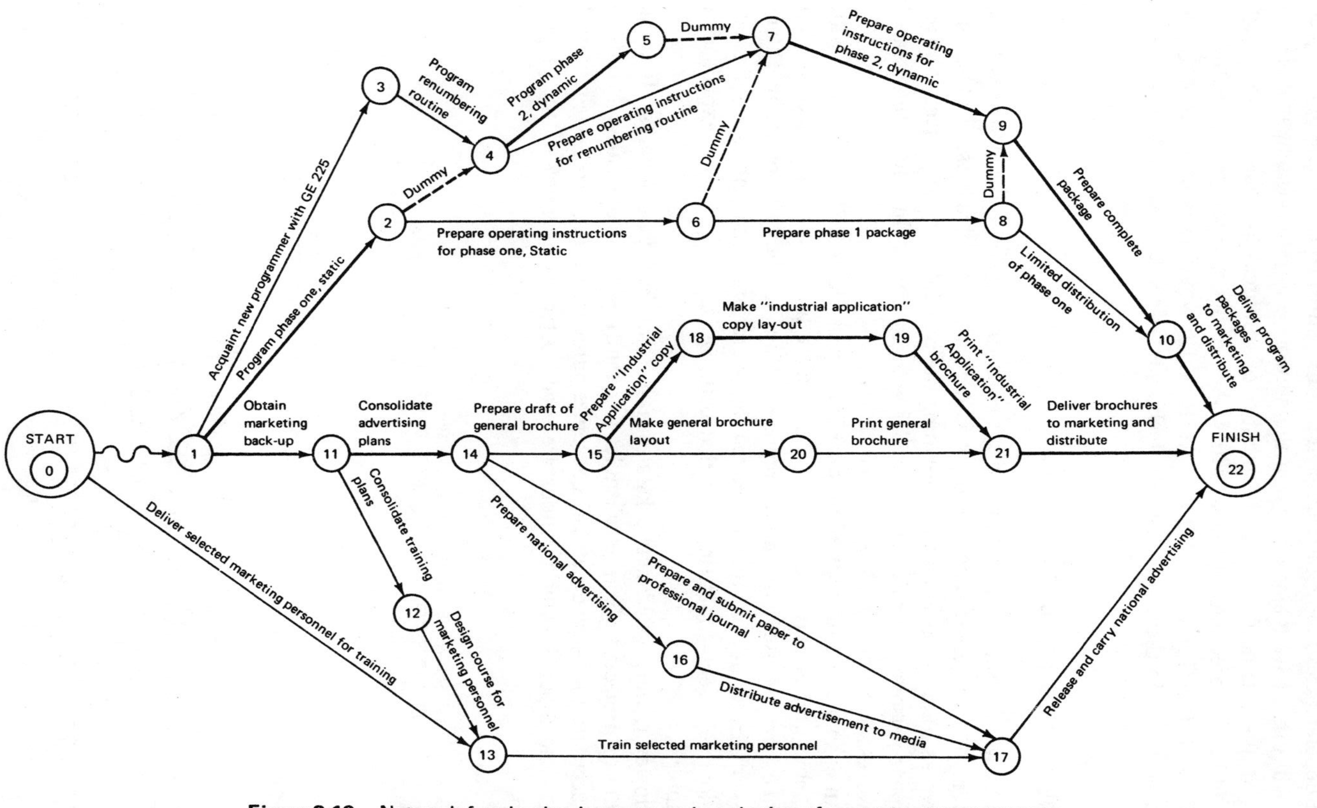


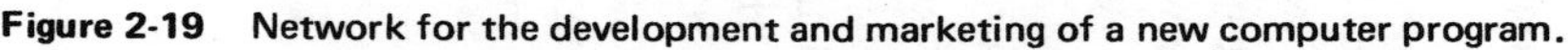
**Figure 2-19** Network for the development and marketing of a new computer program.

firm dependencies represent significant factors in the planning phase of a project and will often have major effects on the network and resulting schedule. However, if there is doubt about the number of resources available or how they should be allocated among the activities, then resource dependencies should be omitted in the first draft of the network, which would then be based only on natural logic. In these situations the techniques of resource allocation discussed in Chapter 7 should be employed.

## TIME-SCALED NETWORKS

If a network is plotted on a time scale, similar to a bar chart, the arrows become vectors that represent time durations as well as project elements. Time-scaled networks have an advantage in presenting the schedule for the project to management personnel or others not familiar with networks. In effect, time-scaled networks are merely extensions of bar charts.

An example of a time-scaled network is given in Figure 2-20. Compare it with the same project network in Figure 2-18. Note that in the time-scaled version dashed lines are used not only to represent a dummy but also represent extensions of arrows beyond their estimated time durations. Thus the dashed-line extensions represent "float time" on that path, which will be further explained in Chapter 4.

However, time-scaled networks also have some key disadvantages. It is the nature of projects to change schedules frequently, almost constantly. Maintaining accurate time-scaled networks becomes burdensome, especially when normal pencil-and-paper drawing methods are used. At each updating of the project

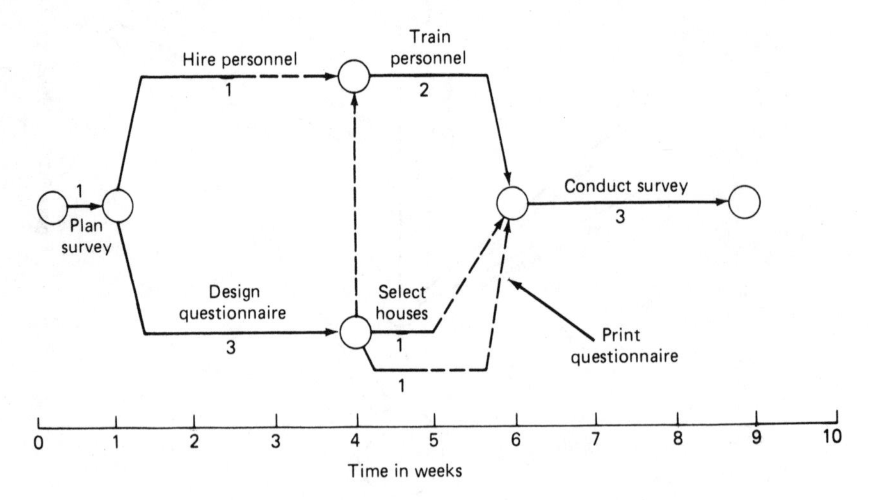


**Figure 2-20** Time-scaled network of survey project.

schedule many future activities have to be redrawn. With networks containing hundreds or thousands of activities, time-scaling becomes impractical. This difficulty can be overcome on smaller networks by the use of special mechanical charting devices. In general, though, time-scaled networks are worthwhile only for occasional presentations for upper levels of management. (See Chapter 6 for further discussion.)

## ACTIVITY-ON-NODE SCHEMES

The graphical method of drawing networks utilized in this chapter is called the *arrow* scheme in this text. This scheme was introduced in the original publications on the Critical Path Method. Since then the arrow scheme has remained one of the most common means of portraying networks. Many or perhaps most computer programs for CPM and PERT are designed to accept the predecessor-successor ($I$ - $J$) event code for activities that are associated with the arrow scheme.

However, the arrow scheme is by no means the only networking procedure, nor is it the most efficient one if we judge efficiency by the number of symbols required to portray a given number of activities in a network. At least three other schemes have been conceived and used widely. In this chapter other schemes are described and evaluated. The individual or organization that has a choice of graphic schemes (that is, not limited by contract or customer preference) should consider all of them and select the one most suitable to his purposes. In most cases, of course, it will be desirable that the organization standardize its networks by using only one of the schemes consistently.

## BASIC NODE SCHEME

The complete reverse of the arrow scheme is the node scheme, in which the nodes represent the activities and the arrows are merely connectors, denoting precedence relationships. The market survey project in the node format is illustrated in Figure 2-21. The principal advantage of the node scheme is that it eliminates the need for special dummies to correct false dependencies. This feature makes the scheme more efficient and, more importantly, easier to learn. In the arrow and event schemes the most difficult aspect is learning to make the proper use of dummies. In the node scheme all the arrows are dummies, in effect, and there are no subtle false dependency problems requiring the use of special dummies.

One of the first proponents of the node scheme and, apparently, its originator was J. W. Fondahl of Stanford University. Professor Fondahl developed a node format in 1958, almost concurrent with the publication of the first PERT and CPM reports (see history of networking methods and references in Chapter 1).

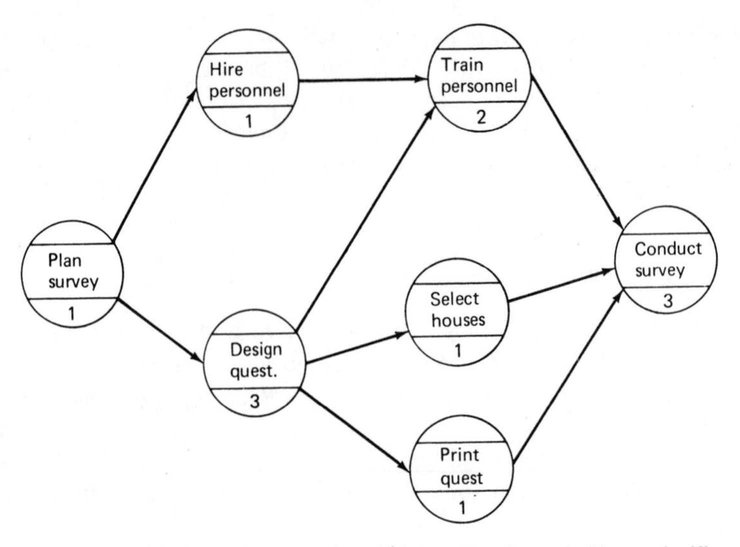


**Figure 2-21** Survey network, node scheme (shape of node symbol is not significant).

However, the PERT and CPM reports received the earliest public attention and the arrow format became the most popular scheme during the first decade of network applications. Most of the computer programs for network analysis would accept only the arrow ($I - J$) notation and logic. Later, in the mid-seventies, commercial computer software that would accept the node scheme began to appear. Since then there has been a noticeable shift in popularity toward the node scheme. An informal survey by the authors in 1981 found that users of networking methods at that time were about equally divided between the arrow and node schemes.

## PRECEDENCE DIAGRAMMING

An extension of the basic node scheme was first mentioned in Fondahl's 1961 report (Chapter 1, Reference 14), in which "lag" values associated with activity relationships and "precedence" matrices were introduced. These concepts gained further notice in the user's manual for an IBM 1440 computer program for network processing, published around 1964. One of the principal authors of that manual was J. David Craig.

As an example of how the precedence rules apply, consider the project network in Figure 2-22. Upon reviewing this first draft of the network, let us assume that the management is dissatisfied with the overall duration of 44 weeks and seeks ways to shorten the time. They find that the times for drawings and procurement are dictated by some complex iron castings, and that the other

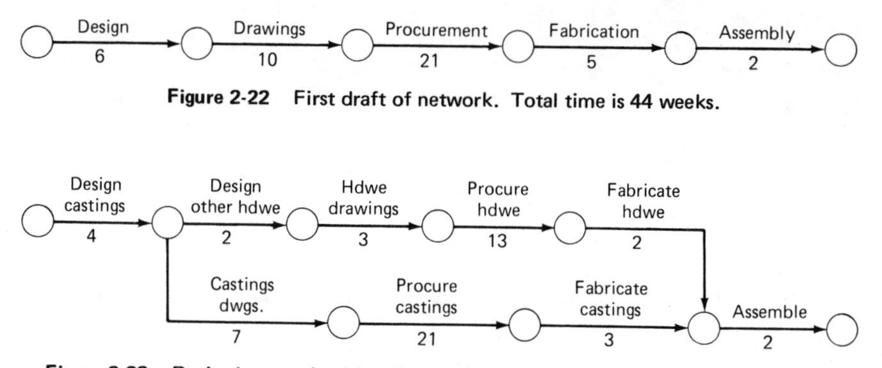


**Figure 2-22** First draft of network. Total time is 44 weeks.

**Figure 2-23** Revised network with split activities. Total time is reduced to 37 weeks.

hardware requires less time to design, prepare drawings, and procure. Thus, all of the activities except the last one (assembly) may be split to form two essentially parallel paths for the castings and other hardware. Figure 2-23 shows this logic, resulting in a total project time estimate of 37 weeks. The revised network in Figure 2-23 is converted to the node scheme in Figure 2-24.

However, assume that under further study it is recognized that the activities are still not as discrete and separate as implied by the revised networks. In particular, management notes that the start of drawings on the castings does not have to wait until all of the castings are designed (4 weeks). Actually the first couple of castings can be designed in the first week and the drawing work can begin with these. Likewise, some of the hardware procurement can begin after the first two weeks of drawing work on the hardware. These facts can affect the overall project schedule and it may be worthwhile to revise the network still further in order to display the more detailed logic. (As an exercise, the reader may sketch such a further revision, using the node scheme.)

In order to represent "overlapping" activities of the kind occurring in this example, modifications of the node scheme were developed. The modifications consist of defining precedence relationships among activities, which are illustrated in Figure 2-25. In addition, one may specify a "lag time" associated with

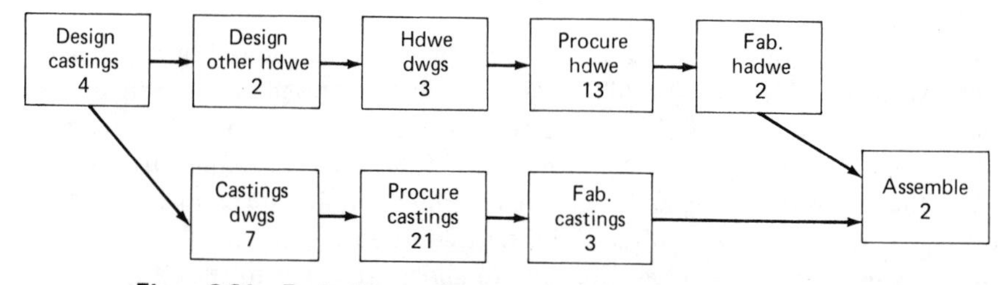


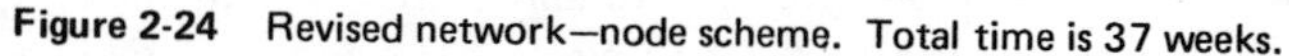

**Figure 2-24** Revised network—node scheme. Total time is 37 weeks.

**Figure 2-25a** Finish-to-start relationship. (Start of B must lag 5 days after the finish of A.)

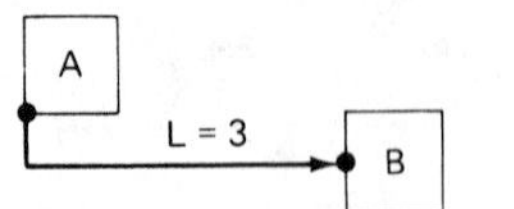


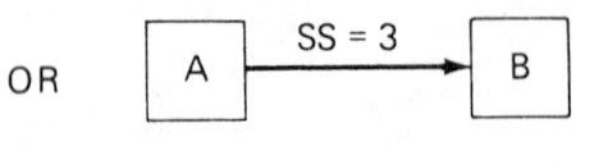


**Figure 2-25b** Start-to-start relationship. (Start of B must lag 3 days after the start of A.)

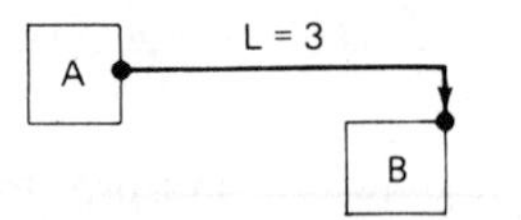


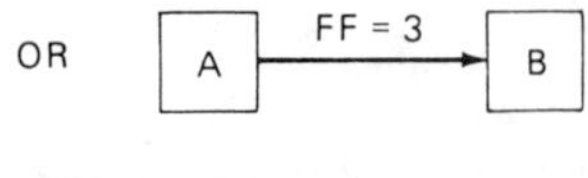


**Figure 2-25c** Finish-to-finish relationship. (Finish of B must lag 3 days after the finish of A.)

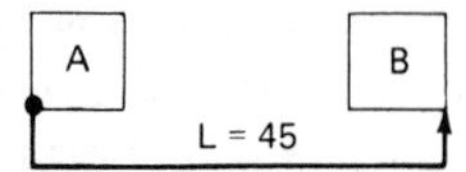


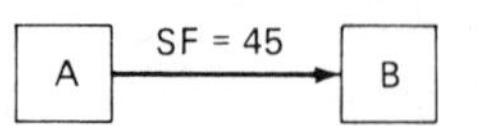


**Figure 2-25d** Start-to-finish relationship. (Finish of B must lag 45 days after the start of A.)

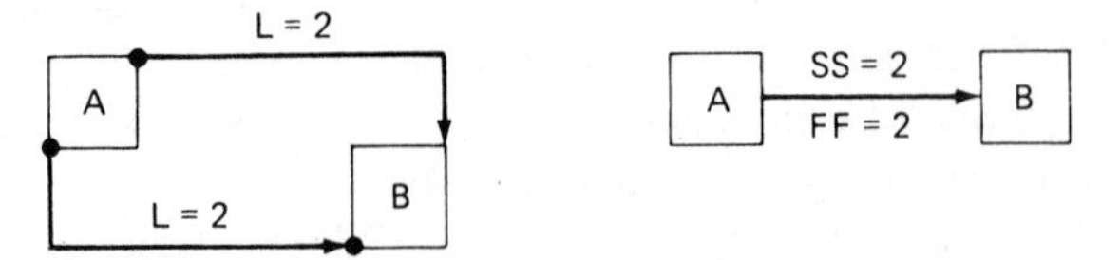


**Figure 2-25e** Composite start-to-start and finish-to-finish relationships. (Start of B must lag 2 days after the start of A, and the finish of B must lag 2 days after the finish of A.)

any of the precedence relationships, which can be used to account for the overlapping times among activities.

The first relationship in Figure 2-25a is the only one allowed in PERT and CPM, with the additional flexibility that a lag time is allowed. For example, the *start* of "concrete forms stripping" (activity B), must lag 5 days after the *finish* of "pour concrete" (activity A). The lag of 5 days is required in this case to allow the concrete to cure and strengthen before the forms are removed. (While the lag times associated with the finish-start constraint are usually zero or positive, it is possible to use a negative time to allow activity B to start a specified number of time units *before* activity A is finished. However, this same result can be achieved without the use of negative lags by using the next type of relationship, start-to-start.)

The second relationship, start-to-start, is also used quite frequently. Suppose activities A and B are "level ground for concrete slab," and "place mechanical pipe," respectively. Then the diagram in Figure 2-25b indicates that "place mechanical pipe" may *start* 3 days after "level ground for concrete slab" *starts.* The 3 day lag here is to allow the grading operation to get a convenient distance ahead of the piping, which is following behind. Now, if the grading should be interrupted for some reason, it is understood that the piping activity would also be interrupted. This type of dependency is not shown explicitly in precedence diagramming, but this does not pose a real problem because field supervisors are adept at handling this type of situation without the aid of project networks.

The third relationship, finish-to-finish, is also frequently used. For example, suppose activity A is "place electrical" (5 days) and activity B is "complete walls" (15 days). Note that it takes 3 days of activity B work to complete the output from one day of activity A. Therefore the *finish* of "complete walls–B" must lag behind the *finish* of "place electrical–A" by 3 days because it will take 3 days of wall completion work to handle the last day of electrical work. This relationship is shown in Figure 2-25c.

The fourth relationship shown in Figure 2-25d, start-to-finish, is not frequently used. An example will be taken from the field of automative design to illustrate that precedence diagramming is not restricted to construction work. Suppose activity A is "design power train" (40 days), and activity B is "design chassis" (30 days). The dependency relationship between A and B might be that the *last* 20 days of the chassis design work depends upon the results that will be obtained from the *first* 25 days of power train design work. Thus, 25 days after the *start* of the power train, 20 days is required to *finish* the chassis design. This implies that the finish of chassis design must lag 25 + 20 = 45 days after the start of the power train design. This is shown in Figure 2-25d by the SF = 45 relationship.

The composite relationship shown in Figure 2-25e is required wherever a series of activities must follow each other. For example, consider the following sequential construction tasks:

A–Erect wall frames
B–Place electrical lines
C–Attach wall boards
D–finish seams on wall boards

Activity B cannot start until 2 days after A has started, and similarly, B cannot finish until 2 days after A has finished. This relationship is shown specifically in Figure 2-25e. The lag of 2 days is chosen from experience as a reasonable separation between the carpenters and electricians. If all four of these tasks work at the same rate (complete same number of walls per day), then this same relation-

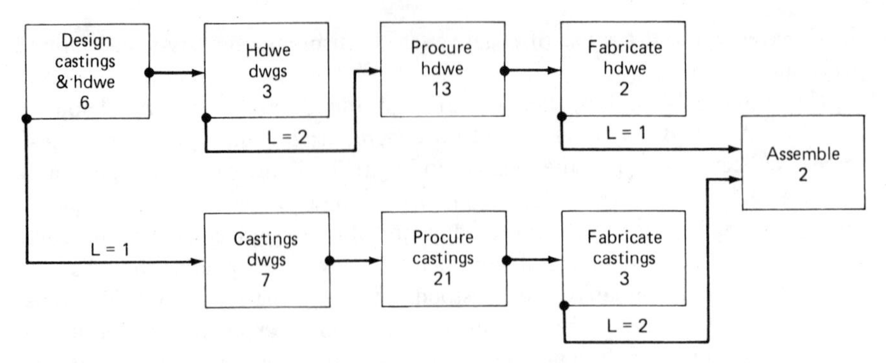


**Figure 2-26** Sample project converted to precedence diagramming network. Total time is reduced to 33 weeks.

ship would hold between activities B and C and between C and D. It is not difficult to visualize the savings in numbers of activities possible with precedence diagramming in networking a construction type of project. Also, it will be seen that representing a task by a single activity, rather than a series of activities, simplifies project time and cost control.

Using these "precedence diagramming" rules in revising the network of Figure 2-24, the management is able to show the overlapping activities in a more accurate way without splitting the activities further, as shown in Figure 2-26. In fact, there is one less node as compared with Figure 2-24, and it can be shown that the project duration is further reduced to 33 weeks.

It is evident that in some situations precedence diagramming can be more efficient than arrow or node schemes. Project managers may also like the fact that certain key activities do not have to be split into two or more separate nodes. In this example, the design work can be retained as a single activity. Advocates of precedence diagramming feel that it is thus easier to understand and less confusing than splitting activities. However, the method also introduces some complexities of its own in the form of connecting arrows with several different definitions, and project time calculations that are not quite as straightforward as in arrow or node networks.

The subject of precedence diagramming is covered further in Chapters 3 and 4.

## SUMMARY

This chapter has been concerned with the translation of the project plan into a series of interconnecting activities and events composing a network model of the plan. A few rules were presented, some being required in order to maintain accuracy and consistency of network interpretation, and others being required by the nature of data-processing procedures. The rules presented relate to the arrow method of networking. The next chapter continues the development of the network, including the addition of activity time estimates and the attainment of the desired level of network detail.

## REFERENCES

1. *GE 225 Application, Critical Path Method Program*, Bulletin CPB 198B, General Electric Computer Deprtment, Phoenix, 1962.
2. Fondahl, J. W., *A Noncomputer Approach to the Critical Path Method for the Construction Industry*, 2nd Ed., Department of Civil Engineering, Stanford University, Stanford, Calif., 1962.

## EXERCISES

1. Review the machinery installation sample network in Figure 2-17 and assume that an activity consisting of "schedule inspector" must precede "inspect machine." Add the activity to the network without causing a false dependency.
2. In Figure 2-27 find at least five errors or unnecessary symbols. State which rule is broken in each case, and suggest how the error might be corrected.

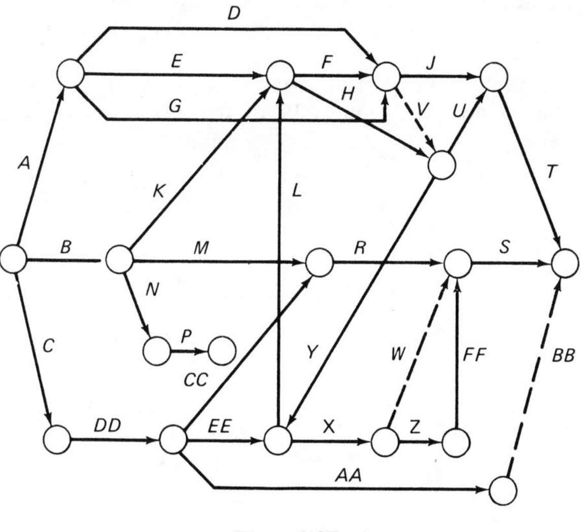


**Figure 2-27**

3. Given the activities and relationships listed below, draw an accurate network with no more than six dummies. Check your results by (1) numbering each event (do not repeat any numbers) and determining that no activity or

dummy has the same pair of identification numbers, (2) making sure that there is only one initial and one terminal event.[2]

| *Activity* | *Predecessor* | *Activity* | *Predecessor* |
|---|---|---|---|
| *A* | — | *I* | *A, B, C, D* |
| *B* | — | *J* | *O, E, N* |
| *C* | — | *K* | *B, C, D* |
| *D* | — | *L* | *K* |
| *E* | *B, C, D* | *M* | *B, C, D* |
| *F* | *A, B, C, D* | *N* | *B, C, D* |
| *G* | *A, B, C, D* | *O* | *A, B, C, D* |
| *H* | *F, G, I* | | |

4. Using the list of activity dependencies given below, draw an accurate and economical (minimizing the use of dummies) network.

| *Activity* Depends on *Activity* | | *Activity* Depends on *Activity* | |
|---|---|---|---|
| *A* | none | *G* | *F* |
| *B* | *A* | *I* | *F* |
| *F* | *A* | *J* | *H* |
| *H* | *A* | *K* | *I* and *J* |
| *C* | *B* | *L* | *G, D,* and *E* |
| *D* | *B* | *M* | *K* |
| *E* | *C* | *N* | *L* and *M* |

5. Figure 2-28 shows a portion of a network for construction of a multi-story building. What do the dummies 29-33 and 34-38 represent? Can you find other dependencies of this type in the network?

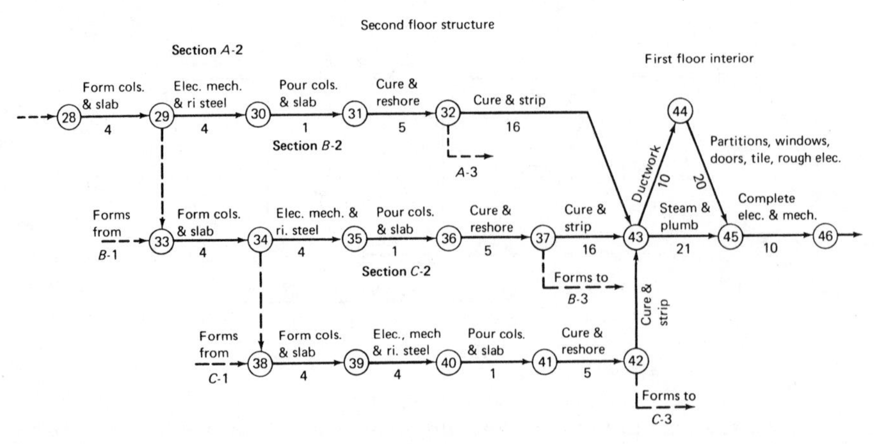


**Figure 2-28**

6. Assume that the following list of activity dependencies is correct. The diagram in Figure 2-29, however, does not represent these dependencies properly. Correct the diagram using only one dummy activity.

| *Activity* | Depends on *Activity* |
|---|---|
| *A* | none |
| *B* | none |
| *C* | *A* |
| *D* | *B* and *G* |
| *E* | *C* and *D* |
| *F* | *D* |
| *G* | *A* |
| *H* | *E* |

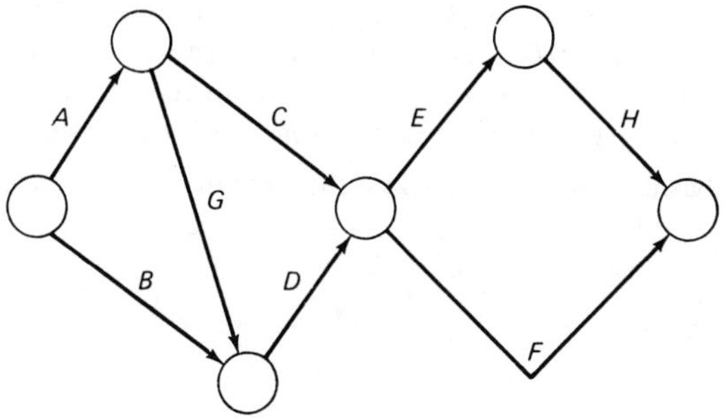

Figure 2-29

7. Draw a network of the following steam pipe maintenance project. The project begins by moving the required material and equipment to the site (5 hours). Then we may erect a scaffold and remove old pipe and valves (3 hours); while this is being done, we may fabricate the new pipe (2 hours). When the old pipe and valves are removed and the new pipe is fabricated, we can place the new pipe (4 hours). However, the new valves can be placed (1 hour) as soon as the old line is removed. Finally, when everything is in place, we can weld and insulate the pipe (5 hours).

8. Consider as a project the servicing of a car at a filling station, including such normal activities as filling the gas tank, checking the oil, etc. List the activities you wish to include, then draw a network of the project. Assume that there are two people available to service the car and that the gas pump has an automatic shut-off valve.

   In carrying out this exercise, remember that the netwrok should reflect technological or physical constraints rather than arbitrary decisions on the order in which the activities are to be carried out. It is therefore suggested that you include time estimates for each activity and then redraw your original network on a time scale to show the actual order of scheduling each activity so as to minimize the time required for two people to service a car.

9. A common error in networking is to place the activities in the network according to the order in which you plan to carry them out, rather than in the order dictated by technological predecessor-successor relationships. Consider the following network.

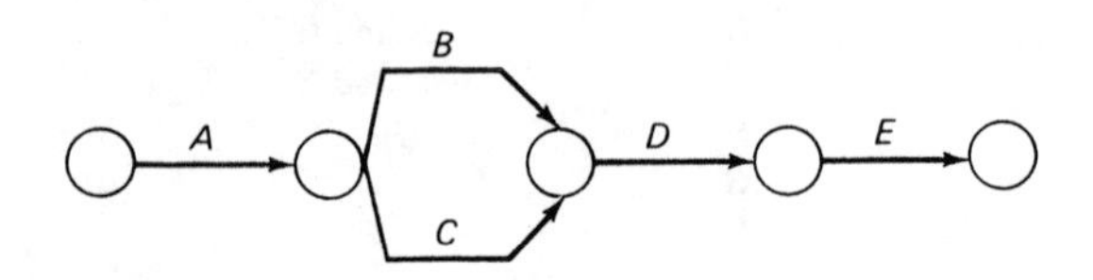


The network was drawn this way because the planner felt that he would not start activity $D$ until $B$ and $C$ were completed, and also that he would not start activity $E$ until $D$ was completed. If the true technological requirements on activity $D$ are that only $A$ must precede $D$, and no other activity depends on activity $D$, how should this network be drawn?

10. Using the basic node scheme, draw the networks described in exercises 3, 4, and 6.

# 3

# TIME ESTIMATES AND LEVEL OF DETAIL

By applying the networking rules presented in Chapter 2, one may develop the first draft of the project network. The basic logic of the plan should be established by the first draft. The next step is to add time estimates to each activity and refine the network as needed to display the desired level of detail.

In practice, the processes of time estimation and network refinement are closely interrelated and are usually accomplished at the same time. For as one begins to make time estimates, it is found that certain activities need to be redefined, condensed into fewer activities, or expanded into more, in order to represent the project accurately and at the desired level of detail.

As mentioned in Chapter 1, there are two methods of applying time estimates: the single estimate method and the three estimate method. This chapter will treat only the single estimate method. The three estimates and their associated statistical treatment are considered in detail in Chapter 9. The three estimate procedure may be useful in projects where the time values for each activity may be uncertain and the estimator wishes to provide a range of time values.

## TIME ESTIMATION IN THE NETWORK CONTEXT

### Accuracy of Estimates

One of the most common first reactions of persons being introduced to critical path methods is that the whole procedure depends upon time estimates made by project personnel. Since these estimates are based upon judgment rather than any "scientific" procedure, it is argued, the resulting CPM or PERT schedules cannot be any better than schedules derived from bar charts or any other method.

While it is true that critical path methods depend upon human time estimates, as all project planning schemes must, there are some significant differences in how the estimates are obtained and in how they are used. To illustrate these differences, consider how estimates for project times and costs are usually derived. The process is similar in most types of industrial or construction projects, so let us take a familiar example, the construction of a house. A builder's estimate for a house may look something like this:

| | *Cost Estimate* | *Time Estimate* |
|---|---|---|
| Clearing and grading | $ 1,920.00 | — |
| Foundations | 5,336.00 | — |
| Framing | 7,290.50 | — |
| ⋮ | ⋮ | ⋮ |
| Total | $221,373.58 | Approximately 3 to 4 Months |

The total cost was developed from detailed, item-by-item estimates. The total time, however, was simply a gross estimate based on the builder's experience with similar projects. Careful attention to costs, of course, is the home builder's key to obtaining contracts and making a profit. He can afford to be less accurate about time estimates because the contracts do not normally have time limitations and the owner is not as concerned about the exact duration of the project.

But suppose the time were a critical factor, as it often is in industrial projects, and suppose the accuracy of the time estimate were made important to the builder's profit, through penalty clauses and other means. How could the builder develop a more accurate estimate of the project duration? We would expect him to break down the job into its time-consuming elements, to obtain good time estimates of each element, and to sequence the elements into a plan that would show which elements must be done in series, which in parallel, etc. In principle we are saying that greater accuracy in time estimation can be developed in a manner similar to the development of accurate cost estimates—through detailed, elemental analysis.

Critical path methods provide a disciplined vehicle for making detailed time estimates, for graphically representing the sequence of project elements, and for computing schedules for each element. Thus the network format effectively utilizes more detail than does a bar chart, which is a particular advantage in the use of time estimates and in schedule analysis. Human knowledge remains the basic ingredient in all methods, but it is how the knowledge is applied that determines the accuracy and power of the results.

**Who Does the Estimating?**

Certainly another key factor in the accuracy of time estimates is who makes them. A general rule in this regard is that the most knowledgeable supervisory person should estimate each activity. This means, for example, that activities that are the primary responsibility of the electrical subcontractor should be estimated by the subcontractor's manager or supervisor most familiar with the job; activities of the research department should be estimated by the research supervisor responsible for and most familiar with the work; and so on. The objective in obtaining time estimates should be to get the most realistic estimates possible.

It is characteristic of network planning to call meetings of all supervisory personnel at the time-estimating stage and to consider each activity for which they are responsible. (It is also desirable for these personnel to participate in preparation of the first network draft. This is not always practical, though, if everyone involved is not familiar with network principles). In addition to the psychological advantages mentioned in Chaper 2, the participation of the key members of the project team has major advantages. Whenever the subcontractors, suppliers, inspectors representing the customer, etc., meet to discuss a project, the discussion will lead to questions of priorities in certain phases of the work, potential interference of work crews, definitions of assignments of engineers, and many other details of planning that might not have been explored until problems arose during the project. These discussions often identify and resolve potential problems before the project begins—rather than tackling them as they actually occur, which can result in the corrective action being expensive or perhaps impossible.

Here again the network merely serves as a cause for calling the meeting and as a detailed agenda. Yet these thorough planning sessions around a network result in what is probably the major benefit of critical path methods as practiced to date. This benefit is the project plan itself, in terms of its validity, its comprehensive scope, and its efficiency in the utilization of time and resources. This is not to say the reader should stop here. There is more to be gained, and many of the more experienced users are applying critical path methods and related procedures to advance the science of project management in a variety of ways. However, it appears that a majority of satisfied users of CPM and PERT have gotten their money's worth out of the initial network planning effort. Following through with the technique in the project control applications, whenever it is done, has provided bonus benefits.[1]

The question of bias or "padding" of time estimates is, naturally, related to who makes the estimates and their motivations. It does not necessarily follow that the most knowledgeable person is also the most objective. It is human nature to try to provide a time estimate that will be accepted as reasonable but will not likely cause embarrassment later. Thus, a certain amount of bias is to be expected in any procedure.

Nevertheless, it is generally felt by CPM users that the network approach tends to help reduce the bias to a manageable level. Again, the increased detail shown by a network plays a useful role. The smaller the work elements, the more difficult it is to hide a padded estimate. Indeed, a certain amount of professional pride is often noticeable in the estimator's attitude, which leads to a degree of optimism in his figures. Another factor favoring realistic estimates is the recognition that biased figures will tend to make the activities involved form the critical path, thus invoking concentrated attention of management and other parties engaged in the project on the group responsible for those activities.

### Research on Time Estimates

Only a limited amount of research on the subject of network time estimation has been published. One of the more interesting studies, by Seelig and Rubin,[2] compared the results of 48 R & D projects, some of which were "PERTed" and some of which were not. The authors concluded that the use of PERT definitely did lead to improvement of schedule performance but had no noticeable effect on technical performance. Furthermore, they concluded that the improvement in schedule performance was primarily a result of *improvement in communication* among the project managers, which was brought about by the use of the network method.

## ESTIMATION METHODS

### When to Add Time Estimates

It is best to complete a rough draft of the total project network before any time estimates are added. This procedure is conducive to concentration on the *logic* of the activity relationships, which must be accurately established. When the draft appears to be complete, the time estimates should be added to each activity. This step will constitute a complete review of the network, and will usually result in a number of modifications based on the diagrammer's new perspective of the total project network.

As soon as the estimates are completed, a simple hand computation of the forward pass should be made (this results in the earliest activity start and finish times, as explained in the next chapter). This is an important step, for it may reveal errors or the need for further refinement before the preparation of the final working draft of the project network.

### Conventional Assumptions

The time estimate to be made for each activity is the mean or average time the activity should take, and the estimate is called the *activity duration*. This term is employed to imply that elapsed time of the activity expressed in units such as working days, rather than a measure of effort expressed in units such as man-days. Units other than working days, such as hours or weeks, may be utilized, provided the unit chosen is used consistently throughout the network. Estimates of activity duration do not include uncontrollable contingencies such as fires, floods, strikes, or legal delays. Nor should safety factors be employed for such contingencies.

In estimating an activity's duration time, the activity should be considered independently of activities preceding or succeeding it. For example, one should not say that a particular activity will take longer than usual because the parts needed for the activity are expected to be delivered late. The delivery should be a separate activity, for which the time estimate should reflect the realistic delivery time.

It is also best to assume a normal level of manpower, equipment, or other resources for each activity. Except for known limitations on resources that cause some activities to be resource dependent (discussed in Chapter 2), do not attempt to account for possible conflicts between activities in parallel that may compete for the same resources. These conflicts will be dealt with later, after the scheduling computations have been made.

### Fixed-Time Activities

Certain project activities require fixed time periods beyond the control of management. Examples include: legal minimums of 30 or 90 days for advance notice for public hearings or other events, advance deadlines of media for receipt of advertising copy, technical minimums such as the time to cure concrete or to transport equipment by ship or barge. Often such activities become part of the critical path and thus are important factors in the total project duration.

What can be done about such fixed-time activities? Of course, the most important step is to recognize these time periods and deadlines early in the planning phase of the project, so that the preparatory steps are scheduled to avoid missing the deadlines. Network planning methods help to identify such deadlines early and to plan properly to meet them.

Another consideration is to make sure that as much as possible can be accomplished in parallel with fixed-time activities rather than in series with them. This may help reduce the total length of time on the paths containing the fixed elements. In some cases hard, creative thinking may even uncover ways to reduce or eliminate "fixed-time" activities. For example, a new product advertising campaign might be restructured to reduce dependence on long lead time monthly publications, while shifting emphasis to weekly publications, radio and television spots, or other media that can accept copy on shorter notice. Considered early enough, even legal and technical limits can be modified sometimes by redesigning the product or the project procedures, if the schedule imperatives warrant such fundamental revisions.

### Accounting for the Weather

In construction projects the weather is one of the greatest sources of scheduling uncertainty. In a single-estimate system, there are two common approaches for taking the weather into account.

The first approach is to omit consideration of the weather when estimating the duration of each activity, and instead, estimate the total effect of weather on the project's duration. For example, suppose a project's duration is computed to be 200 working days. Consideration is now made of the seasons in which the outdoor work will be done, the seasonal temperatures and precipitation in the region, the type of soil, type of construction, and other weather-related factors. It may be estimated that five weeks would be lost because of bad weather. Thus the total project duration would be increased to about 225 working days. However, this approach is no different from the usual method of accounting for the weather in construction estimating. It does not take advantage of the detailed breakdown of activities afforded by the project network.

The second approach involves the consideration of weather effects in making each activity time estimate. In this approach each activity is evaluated as to its weather sensitivity—excavation work being sensitive to rain, concrete work sensitive to freezing, interior plumbing not weather sensitive, etc. Suppose an activity is estimated to require ten man-days, and two men will be placed on the task; the nominal time estimate will be five working days. The weather sensitivity of this activity, the season, and other weather factors may indicate that this activity's estimate duration should be increased about 20 per cent. Thus, the adjusted time estimate is six working days.

The initial scheduling calculations may also provide data useful in dealing with possible weather delays. Activities with relatively large amounts of float may be scheduled early or late in order to avoid bad weather seasons. (The definition and uses of float are treated in Chapter 4.)

The advantage of this detailed approach to weather adjustments is that it applies the adjustments to particular portions of the network, which will result in a more accurate schedule for each activity with reference to calendar dates. A disadvantage is the need to add more notations to the activity descriptions. Activity descriptions, including time estimates, must be clearly understood by all persons expected to work with them. This means that both nominal and weather-adjusted estimates should be noted on each activity; in some cases, it is desirable to add the man-day estimate as well. With these notations, on activity may appear as in Figure 3-1, where *MD* = man days, *WD* = working days. A legend on the network should explain the notations, including the fact that the number standing alone represents the weather-adjusted estimate in working days.

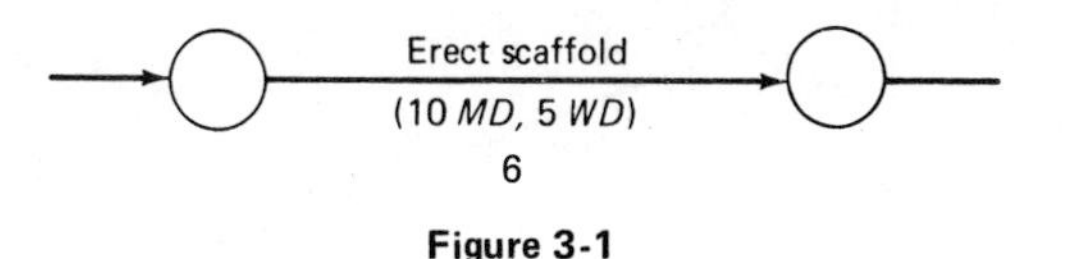


Figure 3-1

### Accounting for Weekend Activities

The use of working days, which is a common time unit in construction projects, results in computations of *project duration* which assume that no activities proceed on weekends and holidays. However, this may be incorrect. For example, concrete may be cured and buildings may be dried out over nonworking days. In such cases, time estimates in working days tend to result in an overestimate of the project duration. When activities of this type are expected to take longer than 5 calendar days, the overestimate can be corrected to a certain extent by a suitable adjustment of the working-day estimate. For example, a curing requirement of 6 or 7 days can be estimated as 5 working days (assuming a five-day work week). A curing requirement of 5 days, however, may actually take 5 working days, although it would likely run over a weekend and thus consume only 3 or 4 working days; in such instances, the estimator should employ the project network to judge the likelihood of curing over the weekends and adjust the working day estimate accordingly.

All time estimates in a network must be based on the same number of working days per week. For activities that will deviate from this standard, adjustments must be made in the time estimates similar to the adjustment for curing activities mentioned above.

Most commercial computer programs for network processing include calendar dating features which automatically convert working-day schedules into calendar dates, including allowances for weekends and holidays.

## ACTIVITY REDEFINITION

As in networking, the proper application of time estimates depends primarily on judgment and experience with critical path techniques. Illustrated here are some of the common networking problems uncovered when one is attempting to make time estimates. In most cases the problems involve activity definitions and the question of detail. Alternative solutions to each problem are discussed.

### Activities in Parallel

In the situation illustrated in Figure 3-2a several interrelated activities may begin and end at approximately the same nodes.

Suppose that upon supplying the time estimates, however, it is realized that *A* cannot begin until *B* has been underway for one day, and will not be completed until one day after *B* is completed. Furthermore, *C* can be accomplished at any

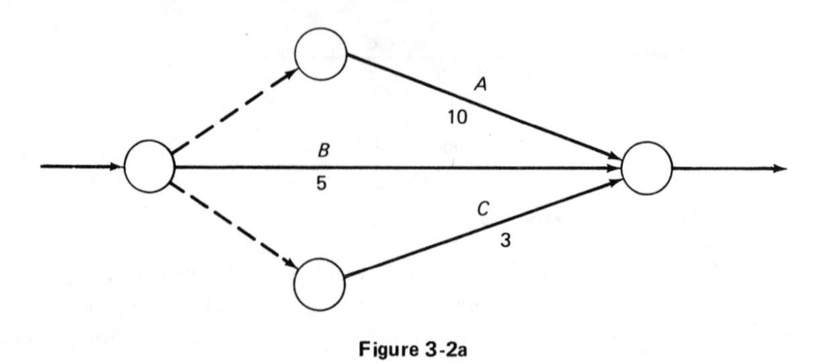


Figure 3-2a

time *B* is going on. A practical illustration of interrelationships of this type is the work prior to pouring a slab of concrete in a building, where the steel, mechanical, and electrical trades may begin and end their work at different times, but during much of the time they are all working in the same area.

To network situations of this type accurately, using the basic arrow scheme, we must redefine the activities. We recall that an activity is any portion of the project that may not begin until other portions are completed. Using the basic arrow scheme, then, and the conditions prescribed above, we may correct Figure 3-2a as shown in Figure 3-2b.

Note the use of percentages to define activities in this illustration. This is occasionally a useful device, but frequent use of percentages is not good practice. They often represent arbitrary definitions which can lead to misunderstandings by subcontractors and others involved in interpreting the network. Where possible, it is better to use physical measures, such as yards of concrete poured, the number of columns formed, the specific items assembled, etc.

Another approach to the correction of the activities-in-parallel problem would be to condense them into a single activity. However, the single activity should have a time estimate representing the total time for the completion of all three activities, a time estimate which is most accurately obtained from a detailed solu-

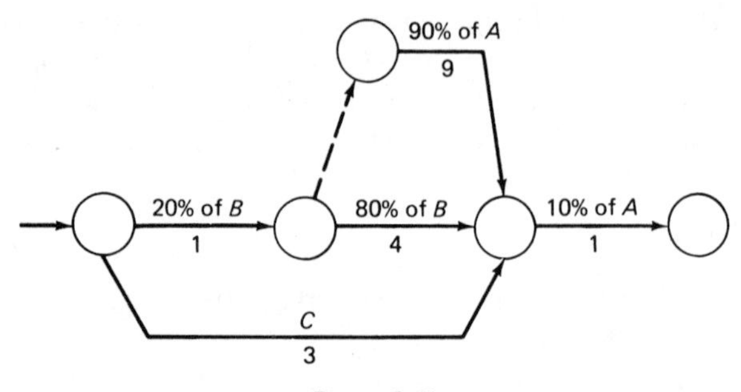


Figure 3-2b

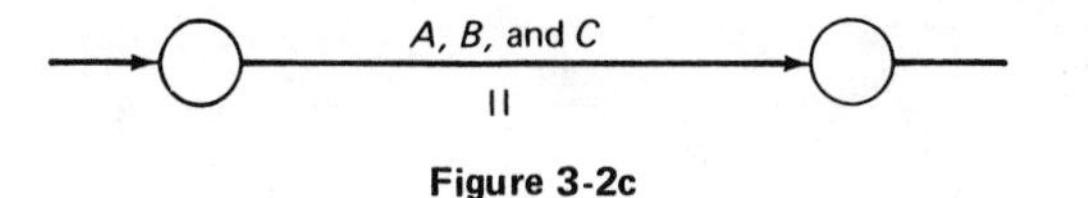


Figure 3-2c

tion as given in Figure 3-2b. From this figure it is clear that the total time required for all three activities is 11 days. Thus the condensed activity may be represented as shown in Figure 3-2c.

Whether 3-2b or 3-2c is the "best" solution to the problem depends on the project, the network objectives, the areas of responsibility involved, and other factors which can be resolved only through the judgment of the project manager. This is essentially the problem of the level of detail, which will be discussed further. It should be noted here, however, that accuracy and detail are directly related, and even when less detail is desired in the final draft of the network, it is often useful in time estimations to sketch certain portions of the network in greater detail.

Some project management systems use the condensation method shown in Figure 3-2c in the computerized model of the project, while the supervisor in the field monitors the actual progress of the work with a hand-processed network in more detail (of the type shown in Figure 3-2b).

These activities-in-parallel situations, or "overlapping activities," occur frequently in construction projects and in certain other kinds of work. The precedence diagramming method introduced briefly in Chapter 2 is designed especially for representing overlapping activities and should be considered when these relationships are expected to occur frequently in the project. To compare the method, let's first illustrate the sample problem in the activity-on-node format. See Figure 3-2d. In this case there is not much difference in network efficiency or clarity; there are still five defined activities.

Now consider the problem represented in a precedence diagram, Figure 3-2e. (To analyze this network, review the precedence relationships defined and illustrated in Figures 2-25 and 2-26.) Note that Figure 3-2e shows only the three

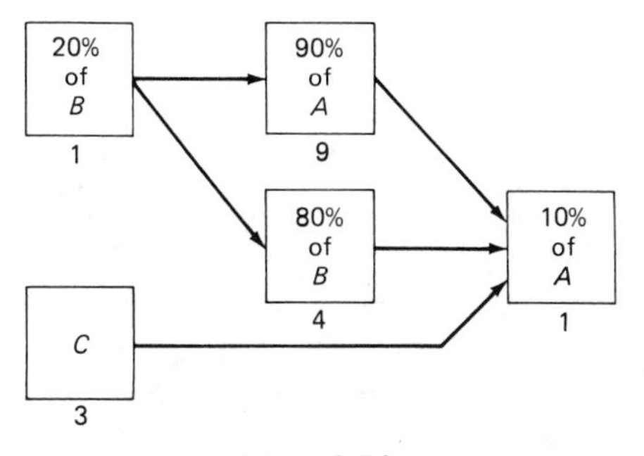


Figure 3-2d

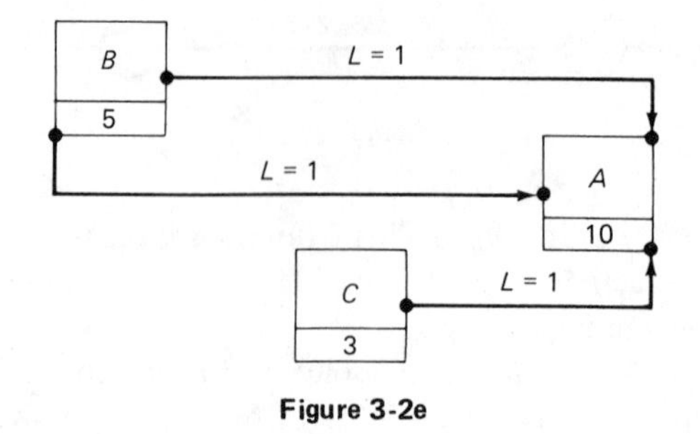


Figure 3-2e

basic activities, and their overlapping relationships are displayed by means of the dependency arrows and lag-time notation. The main feature of the precedence notation is that the activities are not subdivided on the network, which is considered an advantage for readability. However, the dependency relationships are now more involvled. For some people, the precedence diagram may be more difficult to master. (Comparison of scheduling computations for the arrow, node, and precedence methods is presented in the next chapter.)

### Activities in Series

Let us look at another problem of network accuracy, this one arising in a portion of a network in which the activities are drawn in series, as shown in Figure 3-3a. Suppose that upon inspecting this network it was realized that *A* and *B* did not require a total of 17 days. Actually part of *B* could begin at least 2 days before *A* was completed. (This is similar to the previous problem, since we are saying that *A* and *B* are partly concurrent.) An erroneous diagram of this type may be corrected in several ways. One way would be to split *A* at the point that *B* begins (Figure 3-3b). Another way would be to simply absorb the completion of *A* in *B*, making sure that the activity descriptors were clear (Figure 3-3c). A third way, of course, would be to condense *A* and *B* (Figure 3-3d).

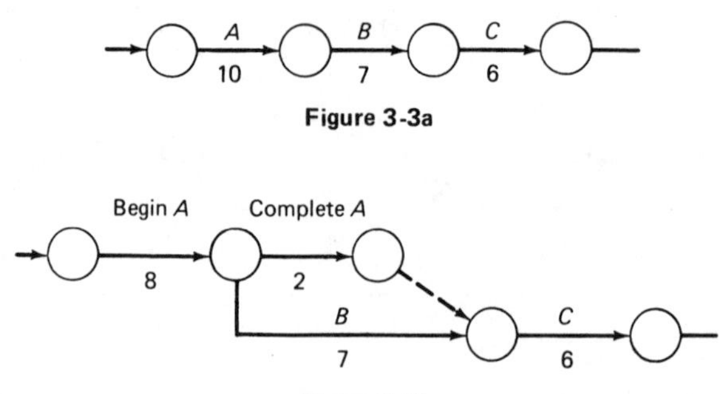


Figure 3-3a

Figure 3-3b

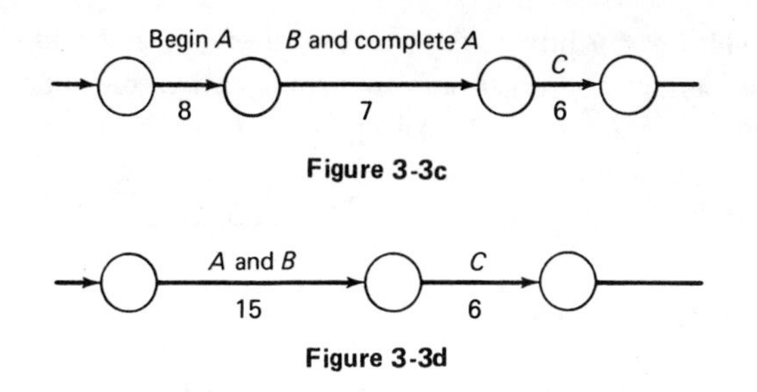


Note that in each of the alternative solutions illustrated in Figures 3-3b, 3-3c, and 3-3d, the total elapsed time for $A$ and $B$ is shown correctly as 15 days. Again, the choice of a solution depends on what one wishes to illustrate and control, and on such factors as the magnitude of the times involved and the feasibility of defining the activity segments clearly.

### Practical Example

The problems discussed above arise repeatedly in practical efforts to draw accurate networks. Therefore, it is worthwhile to review the points made in the form of a small practical example. Consider the project of preparing a technical report. The major activities include the original writing, calculations, and chart sketches done by an engineer, typing and graphics work done by technical support personnel, and final printing and binding done in the reproduction department. The engineer might first draw a network and add time estimates as shown in Figure 3-4a.

The author's reaction to this network might be that 31 days is too long. Her deadline might be only 28 days away. Looking at her report outline, she may

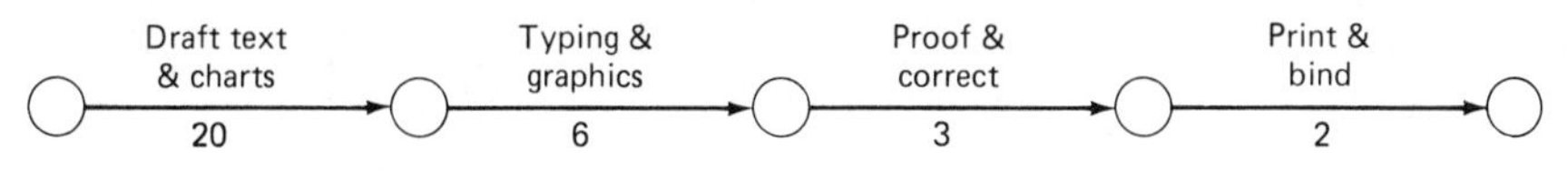

**Figure 3-4a** First network of report project. Total project duration is 31 days.

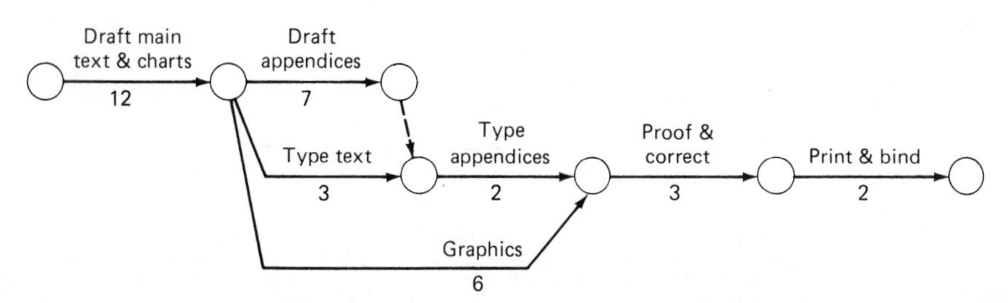


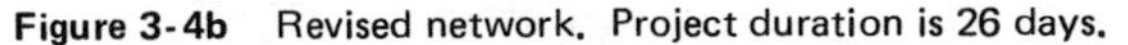

**Figure 3-4b** Revised network. Project duration is 26 days.

note that it divides easily into two parts: (1) the main text and charts, and (2) the technical appendices. Consulting the typing and graphics departments, she finds that they can readily handle the report in those two parts. Thus she revised the network as shown in Figure 3-4b. The resulting total time of 26 days allows for 2 days of errors in time estimates or unexpected delays.

## NETWORK CONDENSATION

In the foregoing illustrations the concepts of condensing and expanding networks are introduced for the purposes of improving accuracy, eliminating excessive detail, and to achieve other objectives in the development of the detailed network. There are also occasions in which it is desired to produce a summary network for review by top management, which calls for the same condensation concept illustrated above, except that it is applied on a broad basis throughout the network, with the purpose of developing a general condensation and summarization of the project plan. In practice this often means that a network of several hundred activities must be reduced to one of a few dozen activities, without distortion of the logic, such that a summary picture of the project may be presented for review by top management, a customer, or other interested audiences. Consequently, some points related to condensation procedures are worthy of attention.

In general, a safe rule of condensation is that groups of activities independent of other activities may be condensed without distorting the network logic. For example, consider Figure 3-5a. In this network there are three independent activity groups that may be condensed, as shown in Figure 3-5b. Note that some

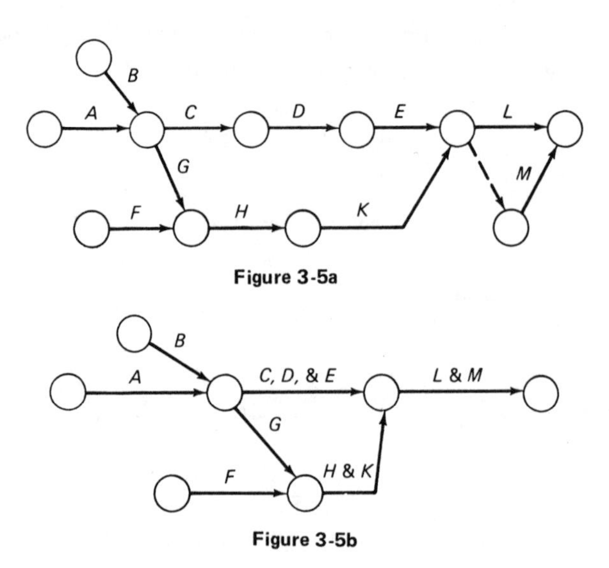


Figure 3-5a



Figure 3-5b

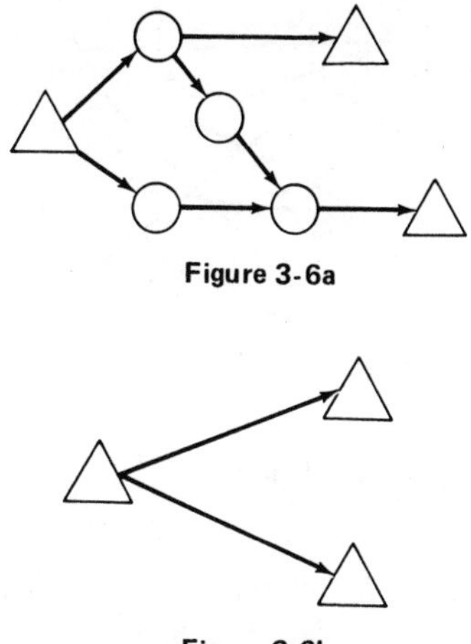

Figure 3-6a

Figure 3-6b

activities in series have been combined, and in one case two activities in parallel, $L$ and $M$, were combined. *But all the dependency relationships in the original diagram still hold.* This is the most important point in condensing networks, for it is very easy to introduce false dependencies. (For a practical example, see Figure 3-9.)

A somewhat different approach to condensation is used by certain computer routines that perform this function. This approach calls for the designation of certain key events in the network which are not to be omited in the condensation procedure. Then all direct and indirect restraints (groups of activities) between each pair of key events are reduced to a single restraint (activity). To illustrate, consider the network of seven activities in Figure 3-6a. The triangles denote the selected key events. Using the condensation process described above, this network would be reduced to the two activities shown in Figure 3-6b.[2]

This particular procedure is vulnerable to the occurrence of "pathological cases," in which the number of activities is not reduced or may even be increased. Using the procedure on the network in Figure 3-7a, for example, pro-

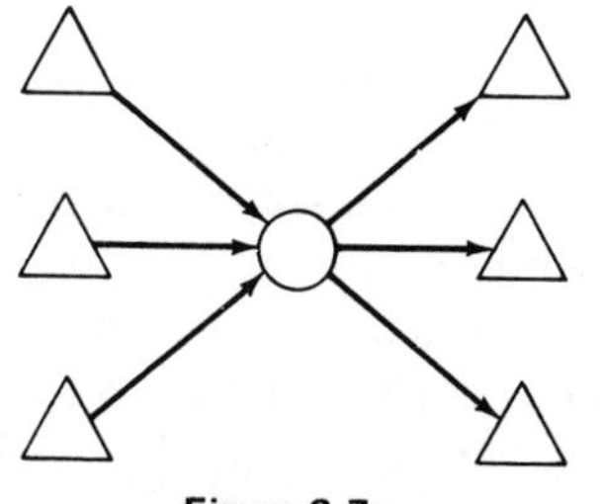

Figure 3-7a

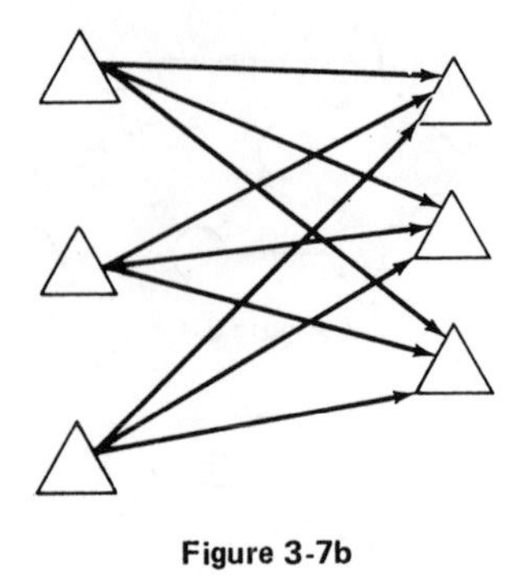

Figure 3-7b

duced the network in Figure 3-7b. However, by selecting key events with this possibility in mind, one can avoid most pathological cases.

## THE LEVEL OF DETAIL

Thus far, comments on the problem of the level of detail in a network have been associated with questions of accuracy and economy of the presentation. There are many other factors involved in determining the most appropriate level. In considering any particular activity or group of activities with regard to expanding, condensing, or eliminating it, the diagrammer may ask himself several questions to guide his decision:

1. Who will use the network, and what are their interests and span of control?
2. Is it feasible to expand the activity into more detail?
3. Are there separate skills, facilities, or areas of responsibility involved in the activity, which could be cause for more detail?
4. Will the accuracy of the logic or the time estimates be affected by more or less detail?

Clearly, these questions are only guides to the subjective decision that must be made in each case. Generally, after working with one or two networks, a person will develop a sense for the appropriate level of detail.

That there are no firm rules that may be followed in determining level of detail is illustrated by the following hypothetical case. The project is the construction of a house. If the network rules of Chapter 2 are followed, one could prepare a complete network, as shown in Figure 3-8a, or one could take a more detailed approach, as indicated in the portion of the network shown in Figure 3-8b.

These appear to be clear examples of too little and too much detail. But suppose the house is one of a hundred identical ones in a large housing project. Three activities per house would thus result in a network of 300 activities, plus

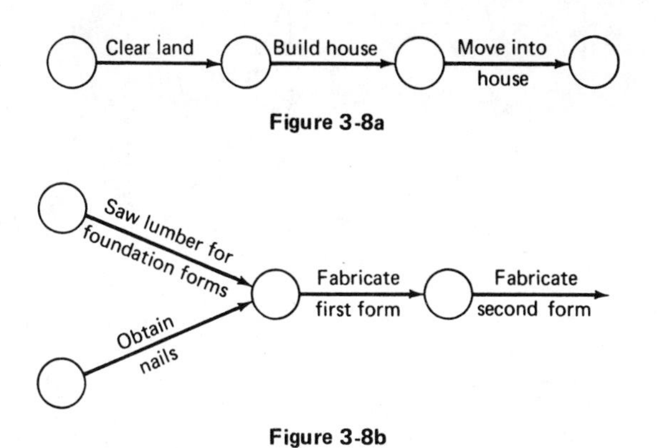


Figure 3-8a

Figure 3-8b

other activities for roads, utilities, etc. Such a broad network may be very useful in analyzing the over-all length of the project, the most desirable sequence of construction, and other problems of general planning. Furthermore, since the house construction in this case is a matter of mass production, it would be worthwhile to work out the construction schedule for one house in considerable detail, for any bottlenecks in the schedule for one house would cause repeated delays in all houses. Thus the detailed network treatment for a typical house might well be justified under these conditions. Under most other conditions the approaches illustrated above would indeed represent a useless extreme on the one hand and an expensive, perhaps impractical extreme on the other.

### Cylical Networks

In the house building case cited above, it was stated that both detailed and condensed networks may be useful if a number of identical houses are to be built. It may be generally stated that whenever a project involves a number of cycles of a group of activities, one should consider (1) developing a detailed network of the group, (2) condensing the detailed network into a summarized version, and (3) using the condensed network in the cycles that compromise the total project network. The purposes of the detailed network are to develop an efficient plan for the group of activities that will be repeated and to derive accurate time estimates for the condensed version. The purpose of the condensed version is network economy, since repetition of the detailed network would be costly in drafting time and would unnecessarily complicate and enlarge the total project network. Project types to which this principle would apply include multistory buildings, bridges, pilot production of a group of missiles, and a series of research experiments.

An example of the application of detailed and condensed networks in a multistory building is shown in Figure 3-9. Here the contractor worked out the de-

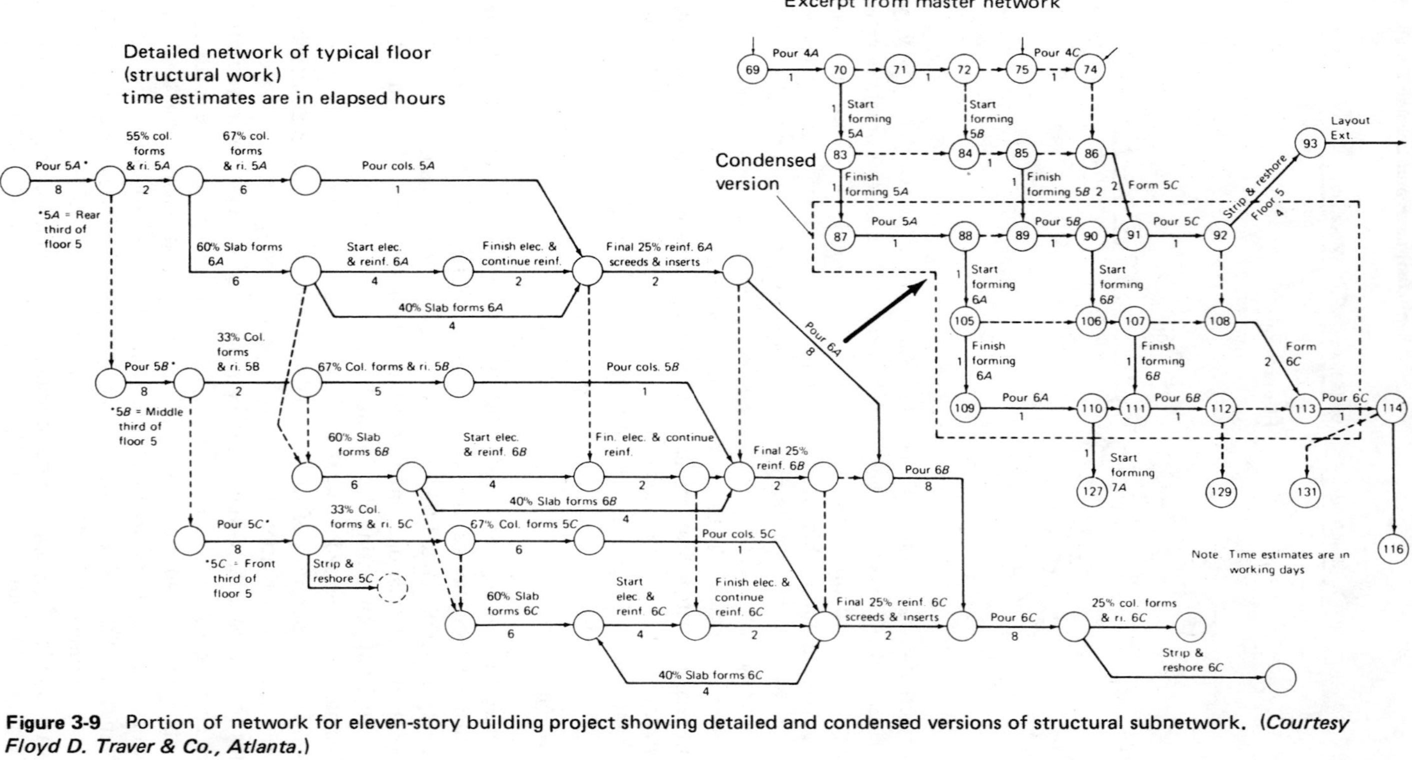


**Figure 3-9** Portion of network for eleven-story building project showing detailed and condensed versions of structural subnetwork. (*Courtesy Floyd D. Traver & Co., Atlanta.*)

tailed network using a time unit of hours in order to balance the crews and minimize delays in the structural work on a floor. Then the condensed version was repeated for each floor of the building in the total project network.

Figure 3-9 also illustrates that large, complex networks are sometimes difficult to read. This is not a trivial matter, since one of the primary purposes of networks is to provide a more readable format for the project plan. An effective approach to improving network readability is to provide large labels for major sections of the network. An example is shown in Figure 3.10.

## ARROW VERSUS PRECEDENCE METHOD

Up to this point we have demonstrated network methods by using the arrow scheme as the basic tool, then showing how the node and precedence schemes compare under certain conditions.

One of the more thorough comparisons of arrow and precedence diagramming was prepared by Keith C. Crandall of the University of California, Berkeley. Professor Crandall developed an example which we reproduce here, courtesy of the author and the publisher.[3]

The basic problem is to pour three consecutive floor slabs in a multifloor building. The problem requirements are as follows:

A. Complete the pour of floor slabs for three floors with the following constraints:
   1. Complete the floors in order so that the crew for each activity is shown to move from one floor to the next, with the exception of pouring which is described below.
   2. Start by placing bottom reinforcing steel. Once adequately underway with bottom steel, start mechanical and electrical rough-ins. Place the top reinforcing mesh once the rough-ins are sufficiently advanced. When complete allow one day for inspection and pour preparation. The pour completes the work on a given floor.

B. Lead/Lag constraints:
   1a. Mechanical rough-in can start once the bottom reinforcing is two days underway.
   1b. Mechanical rough-in can be completed no sooner than one day after the bottom reinforcing.
   2. Electrical rough-in can start three days after bottom reinforcing, yet cannot be completed until two days after bottom reinforcing.
   3. Top reinforcing mesh can start once mechanical and electrical rough-ins are two days underway. Top mesh must be completed two days after the completion of mechanical rough-in and cannot be completed until the same day as electrical rough-in.

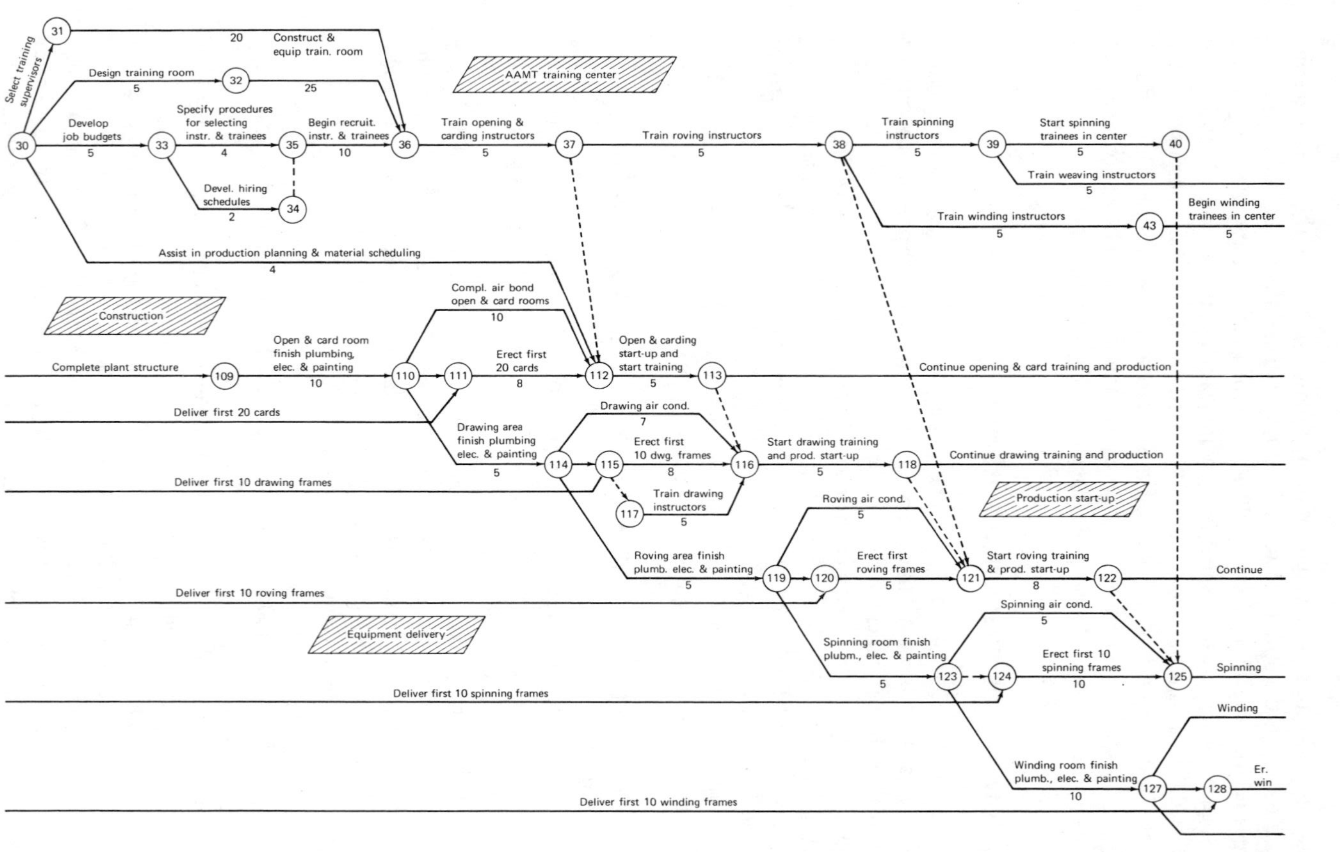


**Figure 3-10** Portion of a network for a new plant start-up project showing labels for major sections of the project. (*Courtesy Textile Industries.*)

4. The pours can be overlapped as long as one day of pour has been completed on the previous floor.

C. Estimated durations for each floor.

| | |
|---|---|
| 1. Bottom reinforcing | 5 days |
| 2. Mechanical rough-in | 10 days |
| 3. Electrical rough-in | 3 days |
| 4. Top mesh | 3 days |
| 5. Delay for inspection | 1 day |
| 6. Pour | 2 days |

Let us first prepare an arrow network of this problem, taking just the activities in the first slab. To illustrate the time relationships more clearly in this case we first use a time-scaled format with a separate arrow for each activity. To show the dependency relationships we add dummy arrows labeled to correspond with the lead/lag constraints given in the problem. See Figure 3-11. The fact that some of the dummy arrows must go backward against time reveals that this time-scaled diagram is not feasible as shown and that several of the activities will need to be delayed (shifted to the right on the network).

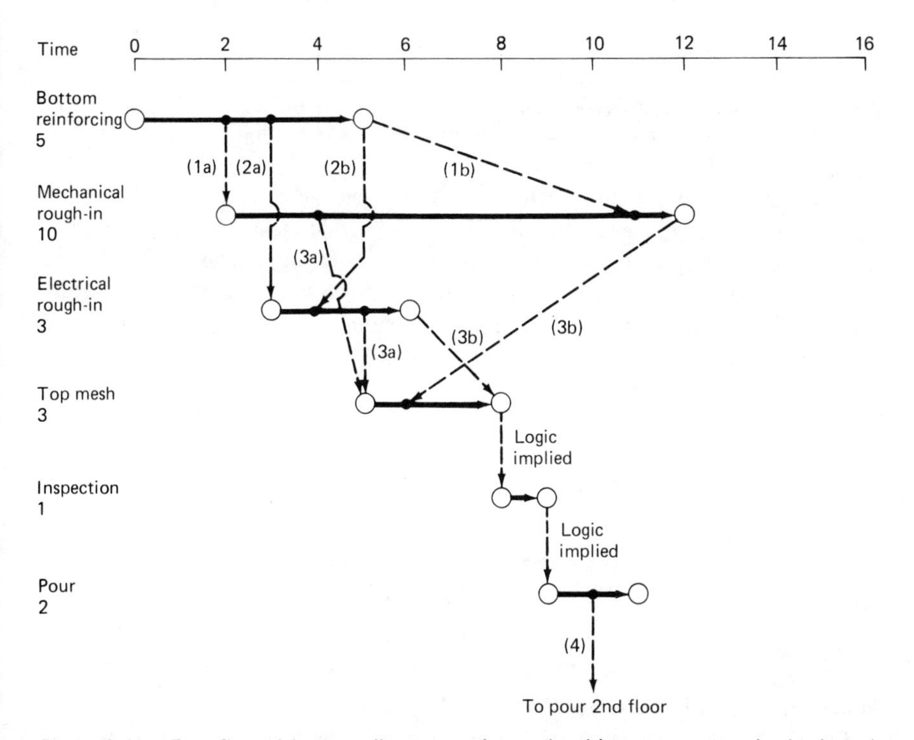


**Figure 3-11** First floor slab arrow diagram on time scale with some arrows going backward against time.

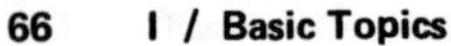

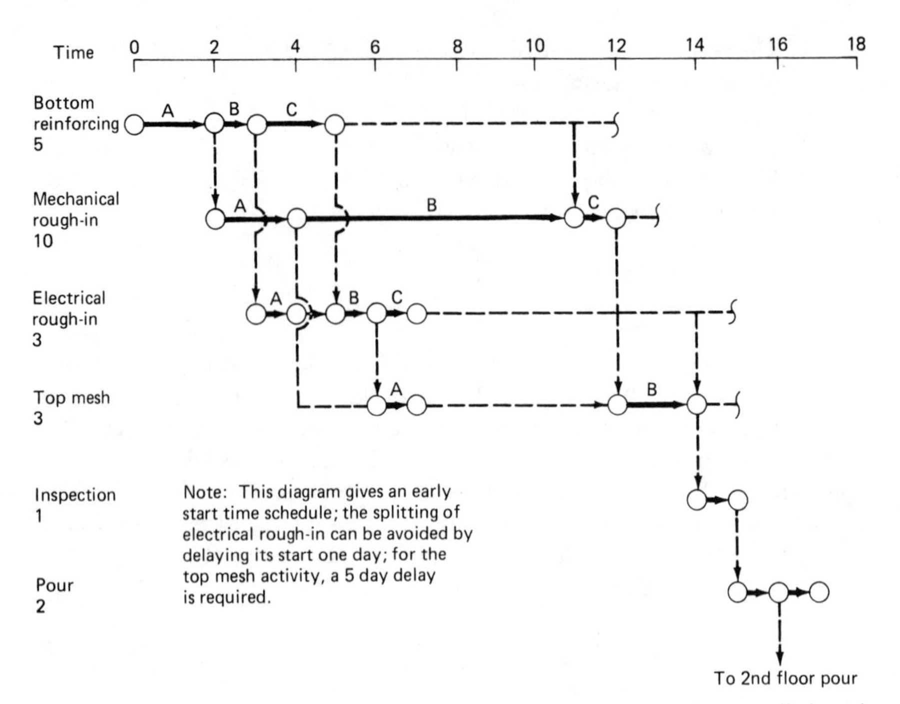


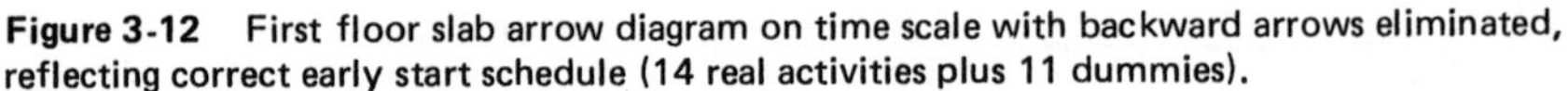
**Figure 3-12** First floor slab arrow diagram on time scale with backward arrows eliminated, reflecting correct early start schedule (14 real activities plus 11 dummies).

Now, let us draw a feasible arrow network, again on a time scale. Using the methods described earlier in this chapter, we subdivide the activities as necessary in order to properly represent the dependency relationships. The result is Figure 3-12. The backward arrows are eliminated and we have a correct early start schedule. However, the activities have been subdivided to the extent that the network has become somewhat complex and tedious to read. It contains 14 real activities and 11 dummies.

Now, let us go back to the original problem statement and network it using precedence diagramming. Figure 3-13 accomplishes the purpose with 5 activities and 9 dependency arrows. In this notational scheme the dependency arrows are coded as follows:

$SS$ = start-to-start constraint
$FF$ = finish-to-finish constraint
$FS$ = finish-to-start delay

The reader should analyze this network for consistency with the problem statement and with the arrow diagram in Figure 3-12.

Clearly the precedence diagram is easier to read with respect to the activities

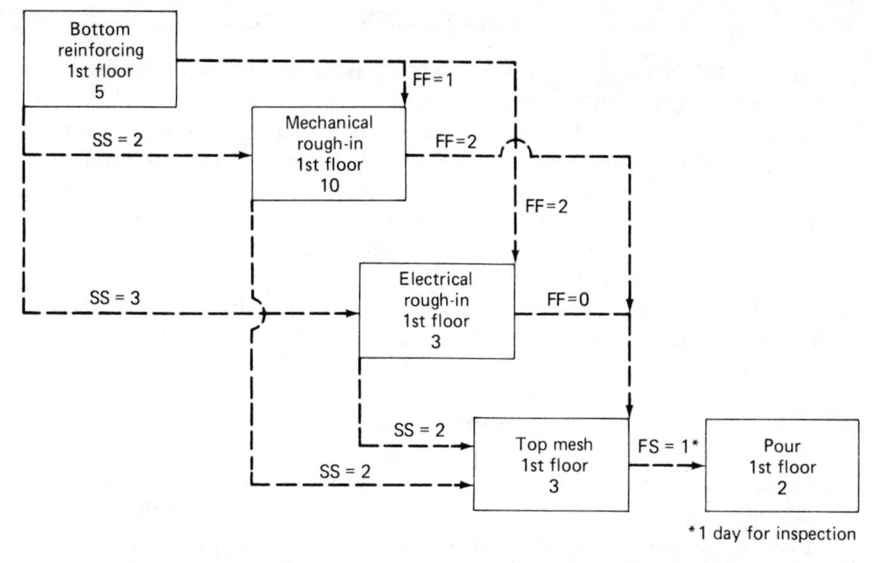


**Figure 3-13** First floor slab precedence diagram. (5 real activities plus 9 dependencies) conventional dependency artwork.

themselves, as compared with the subdivided activities in the arrow diagram of Figure 3-12. However, the dependency relationships in the precedence scheme are more complex and may be more difficult to understand in practice.

In selecting which networking format to employ, one should consider the types of activity relationships involved in the projects being considered, the people who will be reading the networks, and the computational methods available (covered in Chapters 4 and 11).

## SUMMARY

Although the preparation of the network is only the first phase in applications of critical path methods, many users have reported that the greatest benefits from the critical path concepts are derived from this phase alone. They felt that preparing the network caused them to think through the project in a more complete manner than ever before, forcing them to do a more thorough job of advanced planning. However, a great deal of useful information is included in the completed network, and the proper processing and utilization of this information as described in the following chapters can bring important additional benefits not only to the project manager but also to the subcontractors and all other groups engaged in the project effort.

## REFERENCES


1. Davis, E. W., "CPM Use in the Top 400 Construction Firms," *Journal of the Construction Division*, ASCE, March 1974.
2. Seelig, W. D., and I. M. Rubin, "The Effects of PERT in R&D Organizations," published as a Working Paper of the Research Program on the Management of Science and Technology, No. 230-66, Alfred P. Sloan School of Industrial Management, Massachusetts Institute of Technology, December 1966.
3. Crandall, Keith C., "Project Planning with Precedence Lead/Lag Factors," *Project Management Quarterly*, June 1973.
4. Mark, E. J., "How Critical Path Method Controls Piping Installation Progress," *Heating, Piping, and Air Conditioning*, September 1963, pp. 121–126.


## EXERCISES

1. Redraw the network in Figure 3-4b, using an activity-on-node scheme. What advantages and disadvantages of the node scheme do you see in this example?
2. Redraw the detailed network of typical floor shown as part of Figure 3-9, using an activity-on-node scheme. Then condense the network into a node format containing 12 or fewer activities. Compare with the condensed version in Figure 3-9.
3. Condense the network in Figure 3-14 to 10 activities or less, without distorting the dependency relationships.

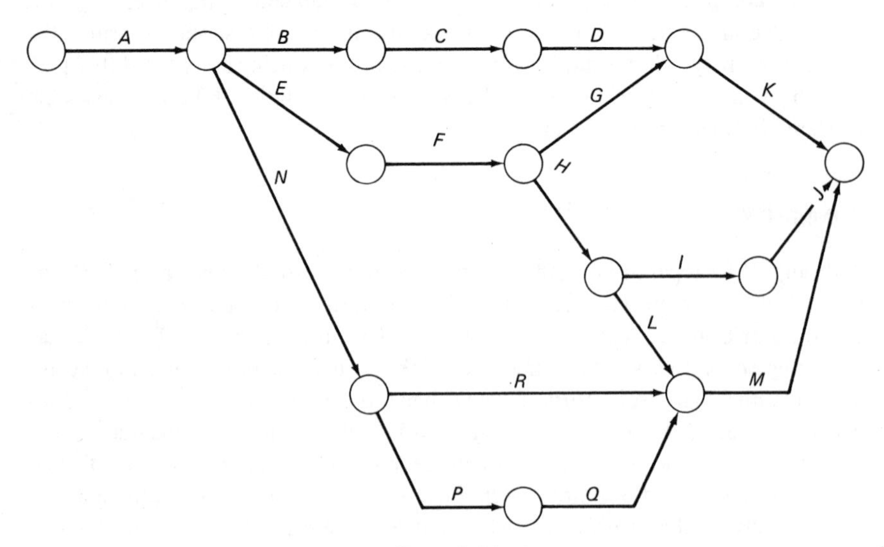


Figure 3-14

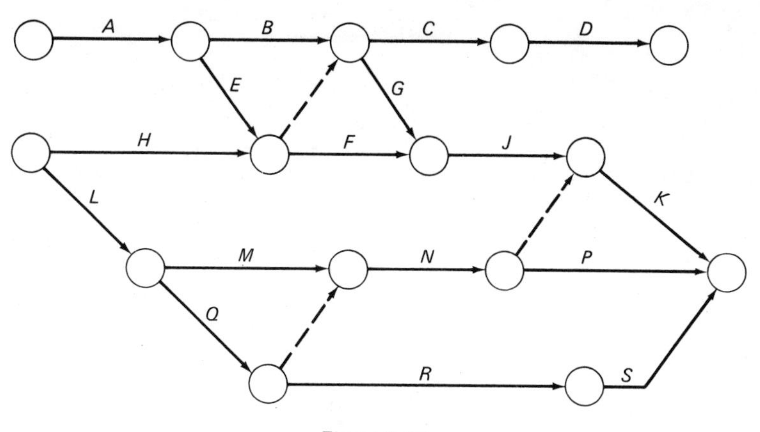


Figure 3-15

4. Condense the network in Figure 3-15 without distorting the logic.
5. Considering Chapters 2 and 3 only, and the initial planning phase of projects (prior to scheduling computations), what planning techniques can reasonably be used to shorten the total project duration?
6. Considering only the material covered in Chapters 2 and 3, list the ways in which network methods can be used to improve communications among project personnel, including management, supervisors, subcontractors, suppliers, and the owner or end-user of the project results.
7. Some persons who supervise part of the work in a project, such as department heads or subcontractors, may inflate their time estimates somewhat in order to more easily stay on schedule and reduce attention and pressure by the general manager of the project. How will the use of network methods and scheduling computations tend to cause this tactic to boomerang?
8. The network shown in Figure 3-16 is an early application of critical path methods by the Pure Oil Company, which was published in 1963.[4] Review the network and answer the following questions:
   a. Assuming a computer was used to process this network, could any of the dummies be eliminated without distorting the logic of the network?
   b. If hand computations are used, which dummies could be eliminated?
   c. What do you suppose the diagrammer accomplishes by the use of the "activity reference number"?
   d. Note that the delivery of certain items, such as truck rack equipment (activities 11-27, 11-28), are included in the network. Why do you suppose the diagrammer did not include the delivery of all items, such as concrete, pumps YP-1 and YP-2 (activities 13-16, 14-15), and insulation (activities 36-37, 77-78), etc?
   e. How could the readability of this network be improved?

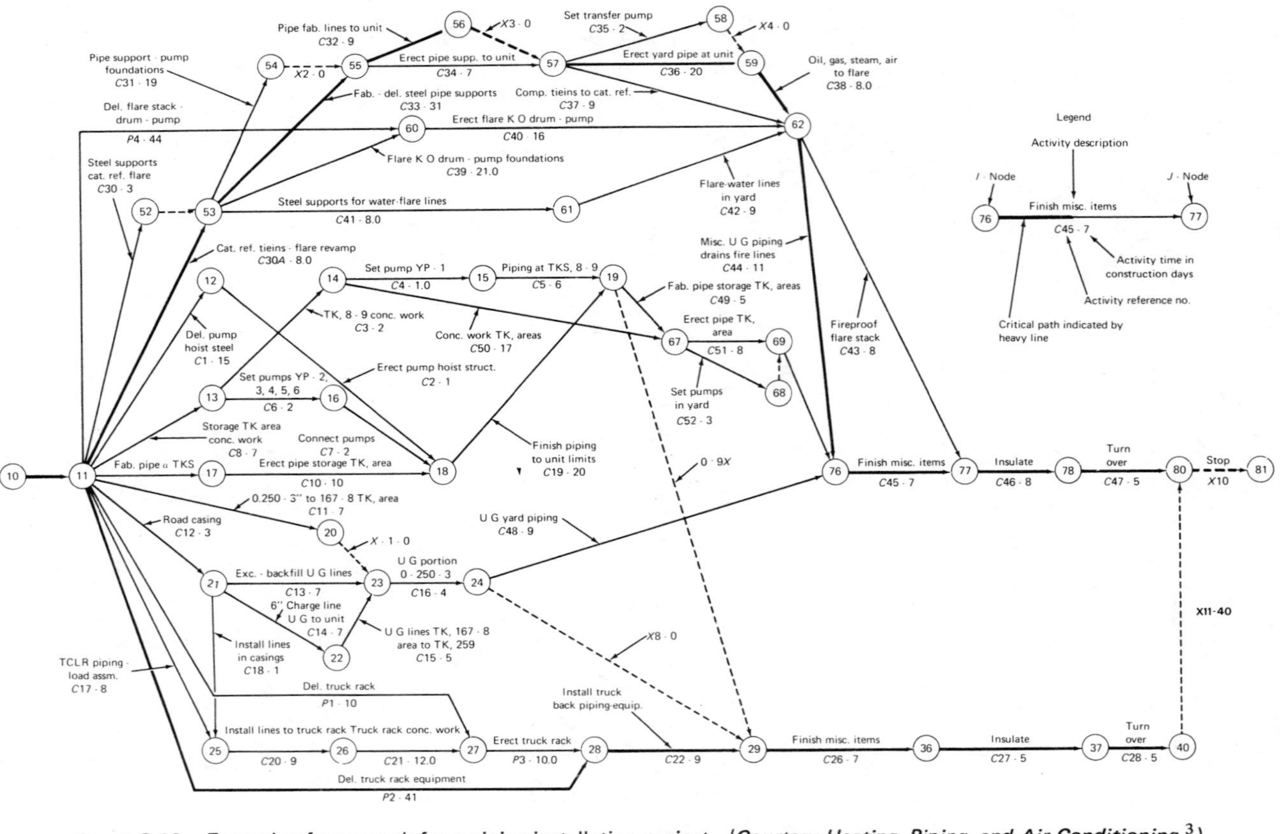


**Figure 3-16** Example of a network for a piping installation project. (*Courtesy Heating, Piping, and Air Conditioning.*[3])

9. Discuss how the level of detail used in a network may be influenced by the purpose and readership of the network. Consider the viewpoints of different levels of management, such as: the project superintendent for a high-rise office building; the president of a television manufacturing firm that is about to introduce a new unit for the retail market; a subcontractor for computer programming in a two-year project to develop a management information system for a major bank.
10. Figures 3-12 and 3-13 are networks of the first slab of construction problem defined by Crandall. Extend the network to include the second and third slabs, using (a) the arrow method and (b) procedence diagramming.

# 4

# BASIC SCHEDULING COMPUTATIONS

At this stage in the application of critical path methods, the project network plan has been completed and the mean performance times have been estimated for each activity. We now consider the questions of how long the project is expected to take and when each activity *may* be scheduled. Answers to these questions are inferred from the network logic and the estimated durations of the individual activities. These estimates may be based on a single time value, as described in Chapter 3, which is basically the original CPM procedure, or it may be based on a system of three time estimates, as described in Chapter 9, which deals with PERT, the statistical approach to project planning. Regardless of which estimation procedure is used, the scheduling computations described in this chapter are the same, since they deal only with the estimates of the mean activity duration time. Since network logic is the underlying basis of these computations, different procedures are required for arrow and precedence diagrams. It will be seen that the more extensive nature of precedence diagramming logic leads to a more involved computational procedure. An understanding of these basic scheduling computations is essential to proper interpretation of computer outputs, and also to enable one to carry them out by hand when the occasion demands. Readers interested in a more rigorous algorithmic formulation of these computational procedures, essential to writing computer programs, are referred to Appendix 4-1 at the end of this chapter.

All basic scheduling computations first involve a forward and then a backward pass through the network. Based on a specified project start time, the *forward pass* computations proceed sequentially from the beginning to the end of the project giving the *earliest (expected) start and finish times* for each activity, and, for arrow diagrams, the *earliest (expected) occurrence time* for each event. The modifier "expected" is sometimes used to remind the reader that these are estimated average occurrence times. The actual times, known only after the various activities are completed, may differ from these expected times because of deviations in the actual and estimated activity performance times.

By the specification of the latest allowable occurrence time for the completion of the project, the *backward pass* computations also proceed sequentially from the end to the beginning of the project. They give the *latest allowable start and finish times* for each activity, and, for arrow diagrams, the *latest allowable occurrence time* for each event. After the forward and backward pass computations are completed, the float (or slack) can be computed for each activity, and the critical and subcritical paths through the network determined.

As mentioned in Chapter 3, it is often appropriate to adopt one working day as the unit of time, so that the network computations are made in working days, beginning with zero as the starting time of the initial project event. The conversion of these computational results to calendar dates merely requires the modification of a calendar wherein the *working days* are numbered consecutively from a prescribed calendar date for the start of the project. This procedure is discussed further in Chapter 11. For convenience, this chapter will use *elapsed* working days for discussion purposes; it should be understood, of course, that other time units than working days may be used with no changes in the computation procedures. In addition, it is assumed at the start that the project begins at time zero and has only one initial and terminal event. These assumptions will be relaxed later in this chapter.

The basic scheduling computations are also termed "time-only" procedures because resource limitations are not explicitly considered, unless of course, they are built into the network logic. The effect of resource limitations on actual activity schedules will be considered in Chapter 7 which deals with this specific problem.

## COMPUTATION NOMENCLATURE—ARROW DIAGRAMS

The following nomenclature will be used in the formulas and discussion which describe the various scheduling computations; for brevity, the modifier "expected" has been omitted from all of these definitions of time and float. Also, these definitions and subsequent formulas will be given in terms of an arbitrary activity designed as $(i - j)$, i.e., an activity with predecessor event $i$, and successor event $j$.

$D_{ij}$ = estimate of the mean duration time for activity $(i - j)$
$E_i$ = earliest occurrence time for event $i$

$L_i$ = latest allowable occurrence time for event $i$
$ES_{ij}$ = earliest start time for activity $(i - j)$
$EF_{ij}$ = earliest finish time for activity $(i - j)$
$LS_{ij}$ = latest allowable start time for activity $(i - j)$
$LF_{ij}$ = latest allowable finish time for activity $(i - j)$
$F_{ij}$ = total path float (or slack) time for activity $(i - j)$
$AF_{ij}$ = activity (free) float (or slack) time for activity $(i - j)$
$T_S$ = scheduled time for the completion of a project or the occurrence of certain key events in a project.

## FORWARD PASS COMPUTATIONS

To compute the *earliest start and finish times* for each activity in the project, the *forward pass* computations are initiated by assigning an arbitrary earliest start time to the (single) initial project event. A value of zero is usually used for this start time since subsequent earliest times can then be interpreted as the project duration up to the point in question. The computations then proceed by assuming that *each activity starts as soon as possible*, i.e., as soon as all of its predecessor activities are completed. These rules are summarized below.

### Forward Pass Rules—Computation of Early Start and Finish Times

RULE 1. The initial project event is assumed to occur at time zero. Letting the initial event be denoted by 1, this can be written as:

$$E_1 = 0$$

RULE 2. All activities are assumed to start as soon as possible, that is, as soon as all of their predecessor activities are completed. For an arbitrary activity $(i - j)$ this can be written as:

$ES_{ij}$ = Maximum of $EF$'s of activities immediately preceding activity $(i - j)$, i.e., all activities ending at node $i$

RULE 3. The early finish time of an activity is merely the sum of its early start time and the estimated activity duration. For an arbitrary activity $(i - j)$ this can be written as:

$$EF_{ij} = ES_{ij} + D_{ij}$$

The above rules are applied to the simple network shown in Figure 4-1a. In the forward pass section, Figure 4-1b, the initial project event 1 is placed at 0 on the time scale according to the first rule. Starting with $E_1 = 0$, the early start time of activity 1-2 is 0, and the early finish time by Rule 3 is merely

$$EF_{1,2} = ES_{1,2} + D_{1,2} = 0 + 2 = 2$$

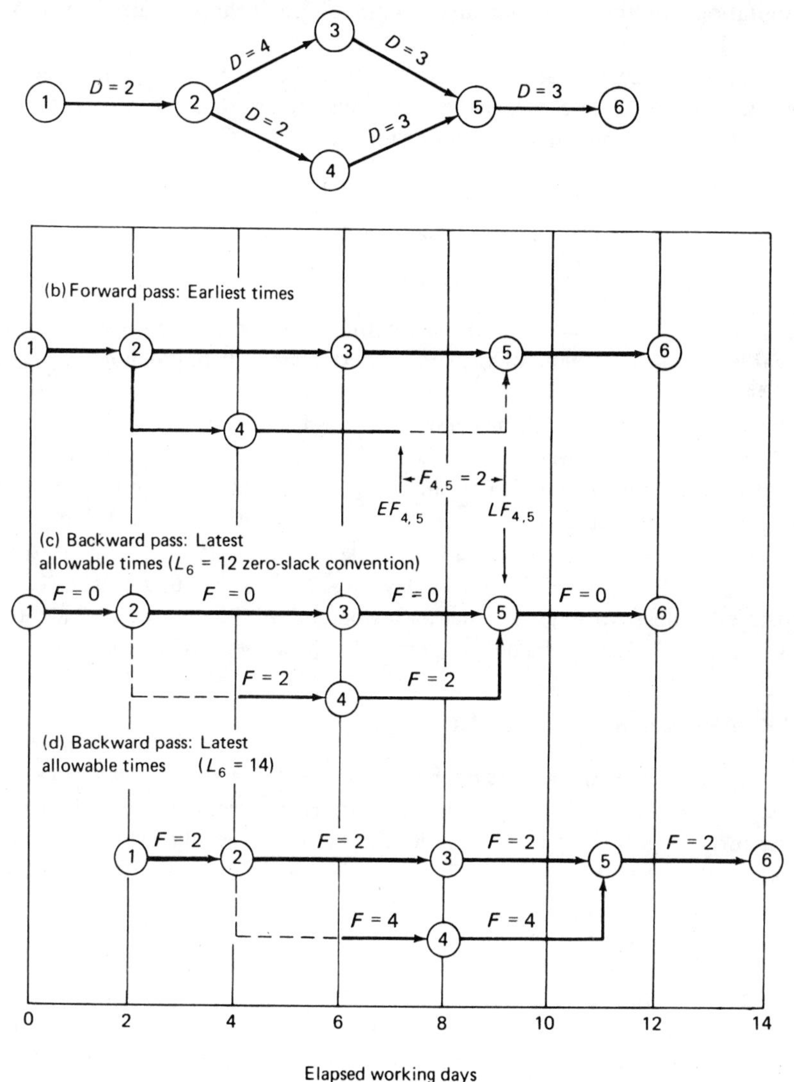


**Figure 4-1** Example of forward and backward pass calculations for a simple network.

The early start and finish times of activities 2-3, 3-5, 2-4, and 4-5 are determined similarly, in that order. Proper sequencing of the activities so that the *EF*'s of all predecessor activities will be available when computing *ES*'s of any follower activity (use of Rule 2) is easily handled in hand computation on networks by working along each path as far as possible, and then "back tracking" to a new path that is then ready for computation. Special procedures required for tabular computations, or the development of computer algorithms is taken up in Appendix 4-1.

The crux of the forward pass computations occurs at the merge event 5, where it is necessary to consider the early finish times for predecessor activities 3-5 and 4-5 to determine the early start time for activity 5-6, i.e.,

$$ES_{5,6} = \text{Maximum of } (EF_{3,5} = 9 \text{ and } EF_{4,5} = 7) = 9$$

Finally, the early finish time of the final network activity is

$$EF_{5,6} = ES_{5,6} + D_{5,6} = 9 + 3 = 12$$

Thus, the early expected finish time for the entire project, corresponding to the earliest occurrence time of the project terminal event 6, is denoted by $E_6 = EF_{5,6} = 12$.

The forward pass network in Figure 4-1b has been drawn to scale on a time base, not only as a convenient means of showing the earliest start and finish times for each activity, but also to show the longest path through the network. This graphical procedure is introduced here for illustrative purposes, but is not practical for routine hand application. In Figure 4-1b, activities 1-2-3-5-6 form the longest path of 2 + 4 + 3 + 3 = 12 days duration. The path consisting of activities 2-4-5 has two days of total path float, as will be discussed below; this path float is indicated by the dashed portion of the activity 4-5 arrow.

## BACKWARD PASS COMPUTATIONS

The purpose of the *backward pass* is to compute the *latest allowable start and finish times* for each activity. These conputations are precisely a "mirror image" of the forward pass computations. First, the term "latest allowable" is used in the sense that the project terminal event must occur on or before some arbitrarily scheduled time, which will be denoted by $T_S$. Thus, the backward pass computations are initiated by specifying a value for $T_S$. If no scheduled date for the completion of the project is specified, then the convention of setting the latest allowable time for the terminal event equal to its earliest time, determined in the forward pass computation, is usually followed, i.e., $L = E$ for the terminal event of the project. This was followed in the initial development of CPM, and will henceforth be referred to as the *zero-slack* convention, or *zero-float* convention.

One result of using this convention is that the float along the critical path(s) is zero, while the float along all other paths is positive. When an arbitrary scheduled date is used for the project terminal event, the float along the critical path may be positive, zero, or negative, depending on whether $T_S$ is greater than, equal to, or less than, respectively, the earliest occurrence time for the terminal event. The zero-float convention has an additional useful property in that the *latest allowable activity finish time gives the time to which the completion of an activity can be delayed without directly causing any increase in the total time to complete the project.*

The zero-float convention is adopted in the illustrative example shown in Figure 4-1c, where the terminal event 6 is placed at time 12, i.e., $L_6 = E_6 = 12$. The latest allowable finish time for activities other than the final activity(s) are then determined from network logic, which dictates that an activity must be completed before its successor activities are started. Thus, the latest allowable finish time for an activity is the smallest, or earliest, of the latest allowable start times of its successor activities. Finally, the latest allowable start time for an activity is merely its latest allowable finish time minus its duration time. These rules are summarized below.

### Backward Pass Rules—Computation of Latest Allowable Start and Finish Times

RULE 1. The latest allowable finish time for the project terminal event ($t$) is set equal to either an arbitrary scheduled completion time for the project, $T_S$, or else equal to its earliest occurrence time computed in the forward pass computations.

$$L_t = T_S \quad \text{or} \quad E_t$$

RULE 2. The latest allowable finish time for an arbitrary activity ($i - j$) is equal to the smallest, or earliest, of the latest allowable start times of its successor activities.

$$LF_{ij} = \text{Minimum of } LS\text{'s of activities directly following activity } (i - j)$$

RULE 3. The latest allowable start time for an arbitrary activity ($i - j$) is merely its latest allowable finish time minus the estimated activity duration time.

$$LS_{ij} = LF_{ij} - D_{ij}$$

These rules are applied in Figure 4-1c labeled Backward Pass. Starting with the final project event 6, we see that according to the zero-float convention, it is placed at time 12 which is the earliest time for event 6 computed in the forward pass calculations.

We next compute the latest allowable start time of activity 5-6 by applying rule 3, i.e., $LS_{5,6} = LF_{5,6} - D_{5,6} = 12 - 3 = 9$. The crux of backward pass computations occurs at the burst event 2. Here, the computation of the latest allowable finish time of activity 1-2 requires consideration of its two successor activities. Applying rule 2 above we obtain

$$LF_{1,2} = \text{Minimum of } (LS_{2,3} = 2 \text{ and } LS_{2,4} = 4) = 2$$

Finally, we obtain $LS_{1,2} = LF_{1,2} - D_{1,2} = 2 - 2 = 0$. This result can be used as a check on the computations when the zero-float convention is followed. If $L_6 = E_6 = 12$ for the terminal event, then $L_1 = E_1 = 0$ must result for the initial event.

## DEFINITION AND INTERPRETATION OF FLOAT (SLACK)

Among the many types of float defined in the literature, two are of most value and are stressed in this text; they are called total path float, or simply path float, and activity free float, or simply activity float. Path float and activity float are also referred to by some authors as total slack and free slack, their definitions being identical to those given below.

### Path Float—Total Float

*Definition:*

Path float, as the name implies, is the total float associated with a path. For a particular path activity, say $ij$, it is equal to the difference between its earliest and latest allowable start *or* finish times. Thus, for activity $(i - j)$, the path float is given by

$$F_{ij} = LS_{ij} - ES_{ij} \quad \text{or} \quad LF_{ij} - EF_{ij}$$

Assume the zero-float convention is followed, i.e., let $L_t = E_t$ for the terminal event. Then the *path float* denotes the amount of time (number of working days) by which the actual completion time of an activity on the path in question can exceed its earliest completion time without affecting the earliest start or occurrence time of any activity or event on the *network critical path*. This is equivalent to not causing any delay in the completion of the project. For example, in Figure 4-1c the *path float* for activities 2-4 and 4-5 is two days. Thus, the slack path (2-4-5) is two days away from becoming critical. Now, suppose activity 2-4 "slips" by starting a day late or taking a day longer to complete than originally estimated. The result is that its completion time occurs at the end of the fifth day instead of the fourth day, and the path float for the remainder of this slack path is thereby reduced by one day. That is, the path float for activity 4-5 is reduced from two to one day. But note that this slippage does not effect the earliest times for any *critical path* activity or event. Also note that if the total slippage along the slack path 2-4-5 exceeds its path float of two days, then the critical path is affected, and the duration of the project is increased accordingly.

There is a subtle difference in the interpretation of *path float* when the zero-float convention of letting $L_t = E_t$ for the terminal event is not followed. Consider Figure 4-1d where $E_t = 12$ but $L_t = T_S = 14$. Here, activity 2-4 has a path float of four days; it can slip up to this amount without causing the project completion to exceed its *scheduled* time of 14 days. However, a slippage of four days will cause the *early* start time of the critical path event 5 to slip by two days, and hence delay the completion of the project, but only up to its scheduled completion time of 14.

### Activity Float–Free Float

Merge point activities that are the last activity on a slack path (activity 4-5 in Fig. 4-1) have what is called *activity* float, sometimes called *free* float. The name follows from the fact that the specific activity is free to use this float without effecting *any* other activity times in the network. This float concept is not widely used, but it will be defined below and discussed briefly.

*Definition:*

Activity float is equal to the earliest start time of the activity's successor activity(s) minus the earliest finish time of the activity in question. Thus, for activity $(i - j)$, the activity float is given as follows, where $j - k$ denotes a successor to the activity in question.

$$AF_{ij} = ES_{jk} - EF_{ij} \quad \text{or} \quad E_j - EF_{ij}$$

Activity float is equal to the amount of time that the activity completion time can be delayed without affecting the earliest start or occurrence time of *any* other activity or event in the network.

In a sense, activity float is "owned" by an individual activity, whereas path or total float is shared by all activities along a slack path.

Again consider the slack path 2-4-5 in Figure 4-1, and assume the zero-float convention is used. The last activity on this path, 4-5, has both a *path float* and an *activity float* of two days. If it slips by an amount of up to two days, no other network activity or event times are affected. However, this is not true for activity 2-4 which has no *activity float.* Here, any slippage is immediately reflected in a corresponding delay in the early start time of its successor activity, 4-5. Thus, when we say that activity 4-5 has an *activity* float and a *path* float of two days, we are assuming that its predecessor will be completed by its *early* finish time.

## CRITICAL PATH IDENTIFICATION

Now that the concept of float has been described, the critical path through a network will be formally defined as follows.

*Definition:*

*The critical path is the one with the least path float. If the zero-float convention of letting* $L_t = E_t$ *for the terminal network event is followed, the critical path will have zero float; otherwise, the float on the critical path may be positive or negative. If the network has single initial and terminal events and no scheduled times are imposed on intermediate network events, then the critical path is also the longest path through the network.*

For the network in Figure 4-1c, the critical path, the one with least path float, is 1-2-3-5-6. It is also the longest path through the network in this case. Its duration is equal to 12 days, and it has zero path float since the convention of letting $L_6 = E_6 = 12$ was followed.

To illustrate the case where the zero float convention is not followed, it has been assumed in Figure 4-1d that the scheduled completion time of the project is 14 working days, i.e., $L_6 = T_S = 14$. In this case, we see that the critical path remains the same, i.e., 1-2-3-5-6, since it is still the path with least path float. However, the float along the critical path is now positive, that is, two days, while the float along the slack path 2-4-5 is now four days. In this case, a slippage up to two days along the critical path will cause the critical path events to slip a corresponding amount. However, the critical path activities will not slip beyond their *latest allowable* start and finish times, and in particular, the project end event 6 will not slip past its scheduled completion time of $T_S = 14$ days.

## OTHER TYPES OF FLOAT

Battersby[1] defines a third type of float called *interfering float*, as the *path float* minus the *activity float*, or total float minus free float. For an arbitrary activity $(i - j)$, this can be written as:

$$ITF_{ij} = F_{ij} - AF_{ij} = L_j - E_j$$

For Figure 4-1c, the activities on the slack path (2-4-5) have the float values shown in Table 4-1. The interpretation follows from the name. Activity 2-4 has both *path* and *interfering* float of two days; if it uses its float, it "interferes," by this amount, with the early times for the down path activity, 4-5. However, activity 4-5 has no *interfering* float, but rather it has both *path* float and *activity* (free) float of two days (assuming, of course, that activity 2-4 is completed at its early finish time). Activity 4-5 can use its *activity* float of two days without "interfering" with *any* other activity or event times in the network.

**Table 4-1. Float Values for Slack Path (2-4-5) in Figure 4-1c**

| | Float Type | | |
|---|---|---|---|
| *Activity* | *Path* | *Activity* | *Interfering* |
| 2-4 | 2 | 0 | 2 |
| 4-5 | 2 | 2 | 0 |

Battersby also defines *independent* float, which occurs only rarely in typical networks. It is the amount of float which an activity will always possess no matter how early or late it or its predecessors and successors are. It is found by assuming a worst case situation. That is, all predecessors end as late as possible and successors start as early as possible. If this time interval exceeds the duration of the activity in question, the excess is called *independent* float. It is illustrated in Figure 4-2, where it can be seen that activity 2-4 has an independent

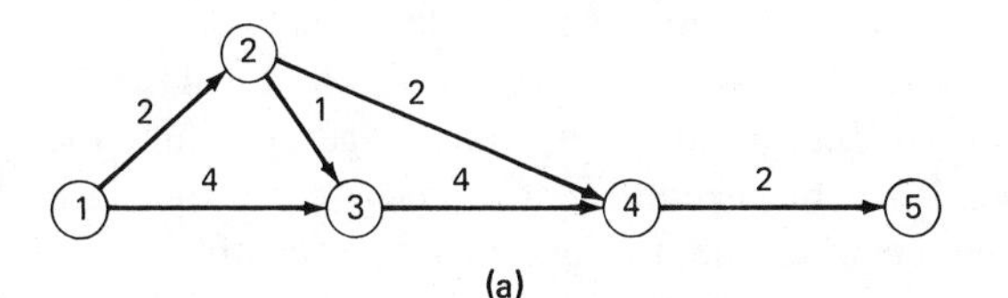


(a)

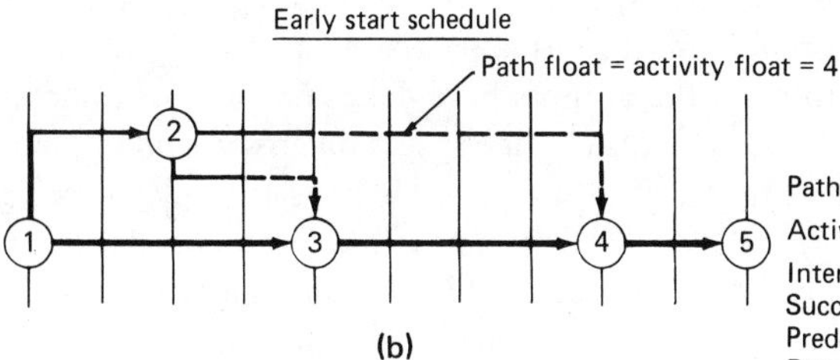


(b)

For activity 2-4

Path float = $L_4 - EF_{2-4} = 8 - 4 = 4$
Activity float = $E_4 - EF_{2-4} = 8 - 4 = 4$
Interfering float = $4 - 4 = 0$
Successor (4-5) ES = 8
Predecessor (1-2) LF = 3
Duration (2-4) = 2
Independent float = $E_4 - L_2 - D_{2-4}$
$= 8 - 3 - 2 = 3$

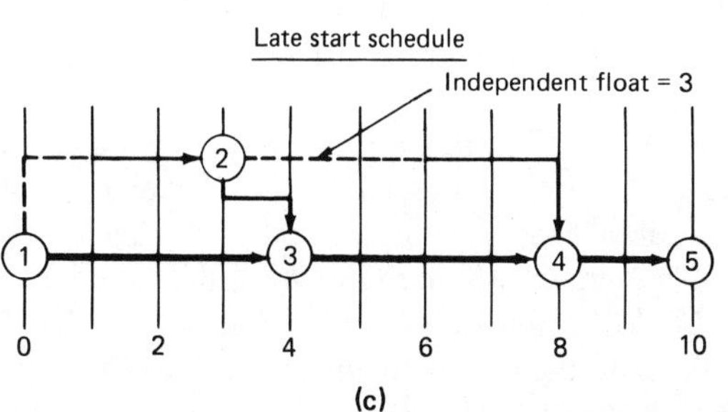


(c)

**Figure 4-2** Network illustrating independent float.

float of 3 days. For an arbitrary activity ($i$ - $j$), this can be written as:

$$IDF_{ij} = \text{Max}\,(0, E_j - L_i - D_{ij})$$

The interpretation is that it has this float "independent" of *any* slippage of predecessor activity 1-2, and any allowable start time of successor activity 4-5. In contrast to this, activity 4-5 in Table 4-1, has no *independent* float and hence its float of 2 days is *not* "independent" of any slippage of its predecessor activity 2-4.

Critical path computer packages always list *path* (or *total*) float for each activity, they occasionally list *activity* (or *free*) float, and to the best of our knowledge, they never list any other forms of float. That is not to say that they have no utility, e.g., in tasks such as resource scheduling, or computerized bar charting. However, these other forms of float are currently judged to be very marginal, and are not utilized in this text.

## USE OF SPECIAL SYMBOLS IN SCHEDULING COMPUTATIONS

Although there are a number of obvious advantages to having the network drawn to scale on a time base as shown in Figure 4-1, the disadvantages of doing this by hand, notably the inflexibility to incorporating network changes, preclude the general use of this procedure. It has been found best in practice not to attach any special significance to the length of the network arrows, but rather to denote the various activity and event times of interest by numerical entries placed directly on the network. As an aid to making the scheduling computations which give these numerical entries, and to display them in an orderly fashion so that they can be easily interpreted, the authors have developed a system incorporating special symbols.[6] These symbols provide spaces on arrows and in event nodes for recording computed times. The spaces are located logically to bring activity earliest finish (and latest allowable start) times close together and thereby facilitate the computation of earliest expected (and latest allowable) event times. These symbols are used throughout this text to make it easier for the reader to follow. In practice, however, the authors omit the enclosures for $LS_{ij}$, $F_{ij}$, and $EF_{ij}$, shown in Figure 4-3a; these numbers are merely written above a single line activity arrow in the appropriate position.

The event identification number is placed in the upper quadrant of the node, as shown in Figure 4-3a. The earliest event occurrence time, $E_i$, which is equal to the earliest start time ($ES_{ij}$) for activity $i$ - $j$, is placed in the left hand quadrant of event $i$. The latest allowable event occurrence time, $L_j$, which is equal to the latest allowable finish time ($LF_{ij}$) of activity $i$ - $j$, is placed in the right hand quadrant of event $j$. The lower quadrant might be used later to note actual event occurrence times if desired. For each activity, the earliest time the activity is expected to be finished, $EF$, is placed in the arrow head and the latest allowable

Reading earliest expected and latest allowable activity start and finish times and float from the special symbols

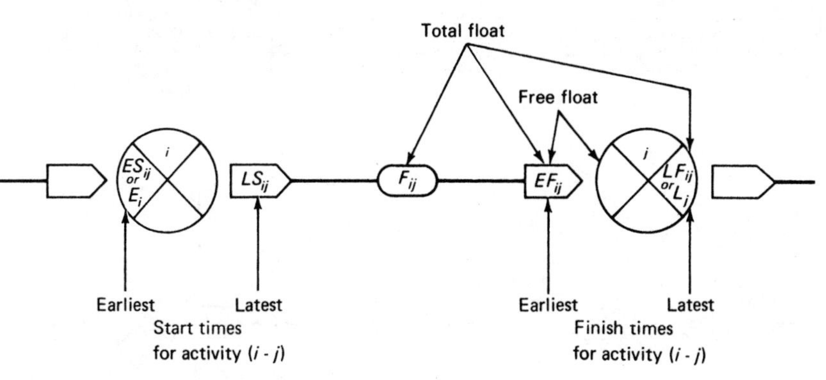


**Figure 4-3a** Key to use and interpretation of special activity and event symbols.

start time, *LS*, in the arrow tail. The estimated duration of the activity along with a description of the activity is placed along the arrow staff. Path float is placed in the bubble along the arrow staff. The lower portion of Figure 4-3a points out how one reads the earliest and latest allowable start and finish times for an activity. The detailed steps involved in carrying out the scheduling computations using these symbols is illustrated in Figures 4-3b and 4-4.

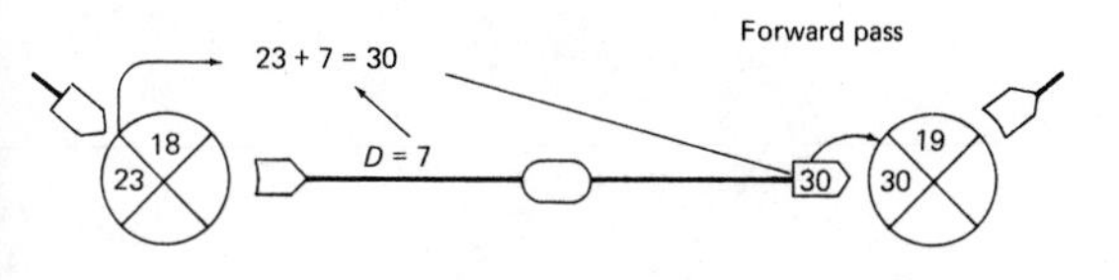


Begin with zero for the earliest start time for the initial project event and compute earliest finish times for all succeeding activities. For a typical activity, place its earliest start time (say, 23 days from project start) in the left quadrant of the event symbol. Then add its duration (7) to the earliest start time to obtain its earliest finish time (30). Write 30 in the arrow head.

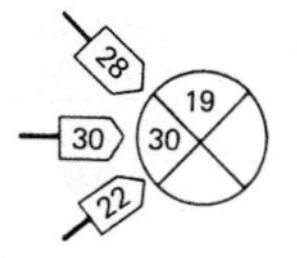


Where activities merge, insert in the left quadrant of the event symbol the largest of the earliest finish times written in the arrowheads of the merging activities.

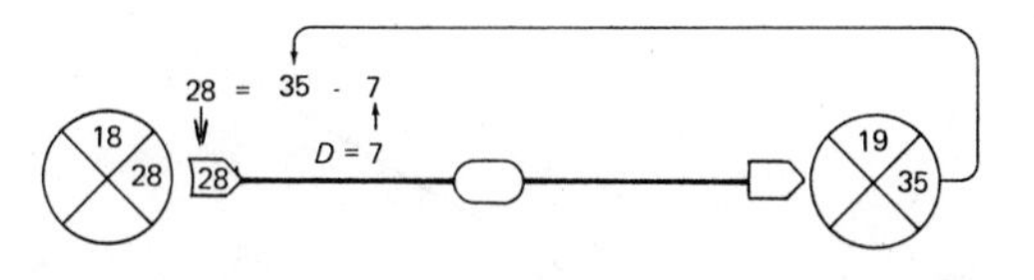


Place the scheduled completion time for the final event in the right quadrant of the project terminal event symbol. For other events, insert instead the latest allowable event occurrence time. For a typical activity, subtract its duration (7) from the latest completion time (35) to obtain the latest allowable activity start time (28). Write 28 in the arrow tail.

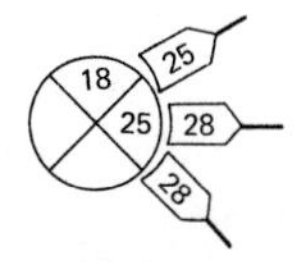


Where two or more activities "burst" from an event, insert in the right quadrant of the event symbol the smallest of the latest allowable activity start times.

**Figure 4-3b** Steps in scheduling computations using special activity and event symbols.

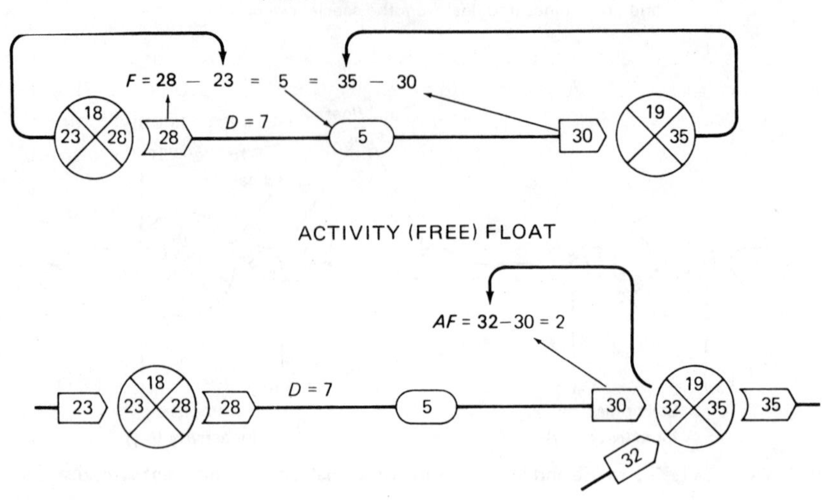


**Figure 4-4** Steps in path float and activity float computations using special activity and event symbols.

## ILLUSTRATIVE NETWORK EMPLOYING SPECIAL SYMBOLS

To illustrate the use of these symbols in making the scheduling computations, a network containing eleven activities is shown in Figure 4-5, which contains the complete forward pass computation. In Figure 4-6 the backward pass and float

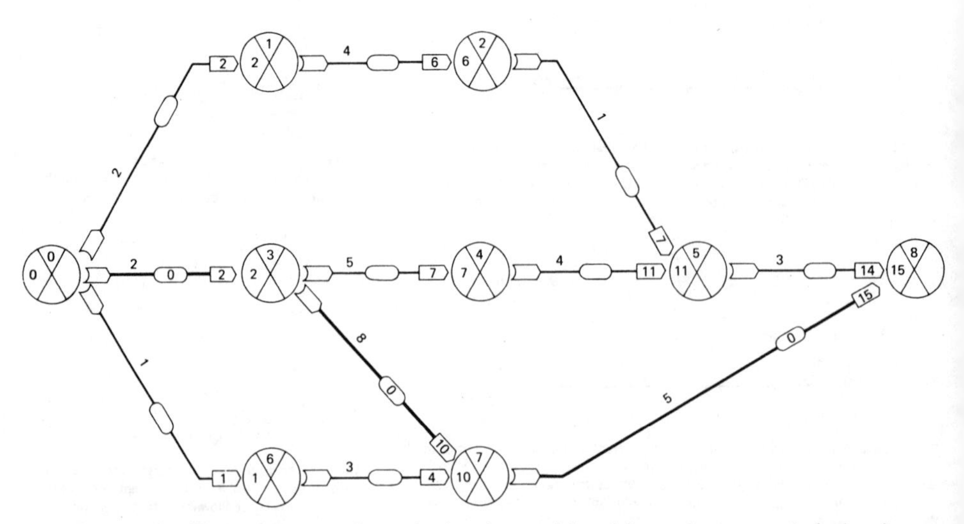

**Figure 4-5** Illustrative network employing the special activity and event symbols showing forward pass computations only.

computations have been added. With a little practice, the computations made on this network can be completed in two to three minutes. With an allowance for an independent check on these computations, the total time is still less than that required to fill out the input forms required to run this problem on a computer. This holds true for networks of any size. The computer is only economical when the network is to be updated periodically, or when a variety of other computations is requested.

In Figure 4-6, the earliest and latest allowable activity start and finish times are clearly displayed to aid in the making of resource allocation checks, the determination of activity schedules, the conducting of time-cost trade-off studies, etc. For example, consider activity 2-5. Its earliest start and finish times are readily observed to be 6 and 7, respectively, and its latest allowable start and finish times are 11 and 12, respectively. The path float is $F_{25} = LF_{25} - EF_{25} = 12 - 7 = 5$ days, while the activity float is only $AF_{25} = E_5 - EF_{25} = 11 - 7 = 4$ days. This activity illustrates quite well the basic difference between path float and activity float, the latter occurring only at the end of a slack path, i.e., at a merge event. If the completion of activity 2-5 is delayed up to 4 days, the amount of its activity float, *no other activity or event time in the network will be affected.* In particular, the earliest expected time for event 5 remains at 11. If the completion of this activity is delayed by an amount exceeding its activity float, but not exceeding its path float, then the earliest expected time for event 5 and the early start time for the following activity 5-8 will be increased. However, no critical path activities or events, such as event 8, will be affected. Finally, if the completion of activity 2-5 is delayed by an amount which exceeds

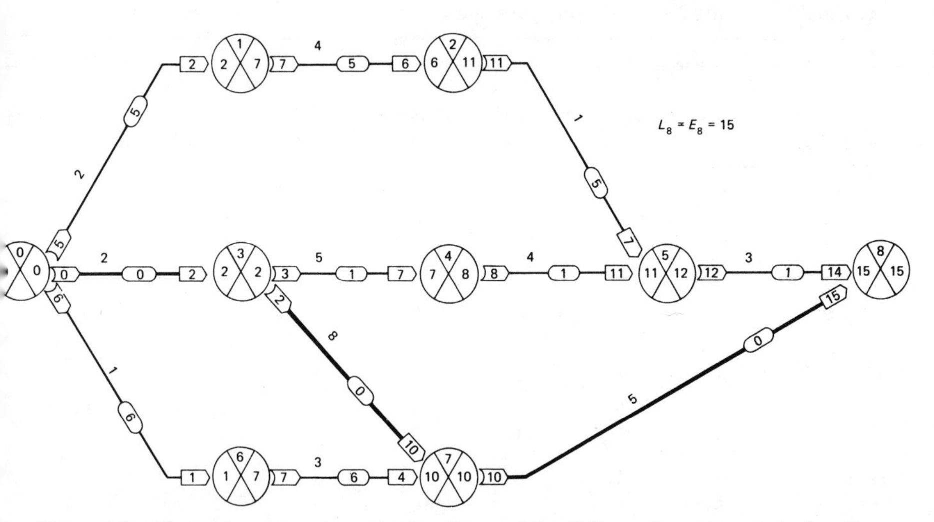


**Figure 4-6** Illustrative network employing the special activity and event symbols showing completed computations.

its path float of 5 days, then the project completion time, i.e., the earliest expected time for event 8, will be increased by a like amount.

According to the previous definition of the critical path, it is made up of activities 0-3-7-8 in the illustrative network shown in Figure 4-6. It has the least amount of path float, which is zero in this case, because the convention of letting $L_8 = E_8 = 15$ was followed. This is also the longest path through the network. In addition to determining the critical path through the network, we can identify various subcritical paths which have varying degrees of path float and hence depart from criticality by varying amounts. These subcritical paths can be found in the following way, which is suggestive of how a computer would handle this problem.

1. Sort the activities in the network by their path float, placing those activities with a common path float in the same group.
2. Order the activities within a group by early start time.
3. Order the groups according to the magnitude of their path float, small values first.

   The first group comprises the critical path(s) and subsequent groups comprise subcritical paths of decreasing criticality.

Application of the above procedure to the network in Figure 4-6 gives the results shown in Table 4-2 below.

**Table 4-2. Listing of Critical and Subcritical Paths by Degree of Criticality for the Network in Figure 4-6**

| | *Earliest* | | *Latest* | | | |
|---|---|---|---|---|---|---|
| *Activity* | *Start Time* | *Finish Time* | *Start Time* | *Finish Time* | *Total Float* | *Criticality* |
| 0-3 | 0 | 2 | 0 | 2 | 0 | |
| 3-7 | 2 | 10 | 2 | 10 | 0 | critical path |
| 7-8 | 10 | 15 | 10 | 15 | 0 | |
| 3-4 | 2 | 7 | 3 | 8 | 1 | |
| 4-5 | 7 | 11 | 8 | 12 | 1 | a "near critical" path |
| 5-8 | 11 | 14 | 12 | 15 | 1 | |
| 0-1 | 0 | 2 | 5 | 7 | 5 | |
| 1-2 | 2 | 6 | 7 | 11 | 5 | third most critical path |
| 2-5 | 6 | 7 | 11 | 12 | 5 | |
| 0-6 | 0 | 1 | 6 | 7 | 6 | path having most float |
| 6-7 | 1 | 4 | 7 | 10 | 6 | |

## CRITICAL PATH FROM FORWARD PASS ONLY

The above procedure for locating the critical path(s) is based on a knowledge of path float, which requires the backward pass for computation. While this procedure is necessary to find the slack along subcritical paths, *the* critical path(s) can be determined from the results of the forward pass only. This is quite useful in the early stages of planning and scheduling a project, when it is desired to determine the expected project duration, and to determine the critical path activities with a minimum of computation. The following steps which make up this procedure are based on the assumption that the forward pass computations have been completed, and the resulting $EF$'s and $E$'s have been recorded on the network.

1. Start with the project final event, which is critical by definition, and proceed backwards through the network.
2. Whenever a merge event is encountered, the critical path(s) follows the activity(s) for which $EF = E$.

To illustrate this procedure, let us trace the critical path of the network shown in Figure 4-5, on which only the forward pass computations have been made. First we start at event 8 for which $E = 15$. The critical path is then along activity 7-8 since $EF = E = 15$ for this activity, while $EF$ is only 14 along the other path, activity 5-8. In this manner, the critical path can be traced next to event 3, and hence to the initial network event 0.

## VARIATIONS OF THE BASIC SCHEDULING COMPUTATIONS

The restrictive assumptions underlying the above basic scheduling computations included the following:

1. The network contained only one initial event, i.e., one event with no predecessor activities.
2. The network contained only one terminal event, i.e., one event with no successor activities.
3. The earliest expected time, $E$, for the initial event was zero.
4. There were no scheduled or directed dates for events other than the network terminal event.

These assumptions were made because they simplify the computational procedure and its subsequent interpretation, and at the same time they do not seriously restrict the usefulness of the procedure. However, occasions will be pointed out where it would be of some value to relax these assumptions; hence, the required modifications in the computational procedures will be taken up here.

It may happen, for example, that the network under study is only a portion of a larger project, or perhaps one of several in a multi-project operation. In this case, it may be that the earliest time for the initial network event does not occur at time zero, but rather at some arbitrary number of time units other than zero. In such a case, this specified earliest occurrence time, $E$, for the initial event is merely used in place of zero, and the forward pass computations are made in the conventional manner.

Similarly, one may wish to specify or direct the latest allowable time for some intermediate network (milestone) event to be a time that is arbitrarily specified as the scheduled or allowed time for the event in question. In this case one uses a scheduled time, $T_S$, in determining the $L$ value for this event, as described in the example below. The important points in this discussion are summarized below.

*Conventions:*

A scheduled time, $T_S$, for an initial project event is interpreted as its earliest expected time, i.e., $T_S = E$ for initial project events.

A scheduled time, $T_S$, for an intermediate (or terminal) project event is interpreted as its latest allowable occurrence time.

## ILLUSTRATIVE NETWORK WITH MULTIPLE INITIAL AND TERMINAL EVENTS

To illustrate how projects with multiple initial and terminal events and scheduled event times are handled, consider the network shown in Figure 4-7. The "main" project in this network has events numbered 101 through 109. This project, which starts with event 101, produces an "end" objective signified by event 109. However, a by-product of this project is that it furnishes an output for a second project whose events are numbered in the 200 series. For example, the objective of the main project might be to develop a new rocket engine, the completion of which is denoted by event 109. Event 210 might signify the delivery of a key component from a second project that has other end objectives. Activity 108-211 might be the preparation and delivery of a report on the testing of this component which took place in the main project. The scheduled occurrence time of event 211 is thus quite important to the second project, and is independent of the main project.

Two events from the second project are pertinent to the main project network shown in Figure 4-7. First, event 210, which initiates an activity preceding event 106, is scheduled to occur 8 units of time after the start of the first project. Hence, if one sets $E = 0$ for event 101, then $E = 8$ for event 210, since the latter

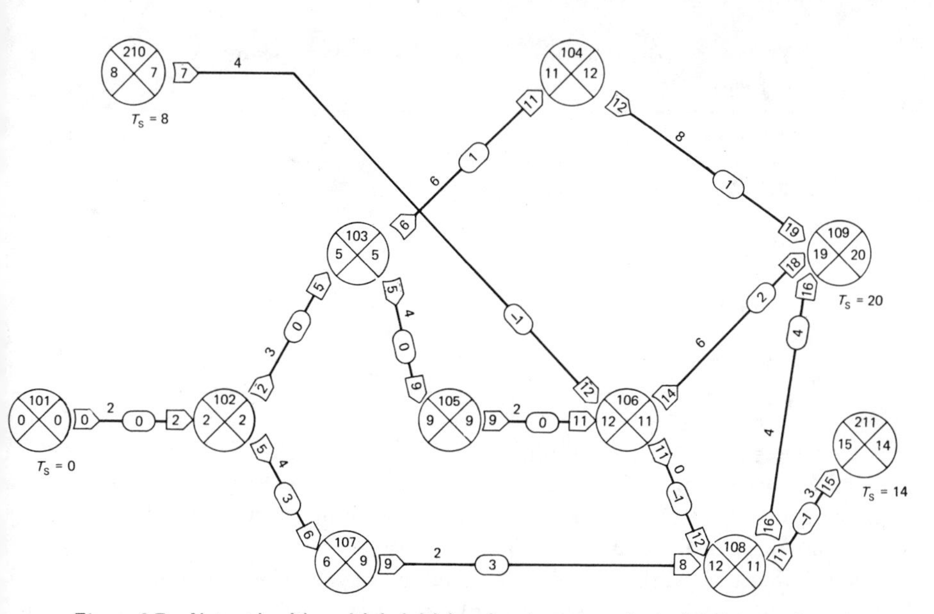


**Figure 4-7** Network with multiple initial and end events and scheduled event times.

is one of the *initial* events in the network. Event 211 of the second project is scheduled to occur 14 units of time after the start of the first project. Hence, since event 211 is a *terminal* event in this network, one sets $T_S = L = 14$ for event 211. Finally, the main project is scheduled to be completed in 20 time units, thus $T_S = L = 20$ for event 109. Thus, the network computations are started with the following times specified.

Event 101: $E = 0$
Event 109: $L = T_S = 20$
Event 210: $E = T_S = 8$
Event 211: $L = T_S = 14$

Since early start times are now specified for both of the initial events in this network, the forward pass computations can be carried out in the conventional manner. These computations indicate that the early finish time for the final event of the main project is 18, i.e., $E = 18$ for event 109. The early finish time for the secondary project is 15, i.e., $E = 15$ for event 211. Thus, the schedule can be met on the main project but not on the secondary project, unless the latter is expedited in some way.

Since all terminal events have scheduled completion times, the backward pass computations are initiated by setting $L = 20$ for event 109, and $L = 14$ for event 211. The critical and subcritical paths through the network can then be determined as given in Table 4-3.

The critical (least float) path through the network has a path float of −1. One

**Table 4-3. Critical and Subcritical Paths for the Network in Figure 4-7**

| *Activities* | *Path Float, F* | *Activities* | *Path Float, F* |
|---|---|---|---|
| 210-106 | −1 | 103-104 | 1 |
| 106-108 | −1 | 104-109 | 1 |
| 108-211 | −1 | 106-109 | 2 |
| 101-102 | 0 | 102-107 | 3 |
| 102-103 | 0 | 107-108 | 3 |
| 103-105 | 0 | 108-109 | 4 |
| 105-106 | 0 | | |

thus expects to be one unit of time late in meeting the scheduled time of 14 for event 211, the termination of the secondary project. With regard to the main project, one notes that all paths leading to the terminal event 109 have path float of at least one time unit.

Multiproject networks can be quite useful in analyzing the effects of one project on another, and thus offer a means of settling disputes which frequently arise in such a situation. For example, it is clear from Figure 4-7 that one way of alleviating the negative slack situation on the critical path would be to move up the schedule for event 210 so that it occurs on or before the late start time of 7 for activity 210-106. In this way the secondary project could help in solving its own scheduling problem. If this could not be done, then the only remaining remedies would be to reduce the duration of either activity 210-106 or 108-211 by one or more time units. It is also clear from Figure 4-7 that the current schedule for event 210, i.e., $T_S = 8$, does not produce the most constraining path to event 109, and hence does not fix the earliest completion time of the main project.

Another method of handling multiple initial and end events is taken up in exercise 4 at the end of this chapter. This procedure is illustrated in Figure 3-16 where two essentially separate projects are tied together by activity 40-80, which has a time estimate of 40 days. In essence, this activity constrains the completion of one project to precede the other by 40 days. Thus, a scheduled date placed on event 80 in one project can force a scheduled date, or latest allowable time, 40 days earlier on event 40 of the second project. This procedure eliminates the second terminal event.

## NETWORKS WITH SCHEDULED OR DIRECTED TIMES ON INTERMEDIATE EVENTS

Scheduled or directed times for intermediate network events are handled in a manner similar to the treatment for terminal events as discussed above. In this case, however, there will be two candidates for the latest allowable time for the event in question. The choice is usually governed by the following convention; however, other conventions may be adopted in certain project management computer packages.

The latest allowable time for an intermediate network event on which a scheduled time, $T_S$, is imposed, is taken as the earlier (smaller) of the scheduled time, $T_S$, and the latest allowable time, $L$, computed in the backward pass.

For example, suppose activity 103-105 in Figure 4-7, which involves earth moving, was scheduled to be completed by time 7 to insure completion prior to the ground freezing. In this case, $L$ for event 105 would be taken as 7, i.e., the smaller of the scheduled time of 7 and the regular backward pass time of 9. The introduction of this scheduled time only affects the path float for activities 101-102-103-105 by reducing it from 0 to −2, a change which is quite important in planning and scheduling these activities. Since this now becomes the path with the least float, it forms the new project critical path. While it starts at the initial project event, it terminates at the intermediate event 105 on which the scheduled time was imposed, without going through the entire project. This is a characteristic result of introducing scheduled or directed times on intermediate events; the computations clearly show up the most constraining scheduled dates in the project.

## NETWORK TIME-STATUS UPDATING PROCEDURE

Updating a network to reflect current status is similar to the problem introduced above in that a project underway is equivalent to a project with multiple start events. After a project has begun, varying portions of each path from the initial project event to the end event will have been completed. According to arrow diagram network logic, all activities preceeding the current status on a particular path must be completed, and no activities succeeding the current status point should have started. By establishing the current status on each path from progress information, the forward pass scheduling computations can then be made in routine fashion. No change in the backward pass computation should be necessary, unless changes are to be incorporated in the logic or time estimates of the uncompleted portion of the network.

To illustrate this updating procedure, consider the network presented in Figure 4-6, which indicates an expected project duration of 15 days. Suppose the project started on Monday and we have just received the Friday evening weekend progress report. Thus, we have just completed the fifth work day on this project, and the progress is as reported in Table 4-4.

The actual activity start and finish times given in Table 4-4 have been written above the arrow tails and heads, respectively, in Figure 4-8. Events that have already occurred have been cross hatched, and activities that are in progress have

**Table 4-4. Status of Project Activities at the End of the Fifth Working Day**

| *Activity* | *Started* | *Finished* |
|---|---|---|
| 0–1 | 1 | 3 |
| 1–2 | 4 | – |
| 0–3 | 0 | 2 |
| 3–7 | 2 | – |
| 0–6 | 2 | 4 |
| 6–7 | 5 | – |

NOTE: all times given are at the *end of* the stated working day.

been so noted by a flag marked 5 to denote that the time of the update is the end of the fifth working day. Only one of the four paths being worked presents a problem, i.e., activity 3-4. At report time, event 3 has occurred, but activity 3-4 has evidently not started. To avoid this problem, it would be desirable to include in progress reports the intended start time of all activities whose predecessor activities have been completed. If this information is not given, then some assumption must be made to complete the update calculations. The usual assumption, which is the one adopted in Figure 4-8, is that the activity will start on the next working day, i.e., at the end of the fifth working day.

Having an actual, or assumed, start time for the "lead" activities on each path in the network, the forward pass calculations are then carried out in the usual manner. The original times are crossed out, with the new updated times written

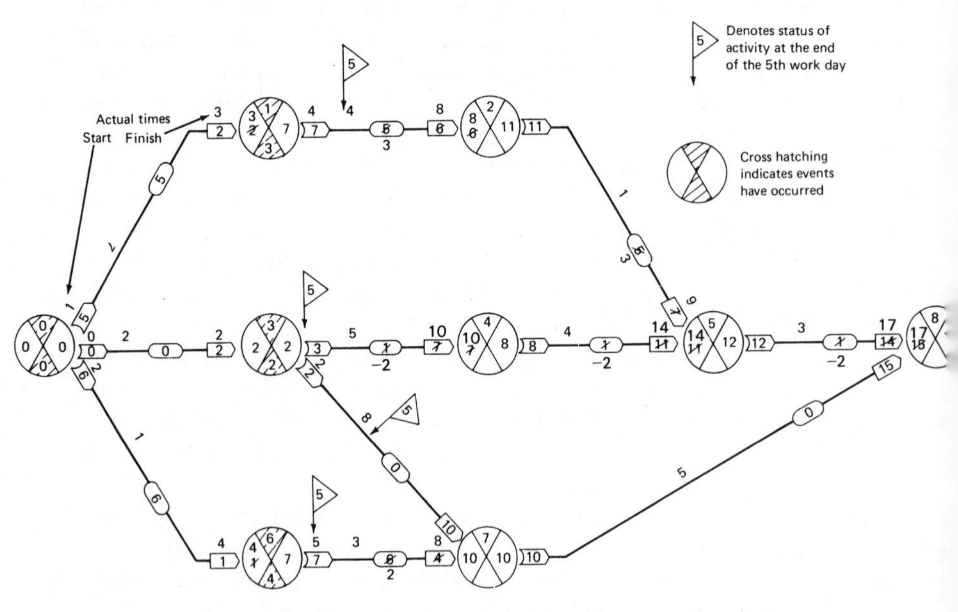


**Figure 4-8** Illustrative network showing time status of project.

nearby. These calculations indicate that the critical path has shifted to activities 3-4-5-8, with a slack of minus two days. Assuming we were scheduled to complete the project in 15 days, the current status indicates we are now two days behind schedule, and appropriate corrective steps are in order.

An alternative updating procedure is as follows; it has the advantage that it can be carried out by a computer program that has no special updating capability. First, add a dummy activity preceding the initial project event, and assign to it a duration value equal to "time now," that is, the number of elapsed working time units from the beginning of the project to the time of the project update. Second, change the duration times of activities that are completed, or partially completed, to the number of time units of work *remaining.* Now, make the forward pass calculations in the conventional manner. It will produce the same results obtained by the method illustrated in Figure 4-8 above.

## ACTIVITY-ON-NODE SCHEDULING COMPUTATIONS

In Chapter 2, the basic node scheme of networking was described. This is the complete reverse of the arrow scheme in which the nodes represent the activities and the arrows are merely connectors. To illustrate the computational procedure for a node diagram, consider the network plan for the market survey project presented in node diagram form in Figure 2-21. (An equivalent arrow diagram was given in Figure 2-18, and a time-scaled version in Figure 2-20.) The basic forward and backward pass scheduling computations for this network are given in Figure 4-9.

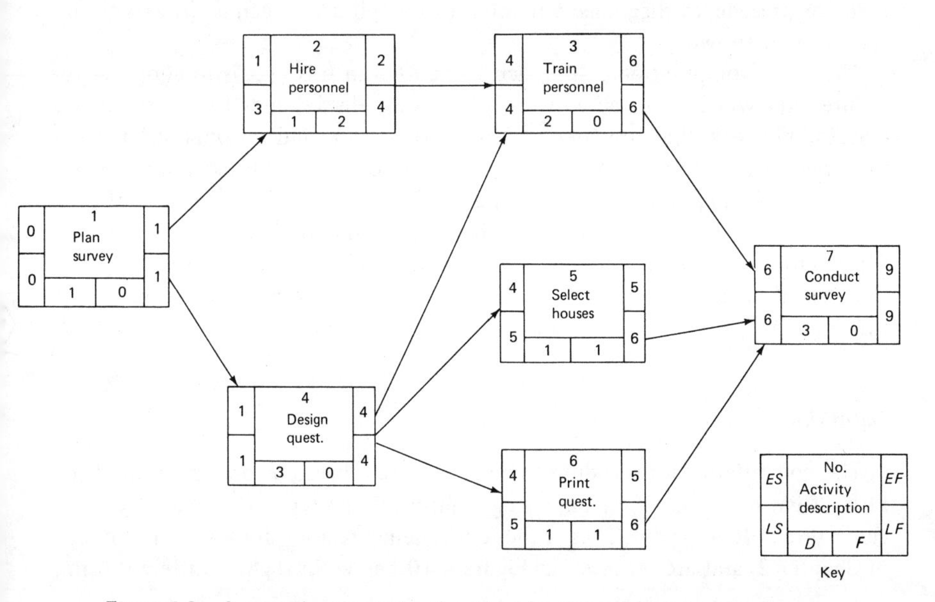


**Figure 4-9** A manual computational method for the activity-on-node scheme.

The node scheme, first promoted by J. W. Fondahl[3] in 1958, lends itself to manual computation as well as the arrow scheme. One symbolic method for manual computation of node networks is illustrated in Figure 4-9. Using the key to the symbol notation, the reader can easily see how the computation is made. For example, the *EF* for node 3 is obtained by adding its times for *ES* and *D*. Thus, $EF = 4 + 2 = 6$. At merge points, such as node 7, the *ES* is selected as the largest of the preceding *EF* times. The backward pass is made similarly.

In comparison with the symbols used in manual computation of arrow networks, the node symbols are somewhat more articulate. All of the numbers associated with an activity are incorporated in the one node symbol for the activity, whereas the arrow symbols contain each activity's data in the predecessor and successor nodes, as well as on the arrow itself.

## PRECEDENCE DIAGRAMMING

An important extension to the original activity-on-node concept appeared around 1964 in the Users Manual for an IBM 1440 computer program.[5] One of the principal authors of the technique was J. David Craig, who referred to the extended node scheme as "precedence diagramming." The computation and interpretation of early/late start/finish times for project activities for this scheme are considerably more complex than those shown above for the basic finish-to-start constraint logic of arrow or node diagrams. For the latter, the computation and interpretation of these times was both simple and unique. This is an important advantage of the PERT/CPM systems. Unfortunately, this does not usually hold for precedence diagrams; a number of complications can arise, as will subsequently be shown.

The basic computational approach to be used in this text is to adopt a procedure that will lead to activity early/late start/finish times for a precedence diagram network that are *identical* to those that would be obtained for the *equivalent* arrow diagram and the conventional forward and backward pass computations. This approach was developed by Keith C. Crandall, Professor of Civil Engineering, University of California, Berkeley, and published in 1973[2] A hand computation version, where activity splitting is allowed, was also supplied by Crandall in private communication in May 1981; it is given in Appendix 4-2 of this chapter.

### *Definitions:*

The computational procedure to be given here is based on an extension of the PERT/CPM network logic from a single finish-to-start type of dependency to include three other types. These other dependency relationships were introduced in Chapter 2, and are repeated in Figure 4-10 below, in slightly different form.

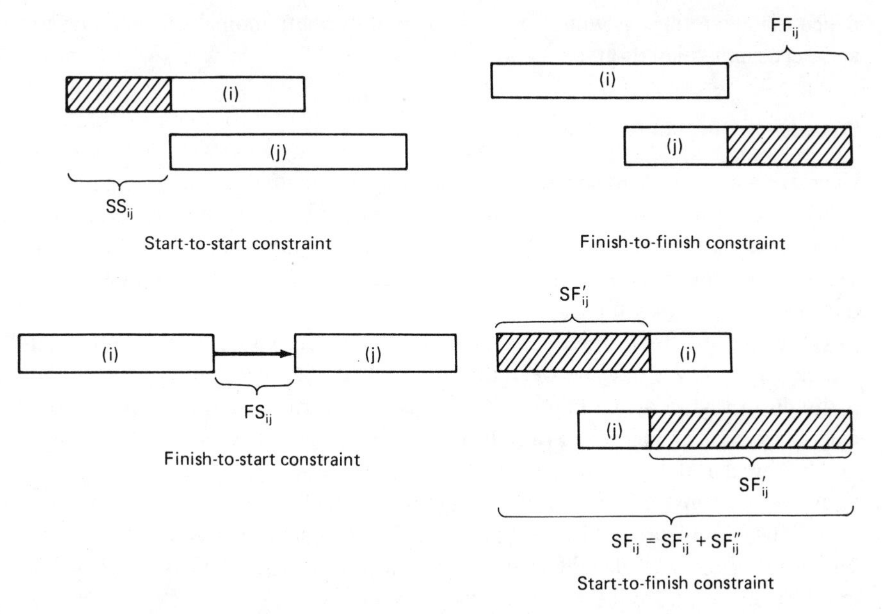


**Figure 4-10** Precedence diagramming constraints with lead/lag times.

The nomenclature and assumptions to be used are as follows:

- $SS_{ij}$ denotes a start-to-start constraint, and is equal to the minimum number of time units that must be complete on the preceding activity ($i$) prior to the start of the successor ($j$).
- $FF_{ij}$ denotes a finish-to-finish constraint, and is equal to the minimum number of time units that must remain to be completed on the successor ($j$) after the completion of the predecessor ($i$).
- $FS_{ij}$ denotes a finish-to-start constraint, and is equal to the minimum number of time units that must transpire from the completion of the predecessor ($i$) prior to the start of the successor ($j$). (Note: This is the sole logic constraint used in PERT/CPM, with $FS_{ij} = 0$).
- $SF_{ij}$ denotes a start-to-finish constraint, and is equal to the minimum number of time units that must transpire from the start of the predecessor ($i$) to the completion of the successor ($j$).
- $ZZ_{ij}$ denotes a frequently used combination of two constraints, i.e., a start-to-start and a finish-to-finish relationship. It is written with the $SS_{ij}$ time units first, followed by the $FF_{ij}$ time units.

The above constraint logic is shown in Figure 4-10. It will be applied in the next section to illustrate the powerful features of the extended logic of precedence diagramming. It will also point up an important anomaly that can occur, and needs an explanation.

### Precedence Diagram Anomalies

Consider a construction subcontract consisting of *Framing* walls, placing *Electrical* conduits, and *Finishing* walls, with the duration of each task estimated to be 10 days, using standard size crews. If the plan is to perform each of these tasks sequentially, the equivalent arrow diagram in Figure 4-11a shows that a project duration of 30 days will result.

To reduce this time, these tasks could be carried out concurrently with a convenient lag of say 2 days between the start and finish of each activity. This plan is shown in Figure 4-11b in precedence diagram notation. The equivalent arrow diagram shown in Figure 4-11c indicates a 14 day project schedule. One important advantage of Figure 4-11b over 4-11c is that each trade is represented by a single activity instead of 2 or 3 subactivities. Also note how the $SS = 2$ and $FF = 2$ lags of Figure 4-11b are built into the equivalent arrow diagram in Figure 4-11c. For example, the first two days of the electrical task in Figure 4-11c must be separated from the remainder of this task to show that 2 days of electrical work must be completed prior to the *start* of the finishing task. Similarly, the last 2 days of electrical work must be separated from the remainder of this task to show that framing must *finish* 2 days before electrical is finished. Thus, the 10 day electrical task must be broken up into 3 subactivities of 2, 6, and 2 days duration, respectively.

So far, precedence diagramming is easy to follow and is parsimonious with activities. But let us see what happens if the duration of the 3 tasks in this project are unbalanced by changing from 10, 10, 10 to 10, 5, and 15 days, respectively. These changes are incorporated in Figures 4-11d and e, along with appropriate new lag times. Note that $SS = 2$ was chosen between framing and electrical to insure that a full days work is ready for electrical before this task is allowed to start. Similarly, $FF = 3$ was chosen between electrical and finishing because the last day of electrical work will require 3 days of finishing work to complete the project. The other lags of 1 day each were chosen as minimal or convenience values needed in each case. These lags define the activity breakdown shown in Figure 4-11e where we see the critical path is the *start* of framing (1-2), then the *start* of electrical (4-5), and, finally, the *totality* of finishing (8-9-10). This is also shown in the precedence diagram, Figure 4-11d, where $ES = LS = 0$ for the *start* of framing, $ES = LS = 2$ for the *start* of electrical, and, finally, $ES = LS = 3$ *and* $EF = LF = 18$ for the totality of finishing. Since the precedence diagram shows each of these tasks in their totality, $EF \neq LF$ even though $ES = LS$ for the framing and electrical tasks. For framing in Figure 4-11d, $LF - EF = 14 - 10 = 4$ days of float, which corresponds to the 4 days of float depicted by activity 7-9 in Figure 4-11e. Similarly, for electrical in Figure 4-11d, $LF - EF = 15 -$

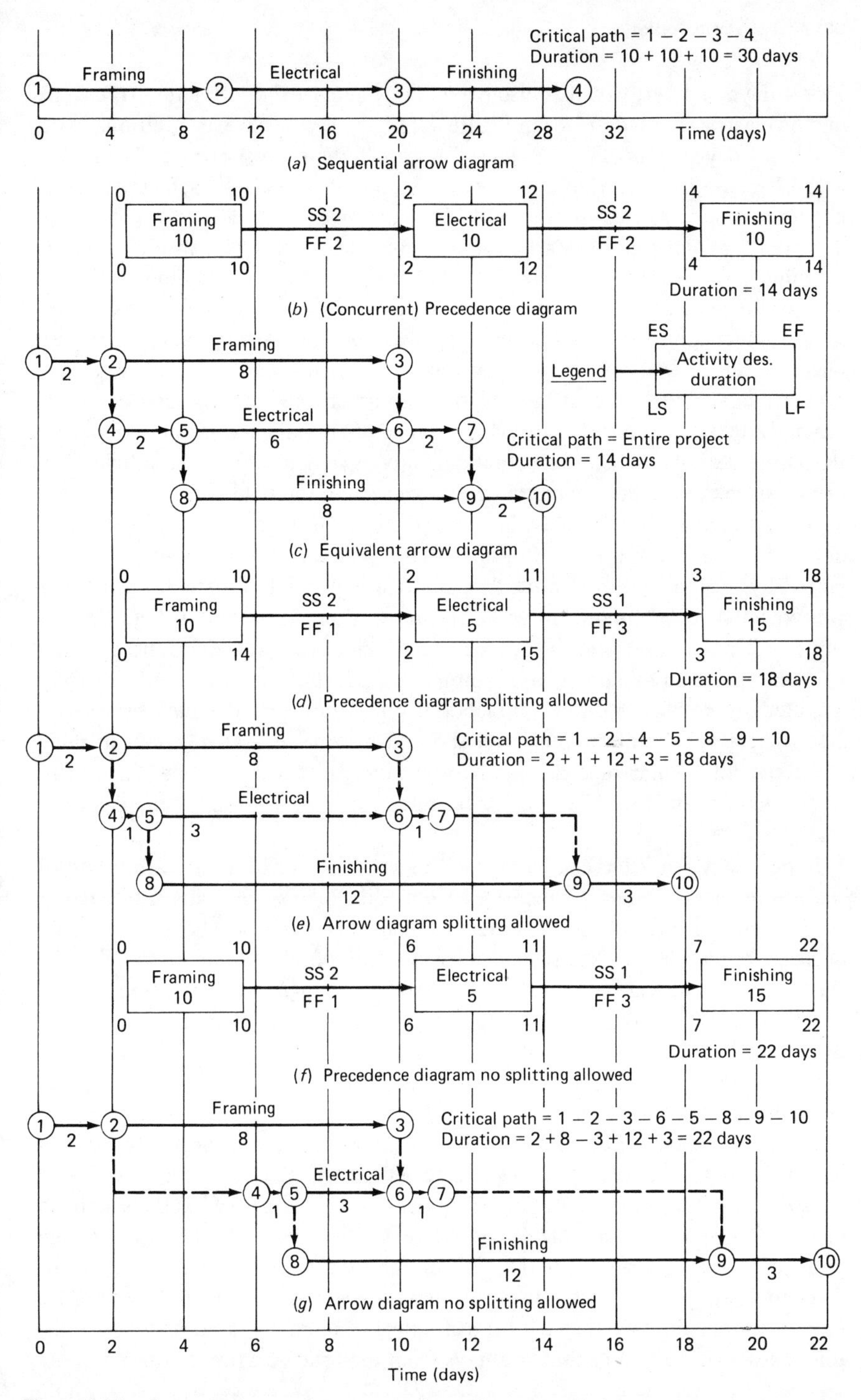


**Figure 4-11** Arrow and precedence networks to illustrate splitting vs. no splitting; activities are shown at their early start times.

11 = 4 days of float, which is also depicted by activity 7-9 in Figure 4-11e. The middle electrical activity (5-6) in Figure 4-11e *appears* to have an additional path float of 4 days, or a total of 8 days. This attribute is not shown at all in Figure 4-11d because it depicts only the beginning and end points of each activity, but not intermediate subactivities such as 5-6. Closer examination will show, however, that any delay in the start of activity 5-6 exceeding 3 days would cause the finishing crew to run out of work, and hence the critical path would be delayed. This problem is shared by both arrow and precedence diagrams, and the user should understand this. It does not, however, present a real problem in the applications since the job foreman generally has no difficulty in the day-to-day management of this type of interrelationship among concurrent activities. It is generally felt that it is not worthwhile to further complicate the networking and the computational scheme to show all interdependencies among activity segments, since these tasks can be routinely managed in the field.

A very important difference between Figures 4-11c and e, other than the 4 day difference in the project durations, lies in the electrical task, which is represented by 3 subactivities in both diagrams. In Figure 4-11c these 3 subactivities are expected to be conducted without interruption. However, in Figure 4-11e this is not possible. Here, the last day of the electrical task (6-7) must follow a 4 day interruption because of the combination effect of successor constraint $SS1$ depicted by activity 5-8, and predecessor constraint $FF1$ depicted by activity 3-6. These constraints require events 5 and 6 to be separated 7 days in time, while the intervening electrical activity requires only 3 days to perform. This forced interruption will henceforth be referred to as *splitting* of the electrical task.

If necessary, *splitting* can be avoided in several ways. First, the duration of the electrical task could be increased from 5 to 9 days. But this is frequently not desirable in projects such as maintenance or construction because it would decrease productivity. The second way to avoid *splitting* would be to delay the start of the electrical task for 4 days, as shown in Figure 4-11g, where it is assumed that activity splitting is not allowed. At first, it may seem that there is no difference between these two alternatives, but this is not so. Reflection on Figure 4-11g shows that delaying the start of the electrical task to avoid *splitting* will delay the start of the finish work, and hence the completion of the project is delayed by 4 days. But increasing the duration of the electrical task will not have this effect. Actually, we have described an anomalous situation where an *increase* of 4 days in the duration of an activity on the critical path (starting 4 days earlier and thus running 4 days longer), will *decrease* the duration of the project by 4 days, from 22 to 18. If you are used to dealing with basic arrow diagram logic ($FS = 0$ logic only), this anomaly will take some getting used to. It results from the fact that the critical path in Figure 4-11g goes "backwards" through activity 5-6, and thus *subtracts* from the total duration of this path. As a result, the project duration shifts in the reverse direction of a shift in the duration of such an activity. That is, the project duration *decreases* when the activity duration *increases*, and *increases* when the activity duration *decreases.* This anomalous situation occurs whenever the critical path *enters* the *completion* of an activity through a *finish* type of constraint (*FF* or *SF*), goes backwards through the activity, and leaves through a *start* type constraint (*SS* or *SF*).

The precedence diagram in Figure 4-11f shows that the entire project is critical, since $ES = LS$ and $EF = LF$ for each task. While it appears that the electrical task has float in Figure 4-11g, this is not true since *splitting* is not allowed. No-splitting is a constraint not explicitly incorporated in the arrow diagram logic.

## Critical Path Characteristics

Wiest[8] describes the anomalous behavior of activity 5-6 in Figure 4-11g picturesquely by stating that this activity is *reverse* critical. Similarly, in Figure 4-11d and e both framing and electrical are called *neutral* critical. They are critical because their $LS = ES$, but they are called *neutral* because their $LF > EF$, and the project duration is independent of the task duration. A task is *neutral* critical when a *pair* of start time constraints result in the critical path entering and exiting from the starting point of the task, or a *pair* of finish time constraints enter and exit from the finish point of a task. These situations could also be referred to as *start* or *finish* critical. In Figure 4-11d and e, the framing and electrical tasks are both *start* critical, while finishing is *normal* critical. That is, a shift in the duration of the finishing task will have a *normal* effect on the project duration, causing it to shift in the same direction. Wiest suggests that precedence diagram computer outputs would be more useful if they identified the way in which tasks are critical. The author suggests that the following nomenclature be considered for this purpose:

*NC* denotes an activity that is *N*ormal *C*ritical; the project duration shifts in the *same* direction as the shift in the duration of a NC-activity.

*RC* denotes an activity that is *R*everse *C*ritical; the project duration shifts in the *reverse* direction as the shift in the duration of a RC-activity.

*BC* denotes an activity that is *Bi-C*ritical; the project duration *increases* as a result of *any* shift in the duration of a BC-activity. (See exercise 13 at the end of this chapter for an example.)

*SC* denotes an activity that is *S*tart *C*ritical; the project duration shifts in the direction of the shift in the *start* time of a SC-activity, but is *neutral* (unaffected) by a shift in the overall duration of the activity.

*FC* denotes an activity that is *F*inish *C*ritical; the project duration shifts in the direction of the shift in the *finish* time of a FC-activity, but is *neutral* (unaffected) by a shift in the overall duration of the activity.

*MNC* denotes an activity whose *M*id portion is *N*ormal *C*ritical.
*MRC* denotes an activity whose *M*id portion is *R*everse *C*ritical.
*MBC* denotes an activity whose *M*id portion is *B*i *C*ritical.

To conclude this discussion, it should be noted that the critical path always starts with a job (or a job start), it ends with a job (or a job finish), and in between it consists of an alternating sequence of jobs and precedence arrows. Although the critical path may pass through a job in any one of the many ways listed above, it *always moves forward* through precedence constraint arrows. Hence, any *increase* (decrease) in the lead-lag times associated with *SS*, *SF*, *FF*, or *FS* constraints on the critical path, will always result in a corresponding *increase* (decrease) in the project duration.

Following the suggestion of stating the nature of the criticality of activities on the critical path, for Figure 4-11d this would consist of the following alternating activities and precedence constraints: Framing (*S*tart *C*ritical–*SC*); *SS*2; electrical (*S*tart *C*ritical–*SC*); *SS*1; finishing (*N*ormal *C*ritical–*NC*). Similarly, for Figure 4-11f it would be: Framing (*NC*); *FF*1; electrical (*RC*); *SS*1; finishing (*NC*). It should be noted here that electrical is labeled reverse critical (*RC*), which puts the manager on notice that any shift in the duration of this activity will shift the duration of the project in the *reverse* direction. As stated above, it is *reverse critical* because its predecessor constraint is a finish type (*FF*1), and its successor constraint is a start type (*SS*1).

## PRECEDENCE DIAGRAMMING COMPUTATIONAL PROCEDURES

Obviously the forward and backward pass computational problem becomes more complex with precedence diagramming, and it calls for establishment of somewhat arbitrary ground rules that were unnecessary with the unique nature of basic arrow diagram logic. In the computational procedures to follow, we will assume that the specified activity durations are fixed, e.g., because of the productivity argument cited above. This assumption can be relaxed, of course, by varying the activity durations of interest, and repeating the calculations. Regarding task splitting, three basic cases will be treated.

*Case 1:* Activity splitting *is not* allowed on any activities.

*Case 2:* Activity splitting *is* allowed on all activities.

*Case 3:* Combination of 1 and 2; activity splitting is permitted only on designated activities.

Figures 4-11g and e represent Cases 1 and 2, respectively. The effect of not allowing splitting (of the electrical task) is a 4 day increase in the project duration. Here, the choice must be made between the (extra) cost of splitting the electrical task, and the cost of a 4 day increase in project duration. Case 3 is provided to allow the project manager to take the possible time (project duration) advantage concomitant with splitting on those activities where it can be tolerated, and to avoid splitting on those activities where it cannot be accommodated.

The computational procedure for Case 1 is reasonably simple and will be described below. The procedure for Case 2 is considerably more complex; it is given in Appendix 4-2. The computational procedure for Case 3 merely amounts to the application of the Case 1 *or* the Case 2 procedure to each activity in turn, depending on whether the activity is designated as one where splitting *is not* allowed, or *is* allowed, respectively.

### Computational Assumptions

The computational procedure for Case 1—No Splitting Allowed, is analogous to the arrow diagram procedure described above. In making the forward pass calculations, one must consider *all* constraints leading into the activity ($j$) in question, i.e., the start time constraints ($SS_{ij}$ and $FS_{ij}$) *as well as* the finish time constraints ($SF_{ij}$ and $FF_{ij}$). For *each* constraint, the early start time for activity ($j$) is computed, and the maximum (latest) of these times then becomes the early start time ($ES_j$) for activity ($j$). Because some project activities may only have finish time constraints, it would be possible for the above procedure to lead to a negative $ES_j$ time, or a time earlier than the specified project start time. For example, referring to Figure 4-12, we see that activity $D$ has no *start* time constraint. If the duration of activity $D$ was 22 (instead of 12), then its early start time would be $EF - D = ES$, or $19 - 22 = -3$ (instead of 7). This would be an erroneous negative value. To prevent the occurrence of this error, an additional time, called the INITIAL TIME, is introduced. It is usually set equal to zero, or else to an arbitrarily specified (nonzero) project scheduled start time, and it overrides the start times computed above if they are all negative, or less (earlier) than the specified project start time.

The backward pass computations follow a similar procedure to find the late finish times for each activity, working backwards along *each* constraint leaving the activity ($i$) in question. In this case, an additional time, called TERMINAL TIME, is required to prevent the occurrence of a late finish time ($LF_i$) *exceeding* the project duration, or the scheduled project completion time. As usual, the project duration is taken as the maximum (latest) of the early finish times computed for each activity in the forward pass computations. For example, this is equal to 42 units in Figure 4-12, which is the largest of all activity early finish times.

The computational procedure given below has the same requirement that prevailed for arrow diagram computations. It requires that the activities are topologically ordered. That is, activities are arranged so that successors to any

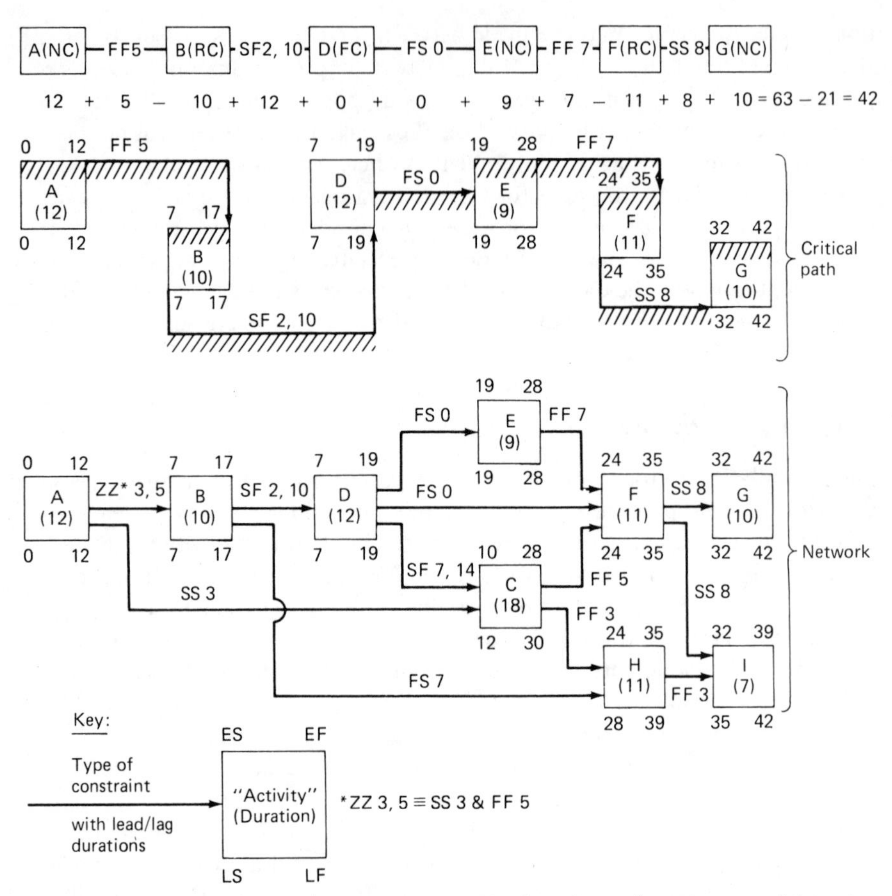


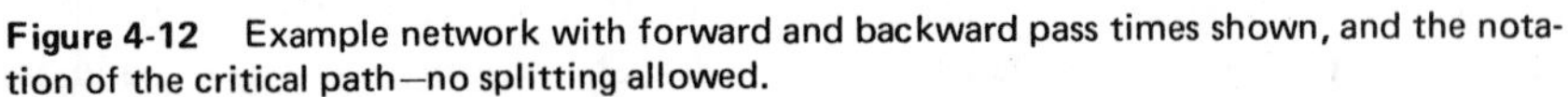
**Figure 4-12** Example network with forward and backward pass times shown, and the notation of the critical path—no splitting allowed.

activity will *always* be found below it in the ordered list. The two step computational procedure is then applied to each activity working the list from the top down. When the computations are performed by hand on a network, this ordering is accomplished automatically by working one path after another, each time going as far as possible. Again, this is the same procedure required to process an arrow diagram.

### Forward Pass Computations—No Splitting Allowed

The following two steps are applied to each project activity, in topological sequence. The term called INITIAL TIME is set equal to zero, or to an arbitrarily specified project scheduled start time.

STEP 1. Compute $ES_j$, the early start time of the activity $(j)$ in question. It is the maximum (latest) of the set of start times which includes the INITIAL TIME, and one start time computed from *each* constraint going to the activity $(j)$ from predecessor activities indexed by $(i)$.

$$ES_j = \underset{\text{all } i}{\text{MAX}} \begin{Bmatrix} \text{INITIAL TIME} \\ EF_i + FS_{ij} \\ ES_i + SS_{ij} \\ EF_i + FF_{ij} - D_j \\ ES_i + SF_{ij} - D_j \end{Bmatrix}$$

STEP 2. $EF_j = ES_j + D_j$

### Backward Pass Computations—No Splitting Allowed

The following two steps are applied to each project activity in the reverse order of the forward pass computations. The term called TERMINAL TIME is set equal to the project duration, or to an arbitrarily specified project scheduled completion time.

STEP 1. Compute $LF_i$ the late finish time of the activity $(i)$ in question. It is the minimum (earliest) of the set of finish times which includes the TERMINAL TIME, and one finish time computed from *each* constraint going from activity $(i)$, to successor activities indexed by $(j)$.

$$LF_i = \underset{\text{all } j}{\text{MIN}} \begin{Bmatrix} \text{TERMINAL TIME} \\ LS_j - FS_{ij} \\ LF_j - FF_{ij} \\ LS_j - SS_{ij} + D_j \\ LF_j - SF_{ij} + D_i \end{Bmatrix}$$

STEP 2. $LS_i = LF_i - D_i$

### Example Problem

To illustrate the application of the above algorithm, a small network consisting of 9 activities with a variety of constraints, is shown in Figure 4-12. The forward pass calculations are as follows, based on the assumption that the project starts at time zero, i.e., INITIAL TIME = 0.

*Activity A*

$$ES_A = \{\text{INITIAL TIME} = 0\} = 0$$

$$EF_A = ES_A + D_A = 0 + 12 = 12$$

*Activity B*

$$ES_B = \max_{A} \begin{Bmatrix} \text{INITIAL TIME} = 0 \\ ES_A + SS_{AB} = 0 + 3 = 3 \\ EF_A + FF_{AB} - D_B = 12 + 5 - 10 = 7 \end{Bmatrix} = 7$$

$$EF_B = ES_B + D_B = 7 + 10 = 17$$

*Activity D*

$$ES_D = \max_{B} \begin{Bmatrix} \text{INITIAL TIME} = 0 \\ ES_B + SF_{BD} - D_D = 7 + (2 + 10) - 12 = 7 \end{Bmatrix} = 7$$

$$EF_D = ES_D + D_D = 7 + 12 = 19$$

*Activity C*

$$ES_C = \max_{A,D} \begin{Bmatrix} \text{INITIAL TIME} = 0 \\ ES_A + SS_{AC} = 0 + 3 = 3 \\ ES_D + SF_{DC} - D_C = 7 + (7 + 14) - 18 = 10 \end{Bmatrix} = 10$$

$$EF_C = ES_C + D_C = 10 + 18 = 28$$

etc.

The backward pass calculations are as follows, wherein the TERMINAL TIME is set equal to the project duration, determined from the forward pass calculations to be 42, i.e., the *EF* time for the last critical path activity G.

*Activity G*

$$LF_G = \{\text{TERMINAL TIME} = 42\} = 42$$

$$LS_G = LF_G - D_G = 42 - 10 = 32$$

*Activity I*

$$LF_I = \{\text{TERMINAL TIME} = 42\} = 42$$

$$LS_I = LF_I - D_I = 42 - 7 = 35$$

*Activity H*

$$LF_{\mathrm{H}} = \underset{\mathrm{I}}{\mathrm{MIN}} \begin{Bmatrix} \text{TERMINAL TIME} = 42 \\ LF_{\mathrm{I}} - FF_{\mathrm{HI}} = 42 - 3 = 39 \end{Bmatrix} = 39$$

$$LS_{\mathrm{H}} = LF_{\mathrm{H}} - D_{\mathrm{H}} = 39 - 11 = 28$$

*Activity F*

$$LF_{\mathrm{F}} = \underset{\mathrm{G,I}}{\mathrm{MIN}} \begin{Bmatrix} \text{TERMINAL TIME} = 42 \\ LS_{\mathrm{G}} - SS_{\mathrm{FG}} + D_{\mathrm{F}} = 32 - 8 + 11 = 35 \\ LS_{\mathrm{I}} - SS_{\mathrm{FI}} + D_{\mathrm{F}} = 35 - 8 + 11 = 38 \end{Bmatrix} = 35$$

$$LS_{\mathrm{F}} = LF_{\mathrm{F}} - D_{\mathrm{F}} = 35 - 11 = 24$$

etc.

From the computational results shown in Figure 4-12, the critical path consists of activities A-B-D-E-F-G. The nature of the criticality of each activity is indicated at the top of Figure 4-12, along with the critical constraints between each pair of activities. Activities A, E, and G are *normal* critical, activities B and F are *reverse* critical (noted by the reverse direction cross-hatching), and activity D is only *finish* time critical. The duration of the critical path, 42, is also noted, with the net contributions of the activity durations being (12 - 10 + 0 + 9 - 11 + 10) = 10 and the contributions of the constraints being (5 + 12 + 0 + 7 + 8) = 32, for a total of 42 time units. The early/late start/finish times for each activity have the conventional interpretations. For example, for the critical activity E, both the early and late start/finish times are 19 and 28; the activity has no slack. But for activity H, the early start/finish times are 24 and 35, while the late start/finish times are 28 and 39. In this case, the activity has 4 units of activity slack or free slack, because the completion of activity H can be delayed up to 4 units without affecting the slack on its successor activity I.

To illustrate the use of computers for these computations, the completed basic scheduling computations for this network are shown in Figure 4-13, using a program written by K. C. Crandall. The upper portion of this figure gives times in elapsed working days, while the lower portion is given in calendar dates. The latter assumes that start times denote the beginning of the day, and finish times the end of the day. For example, activity A has a duration of 12 days, corresponding to an early start date of 3 August and an early finish date of 14 August. This computer run assumes, of course, a 7 day work week. The same information is very conveniently displayed in bar-chart form in Figure 4-14. The upper portion of this figure gives the early-start activity schedule, while the lower portion gives the late-start schedule. A study of this figure will reveal that

```
OWNER: MODER TEXT - THIRD EDITION                              CONTRACTOR: I AM AN EXCELLENT CONTRACTOR

COMPUTER SOLUTION SHOWING THE 'NON SPLITTING' OPTION CALCULATION              UPDATE                             NET MOD    1
FOR THE THIRD EDITION OF THE MODER TEXT ON 'CPM & PERT'.                      CALENDAR      3AUG81 16:55:11     RUN NO.    1
TEST CALCULATED ON PACKAGE BY K C CRANDALL                                    MODIFICATION  3AUG81 17:04:38
                                                                              CALCULATION   3AUG81 17:17:11

ACTIVITY SCHEDULE LISTING -                                                                                      PAGE    1
```

| | LABL | ACTIVITY DESCRIPTION | DURATION | START EARLY | START LATE | FINISH EARLY | FINISH LATE | FLOAT TOTAL | FLOAT FREE |
|---|---|---|---|---|---|---|---|---|---|
| * | AAAA | THIS IS ACTIVITY 'A' | 12 | 0 | 0 | 12 | 12 | 0 | 0 |
| * | BBBB | THIS IS ACTIVITY 'B' | 10 | 7 | 7 | 17 | 17 | 0 | 0 |
| | CCCC | THIS IS ACTIVITY 'C' | 18 | 10 | 12 | 28 | 30 | 2 | 2 |
| * | DDDD | THIS IS ACTIVITY 'D' | 12 | 7 | 7 | 19 | 19 | 0 | 0 |
| * | EEEE | THIS IS ACTIVITY 'E' | 9 | 19 | 19 | 28 | 28 | 0 | 0 |
| * | FFFF | THIS IS ACTIVITY 'F' | 11 | 24 | 24 | 35 | 35 | 0 | 0 |
| * | GGGG | THIS IS ACTIVITY 'G' | 10 | 32 | 32 | 42 | 42 | 0 | 0 |
| | HHHH | THIS IS ACTIVITY 'H' | 11 | 24 | 28 | 35 | 39 | 4 | 4 |
| | IIII | THIS IS ACTIVITY 'I' | 7 | 32 | 35 | 39 | 42 | 3 | 3 |

```
ACTIVITY SCHEDULE LISTING -                                                                                      PAGE    1
```

| | LABL | ACTIVITY DESCRIPTION | DURATION | START EARLY | START LATE | FINISH EARLY | FINISH LATE | FLOAT TOTAL | FLOAT FREE |
|---|---|---|---|---|---|---|---|---|---|
| * | AAAA | THIS IS ACTIVITY 'A' | 12 | 3 AUG 81 | 3 AUG 81 | 14 AUG 81 | 14 AUG 81 | 0 | 0 |
| * | BBBB | THIS IS ACTIVITY 'B' | 10 | 10 AUG 81 | 10 AUG 81 | 19 AUG 81 | 19 AUG 81 | 0 | 0 |
| | CCCC | THIS IS ACTIVITY 'C' | 18 | 13 AUG 81 | 15 AUG 81 | 30 AUG 81 | 1 SEP 81 | 2 | 2 |
| * | DDDD | THIS IS ACTIVITY 'D' | 12 | 10 AUG 81 | 10 AUG 81 | 21 AUG 81 | 21 AUG 81 | 0 | 0 |
| * | EEEE | THIS IS ACTIVITY 'E' | 9 | 22 AUG 81 | 22 AUG 81 | 30 AUG 81 | 30 AUG 81 | 0 | 0 |
| * | FFFF | THIS IS ACTIVITY 'F' | 11 | 27 AUG 81 | 27 AUG 81 | 6 SEP 81 | 6 SEP 81 | 0 | 0 |
| * | GGGG | THIS IS ACTIVITY 'G' | 10 | 4 SEP 81 | 4 SEP 81 | 13 SEP 81 | 13 SEP 81 | 0 | 0 |
| | HHHH | THIS IS ACTIVITY 'H' | 11 | 27 AUG 81 | 31 AUG 81 | 6 SEP 81 | 10 SEP 81 | 4 | 4 |
| | IIII | THIS IS ACTIVITY 'I' | 7 | 4 SEP 81 | 7 SEP 81 | 10 SEP 81 | 13 SEP 81 | 3 | 3 |

**Figure 4-13** Time and date computer outputs for the basic scheduling computations on the network shown in Figure 4-12—no splitting allowed.

```
OWNER: MODER TEXT - THIRD EDITION                          CONTRACTOR: I AM AN EXCELLENT CONTRACTOR

COMPUTER SOLUTION SHOWING THE 'NON SPLITTING' OPTION CALCULATION            UPDATE                             NET MOD     1
FOR THE THIRD EDITION OF THE MODER TEXT ON 'CPM & PERT'.                    CALENDAR      3AUG81 16:55:11      RUN NO.     1
TEST CALCULATED ON PACKAGE BY K C CRANDALL                                  MODIFICATION  3AUG81 17:04:38
                                                                            CALCULATION   3AUG81 17:17:11

                                    * * * CPM EARLY START ACTIVITY BARCHART * * *                           PAGE:  1-  1

  SYMBOLS USED ARE: <*> CRITICAL PATH ; <O> WORK DAY ; <-> TOTAL FLOAT ; <.> HOLIDAY-WEEKEND ; <^> UPDATE DATE

                                      AUG 81                        SEP 81
                                             1         2         3         1
                                      34567890123456789012345678901123456789012 3
                                      MTWTFSSMTWTFSSMTWTFSSMTWTFSSMTWTFSSMTWTFSS
----------------------------------------------------------------------------------------------------------------------------
LABL DESCRIPTION           CALNDR DAYS ->      10        20        30        40        50        60        70        80        90
----------------------------------------------------------------------------------------------------------------------------
AAAA THIS IS ACTIVITY 'A'             ************
BBBB THIS IS ACTIVITY 'B'                    **********
CCCC THIS IS ACTIVITY 'C'                       000000000000000000--
DDDD THIS IS ACTIVITY 'D'                    ************
EEEE THIS IS ACTIVITY 'E'                                *********
FFFF THIS IS ACTIVITY 'F'                                    ***********
GGGG THIS IS ACTIVITY 'G'                                            **********
HHHH THIS IS ACTIVITY 'H'                                    00000000000----
IIII THIS IS ACTIVITY 'I'                                            0000000---
                                    * * * CPM  LATE START ACTIVITY BARCHART * * *                           PAGE:  1-  1

  SYMBOLS USED ARE: <*> CRITICAL PATH ; <O> WORK DAY ; <-> TOTAL FLOAT ; <.> HOLIDAY-WEEKEND ; <^> UPDATE DATE

                                      AUG 81                        SEP 81
                                             1         2         3         1
                                      34567890123456789012345678901123456789012 3
                                      MTWTFSSMTWTFSSMTWTFSSMTWTFSSMTWTFSSMTWTFSS
----------------------------------------------------------------------------------------------------------------------------
LABL DESCRIPTION           CALNDR DAYS ->      10        20        30        40        50        60        70        80        90
----------------------------------------------------------------------------------------------------------------------------
AAAA THIS IS ACTIVITY 'A'             ************
BBBB THIS IS ACTIVITY 'B'                    **********
CCCC THIS IS ACTIVITY 'C'                         ******************
DDDD THIS IS ACTIVITY 'D'                    ************
EEEE THIS IS ACTIVITY 'E'                                *********
FFFF THIS IS ACTIVITY 'F'                                    ***********
GGGG THIS IS ACTIVITY 'G'                                            **********
HHHH THIS IS ACTIVITY 'H'                                        ***********
IIII THIS IS ACTIVITY 'I'                                                *******
```

**Figure 4-14** Bar-chart computer outputs for the basic scheduling computations on the network shown in Figure 4-12—no splitting allowed.

the information given is identical to Figure 4-13, but in graphical rather than numerical form.

## DISCUSSION OF COMPUTATIONS WITH SPLITTING ALLOWED

The Case 2–Splitting Allowed computational procedure is more complex than Case 1, and for this reason the algorithm is deferred to Appendix 4-2. This procedure assumes that activities can be split whenever the *combination* of constraints associated with the activity in question result in early finish and early start (or late finish and late start) times whose difference exceeds the activity duration. When this occurs, the activity is assumed to split so that it *preserves the continuity of work flow.* For example, in Figure 4-11e above, the electrical task must split as shown: 4 days of electrical, 4 days of interruption, and then the final 1 day of electrical. Note how this split preserves the continuity of work flow. Just as framing completes its duration of 10 days, electrical *recommences* for its final 1 day of its overall duration of 5 days. A different split is assumed in the backward pass to preserve flow in the reverse direction. That is, the first portion of electrical would have a duration of 1 day and the last portion of 4 days. This preserves flow moving backwards through finishing, and then the 1 day initial portion of electrical. An important result of this procedure is that the early/late start/finish times will be identical to those that would be obtained for an equivalent arrow diagram. (The reader may find it helpful to redraw the arrow diagram shown in Figure 4-11e and perform the basic arrow diagram forward and backward pass calculations to verify this assertion.) It is important to note, however, that these splitting rules are arbitrary. The actual activity split can be chosen later by the "foreman" to be at any acceptable point within the range of the start and finish times computed for the job in question. It should also be noted that other splitting rules, e.g., split all activities in the middle, may give different project durations, longer or shorter, from that which results from the rule adopted here. Without referring further to the algorithm, the results of its application will be given here to illustrate how it would be used when a computer is utilized to perform the computations.

Utilizing the same network shown in Figure 4-12 where no splitting was allowed, the results of allowing splitting are given in Figure 4-15. To show that these results are identical to those obtained for the equivalent arrow diagram, Figure 4-16 is given with early/late start/finish times and (total) path float times noted. To compare these two figures, recall that the precedence diagram treats each task (activity) in totality. For example, activity F is broken into 4 subactivities in Figure 4-16. The early start time of the first subactivity is 15, and the early finish time of the last one is 31. These are the only two times shown for activity F in Figure 4-15. Similarly, the latest start/finish times are 20 and 38, respectively. A special situation occurs here in that activity F is on the criti-

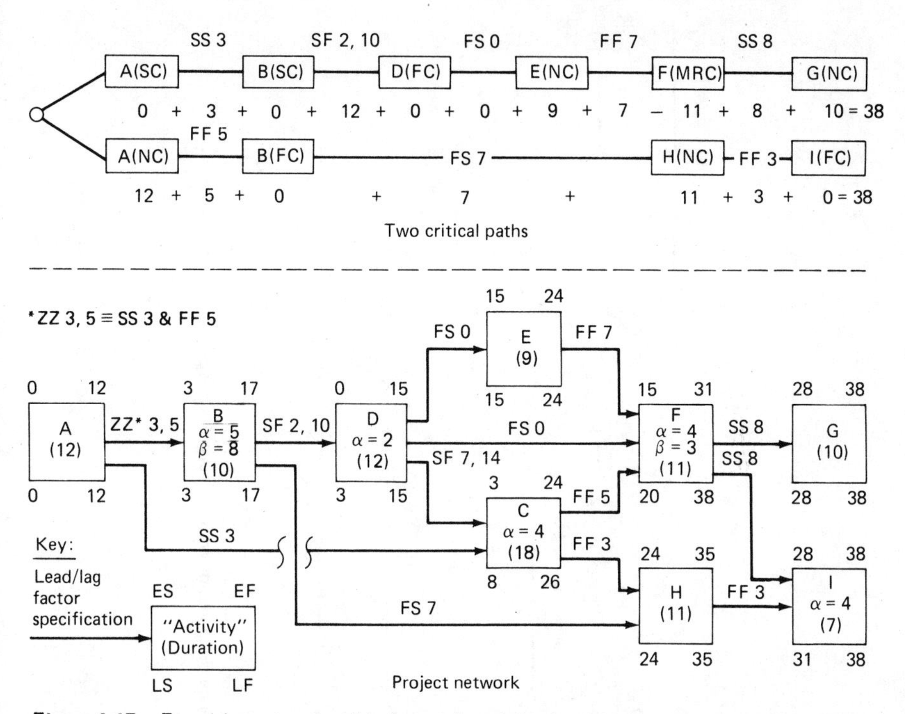


**Figure 4-15** Example network with forward and backward pass times shown—splitting allowed.

cal path, yet its start time has 5 units (20-15) of float, and its finish time has 7 units (38-31) of float. The critical nature of activity F lies in the fact that its mid-subactivities (8-10 and 10-15 in Figure 4-16) are critical. This fact is noted at the top of Figure 4-15, where the critical path is shown to contain F(*MRC*), i.e., activity F has a *m*idportion that is *r*everse *c*ritical. The latter occurs because its predecessor constrains its finish time (*FF*7) and its successor constrains its start time (*SS*8).

Comparing Figures 4-12 and 4-15, we note that the project duration is 4 units less (42-38) when splitting is allowed. This results from the fact that activity B is *required* to split, because its *late start time plus its duration is less* (*earlier*) than its *early finish time.* That is,

$$LS_B + D_B < EF_B \qquad \text{or} \qquad 3 + 10 < 17.$$

Thus, there must be a 4 unit idle time between the two segments of activity B, and this results in a 4 unit reduction in the project duration. This reduction occurs because the splitting of activity B allows it to start 4 units earlier, and hence its successor D can finish 4 units earlier, since these two activities are connected by a critical start finish constraint (*SF*2, 10). Examination of all other activities

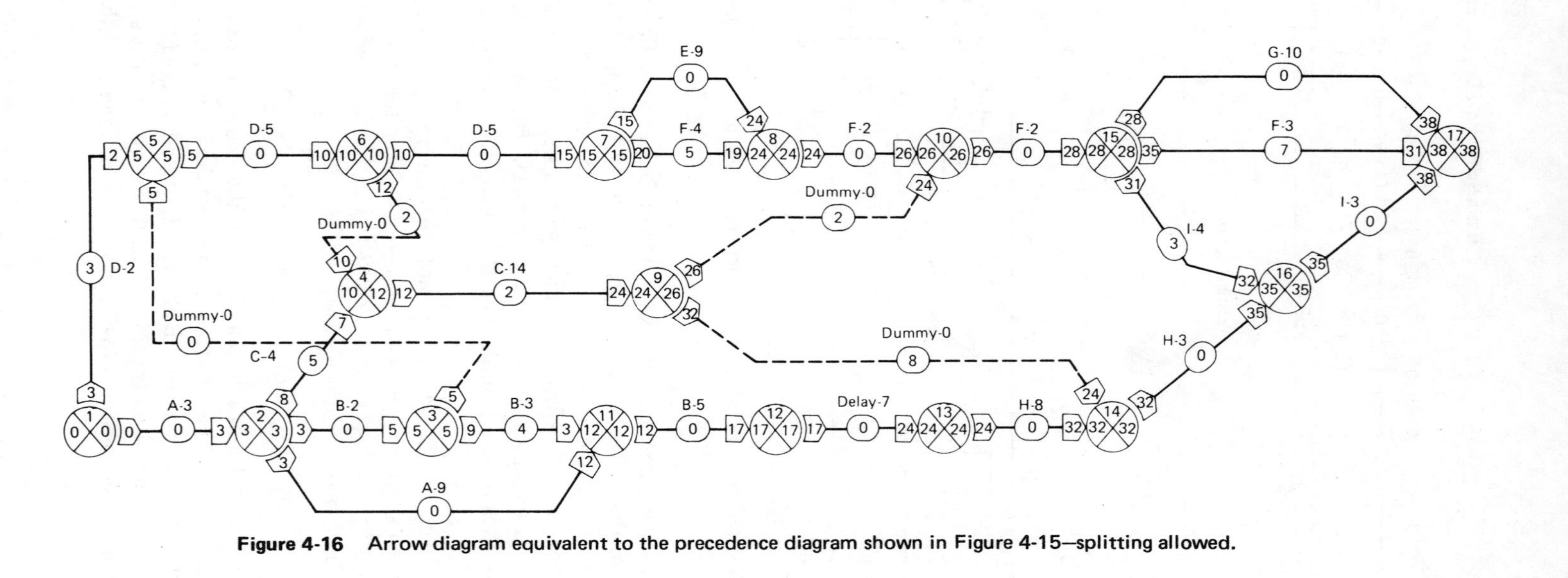


**Figure 4-16** Arrow diagram equivalent to the precedence diagram shown in Figure 4-15—splitting allowed.

indicates that none are *required* to split because, in each case, their late start times plus their durations equals or exceeds the early finish time ($LS + D \geqq EF$).

Another difference in Figures 4-12 and 15 is found in their critical paths. In Figure 4-12, it is a single path, with a duration of 42 units. In Figure 4-15, there are two critical paths, each having a duration of 38 units. These paths are shown at the top of Figure 4-15.

To illustrate the use of computers for these splitting allowed computations, the completed basic scheduling computations for this network are shown in Figure 4-17. The upper portion of this figure essentially gives the results shown in Figure 4-15 in elapsed working days, and in the lower portion in calendar dates. The results given in Figures 4-15 and 4-17 are identical. From the above discussion, it is clear that the equivalent arrow diagram in Figure 4-16, contains more information about the subactivity schedules than is shown in either Figures 4-15 or 4-17. To recapture this information, the precedence diagram computations are given in bar-chart form in Figure 4-18. Note how the two portions of this figure give identical activity breakdowns, and early/late start/finish and float times. For example, the peculiar critical nature of activity F was discussed above. In Figure 4-16, note that the middle 4 units of activity F are critical (zero float) with start/finish times of 24 and 28. This is shown in the upper portion of Figure 4-18 by the critical symbol (*) placed on days 25, 26, 27 and 28. Also, the last 3 units of F are noted with the noncritical symbol (0), followed by 7 units of slack noted by the slack symbol (-), exactly as shown on the arrow diagram in Figure 4-16. To recapitulate for activity F with splitting allowed, the bar chart indicates that its *first four* units can be started as early as the 16th day, and they have 5 days of float. The middle 4 days are critical and must be performed on days 25 through 28. Finally, the last 3 units have an early start on the 29th day, with 7 days of float. This figure does indeed contain a large amount of information with the advantage of having all subactivities shown by a single bar (line). A longer time will be required to instruct users how to interpret this chart, however, the benefit/cost ratio appears to favor its more widespread use in the future.

## SUMMARY OF BASIC SCHEDULING COMPUTATIONS

The basic scheduling computations have been defined for arrow and node diagrams as the computation of the earliest and latest start and finish times of each activity; the computation of path float then follows immediately. In this chapter a simple procedure has been presented to make these computations directly on the network arrow diagram using special symbols.

As an alternative, one may make them in a tabular manner, separate from the network. A tabular procedure is given in Appendix 4-1. In comparison with computations on the network, tabular procedures are somewhat tedious and are

OWNER: MODER TEXT - THIRD EDITION CONTRACTOR: I AM AN EXCELLENT CONTRACTOR

COMPUTER SOLUTION SHOWING THE 'SPLITTING' OPTION CALCULATION
FOR THE THIRD EDITION OF THE MODER TEXT ON 'CPM & PERT'.
TEST CALCULATED ON PACKAGE BY K C CRANDALL

UPDATE
CALENDAR 3AUG81 16:55:11
MODIFICATION 3AUG81 17:04:38
CALCULATION 3AUG81 17:42:38

NET MOD 1
RUN NO. 1

ACTIVITY SCHEDULE LISTING - PAGE 1

| | LABL | ACTIVITY DESCRIPTION | DURATION | START EARLY | START LATE | FINISH EARLY | FINISH LATE | FLOAT TOTAL | FLOAT FREE |
|---|---|---|---|---|---|---|---|---|---|
| * | AAAA | THIS IS ACTIVITY 'A' | 12 | 0 | 0 | 12 | 12 | 0 | 0 |
| * | BBBB | THIS IS ACTIVITY 'B' | 10 | 3 | 3 | 17 | 17 | 0 | 0 |
| | CCCC | THIS IS ACTIVITY 'C' | 18 | 3 | 8 | 24 | 26 | 2 | 2 |
| * | DDDD | THIS IS ACTIVITY 'D' | 12 | 0 | 3 | 15 | 15 | 0 | 0 |
| * | EEEE | THIS IS ACTIVITY 'E' | 9 | 15 | 15 | 24 | 24 | 0 | 0 |
| * | FFFF | THIS IS ACTIVITY 'F' | 11 | 15 | 20 | 31 | 38 | 0 | 0 |
| * | GGGG | THIS IS ACTIVITY 'G' | 10 | 28 | 28 | 38 | 38 | 0 | 0 |
| * | HHHH | THIS IS ACTIVITY 'H' | 11 | 24 | 24 | 35 | 35 | 0 | 0 |
| * | IIII | THIS IS ACTIVITY 'I' | 7 | 28 | 31 | 38 | 38 | 0 | 0 |

ACTIVITY SCHEDULE LISTING - PAGE 1

| | LABL | ACTIVITY DESCRIPTION | DURATION | START EARLY | START LATE | FINISH EARLY | FINISH LATE | FLOAT TOTAL | FLOAT FREE |
|---|---|---|---|---|---|---|---|---|---|
| * | AAAA | THIS IS ACTIVITY 'A' | 12 | 3 AUG 81 | 3 AUG 81 | 14 AUG 81 | 14 AUG 81 | 0 | 0 |
| * | BBBB | THIS IS ACTIVITY 'B' | 10 | 6 AUG 81 | 6 AUG 81 | 19 AUG 81 | 19 AUG 81 | 0 | 0 |
| | CCCC | THIS IS ACTIVITY 'C' | 18 | 6 AUG 81 | 11 AUG 81 | 26 AUG 81 | 28 AUG 81 | 2 | 2 |
| * | DDDD | THIS IS ACTIVITY 'D' | 12 | 3 AUG 81 | 6 AUG 81 | 17 AUG 81 | 17 AUG 81 | 0 | 0 |
| * | EEEE | THIS IS ACTIVITY 'E' | 9 | 18 AUG 81 | 18 AUG 81 | 26 AUG 81 | 26 AUG 81 | 0 | 0 |
| * | FFFF | THIS IS ACTIVITY 'F' | 11 | 18 AUG 81 | 23 AUG 81 | 2 SEP 81 | 9 SEP 81 | 0 | 0 |
| * | GGGG | THIS IS ACTIVITY 'G' | 10 | 31 AUG 81 | 31 AUG 81 | 9 SEP 81 | 9 SEP 81 | 0 | 0 |
| * | HHHH | THIS IS ACTIVITY 'H' | 11 | 27 AUG 81 | 27 AUG 81 | 6 SEP 81 | 6 SEP 81 | 0 | 0 |
| * | IIII | THIS IS ACTIVITY 'I' | 7 | 31 AUG 81 | 3 SEP 81 | 9 SEP 81 | 9 SEP 81 | 0 | 0 |

**Figure 4-17** Time and date computer outputs for the basic scheduling computations on the network shown in Figure 4-15—splitting allowed.

```
OWNER: MODER TEXT - THIRD EDITION                           CONTRACTOR: I AM AN EXCELLENT CONTRACTOR

COMPUTER SOLUTION SHOWING THE 'SPLITTING' OPTION CALCULATION                 UPDATE                                NET MOD    1
FOR THE THIRD EDITION OF THE MODER TEXT ON 'CPM & PERT'.                     CALENDAR      3AUG81 16:55:11   RUN NO.    1
TEST CALCULATED ON PACKAGE BY K C CRANDALL                                   MODIFICATION  3AUG81 17:04:38
                                                                             CALCULATION   3AUG81 17:42:38

                                       * * * CPM EARLY START ACTIVITY BARCHART * * *                                  PAGE:  1-  1

    SYMBOLS USED ARE: <*> CRITICAL PATH ; <O> WORK DAY ; <-> TOTAL FLOAT ; <.> HOLIDAY-WEEKEND ; <^> UPDATE DATE

                                       AUG 81                         SEP 81
                                              1         2         3
                                       345678901234567890123456789011234567890
                                       MTWTFSSMTWTFSSMTWTFSSMTWTFSSMTWTFSSMTW
-------------------------------------------------------------------------------------------------------------------------------------
LABL DESCRIPTION         CALNDR DAYS ->      10        20        30        40        50        60        70        80        90
-------------------------------------------------------------------------------------------------------------------------------------
AAAA THIS IS ACTIVITY 'A'              ************
BBBB THIS IS ACTIVITY 'B'                 **000----*****
CCCC THIS IS ACTIVITY 'C'                 0000---0000000000000--
DDDD THIS IS ACTIVITY 'D'              00---**********
EEEE THIS IS ACTIVITY 'E'                             *********
FFFF THIS IS ACTIVITY 'F'                             0000-----****000-------
GGGG THIS IS ACTIVITY 'G'                                          **********
HHHH THIS IS ACTIVITY 'H'                                      ***********
IIII THIS IS ACTIVITY 'I'                                          0000---***
                                       * * * CPM  LATE START ACTIVITY BARCHART * * *                                  PAGE:  1-  1

    SYMBOLS USED ARE: <*> CRITICAL PATH ; <O> WORK DAY ; <-> TOTAL FLOAT ; <.> HOLIDAY-WEEKEND ; <^> UPDATE DATE

                                       AUG 81                         SEP 81
                                              1         2         3
                                       345678901234567890123456789011234567890
                                       MTWTFSSMTWTFSSMTWTFSSMTWTFSSMTWTFSSMTW
-------------------------------------------------------------------------------------------------------------------------------------
LABL DESCRIPTION         CALNDR DAYS ->      10        20        30        40        50        60        70        80        90
-------------------------------------------------------------------------------------------------------------------------------------
AAAA THIS IS ACTIVITY 'A'              ************
BBBB THIS IS ACTIVITY 'B'                 **----********
CCCC THIS IS ACTIVITY 'C'                      ******************
DDDD THIS IS ACTIVITY 'D'                 ************
EEEE THIS IS ACTIVITY 'E'                             *********
FFFF THIS IS ACTIVITY 'F'                                  ********-------***
GGGG THIS IS ACTIVITY 'G'                                          **********
HHHH THIS IS ACTIVITY 'H'                                      ***********
IIII THIS IS ACTIVITY 'I'                                             *******
```

**Figure 4-18** Bar-chart computer outputs for the basic scheduling computations on the network shown in Figure 4-15—splitting allowed.

less efficient than computations made directly on the network. A tabular procedure is included in this text primarily because of its value in helping to understand the logic of computer procedures used to carry out the basic scheduling computations. Anyone planning to program a computer for this purpose should study these tabular procedures.

The computational problems resulting from the introduction of multiple initial and terminal network events, or the introduction of scheduled times on key milestone events have been treated in this chapter. Their effects on the computational procedures are trivial; however, occasions may arise where they can be profitably applied. The problem of network time status updating was also considered. Again this results in a trivial modification of the basic scheduling computations, but it is an important procedure in the control phase of project management. Finally, the computational procedures associated with precedence diagrams is given, for the cases where splitting of activities is or is not allowed. In this procedure, activities are treated in their totality; early and late start times are given for the beginning segment, and early and late finish times for the ending segment of each activity.

After having developed arrow and precedence diagram procedures in this chapter, it is clear that each has some significant advantages over the other. The uniqueness and simplicity of the arrow or node diagram are very desirable properties. However, the compactness of the precedence diagram, 9 activities in Figure 4-15 vs. 25 activities in Figure 4-16, is also a significant advantage for precedence diagrams. The latter advantage will be greatest in construction type projects where concurrency of activities is the rule rather than the exception, and least in some developmental type projects where concurrency tends to be the exception. On this basis, one might conjecture that the use of precedence diagramming will undoubtedly grow in the 1980s, but not to the point of complete replacement of the arrow diagram.

## REFERENCES


1. Battersby, A., *Network Analysis for Planning and Scheduling*, St. Martins Press, Inc., New York, 1964, Chapter 3.
2. Crandall, K. C., "Project Planning With Precedence Lead/Lag Factors," *Project Management Quarterly*, Vol. 4, No. 3, Sept. 1973, pp. 18–27.
3. Fondahl, J. W., *A Noncomputer Approach to the Critical Path Method for the Construction Industry*, Dept. of Civil Eng., Stanford University, Stanford, Calif., 1st Ed., 1961, 2nd Ed., 1962.
4. Fulkerson, D. R., "Expected Critical Path Lengths in PERT Networks," *Operations Research*, Vol. 10, No. 6, Nov.–Dec. 1962, pp. 808–817.
5. IBM, *Project Management System, Application Description Manual* (H20-0210), IBM, 1968.
6. Moder, J. J., "How to Do CPM Scheduling Without a Computer," *Engineering News-Record*, March 14, 1963, pp. 30–34.

7. Montalbano, M., "High-Speed Calculation of Critical Paths of Large Networks," IBM Systems Research and Development Center *Technical Report*, Palo Alto, California, undated report, about 1963.
8. Weist, Jerry D., "Precedence Diagramming Methods: Some Unusual Characteristics and Their Implications for Project Managers," *Journal of Operations Management*, Vol. 1, No. 3, February 1981, pp. 121–130.

## EXERCISES

1. a. In Figure 4-6, suppose an activity 7-5 must be added to the network which requires 1 time unit to carry out. Will this change any of the times computed in the basic scheduling computations?
   b. What time value for activity 7-5 would cause it to just become critical?
   c. Suppose the project represented by Figure 4-6 is the maintenance of a chemical pipeline in which activity 0-6 represents the deactivation of the line. To minimize the time the line is out of service, when would you schedule this activity?
2. Redraw the network shown in Figure 4-7 and perform the network computations using the following scheduled times.
   a. The main project is scheduled to start at time zero.
   b. The activities leading to event 210 are scheduled to be completed 12 days after the start of the main project.
   c. The scheduled time for the completion of activity 108–211 is 20 days after the start of the main project.
   d. The scheduled time for the completion of the main project (event 109) is also 20 days.
3. a. In exercise 2 what is the critical path?
   b. In exercise 2 what is the effect of assigning a scheduled time of 16 to event 106, or a time of 12 to event 106?
4. It is possible to modify the network in Figure 4-7 so that the correct basic scheduling computations can be carried out by the simple procedure described at the beginning of this chapter. The required modifications are in the form of dummy type activities with suitable time estimates. For example, the addition of an activity 101–210 with a time estimate of 8 would eliminate the multiple initial events. If the main project had a scheduled completion time of 17, i.e., $L_{109} = 17$, what activity and time estimate would eliminate the multiple end events?
5. A reactor and storage tank are interconnected by a 3″ insulated process line that needs periodic replacement. There are valves along the lines and at the terminals and these need replacing as well. No pipe and valves are in stock. Accurate, as built, drawings exist and are available. The line is overhead and requires scaffolding. Pipe sections can be shop fabricated at the plant. Adequate craft labor is available.

   You are the maintenance and construction superintendent responsible for this project. The works engineer has requested your plan and schedule for a review with the operating supervision. The plant methods and standards section has furnished the following data. The precedents for each activity have been determined from a familiarity with similar projects.

| *Symbol* | *Activity Description* | *Time (Hrs.)* | *Precedents* |
|---|---|---|---|
| *A* | Develop required material list | 8 | — |
| *B* | Procure pipe | 200 | *A* |
| *C* | Erect scaffold | 12 | — |
| *D* | Remove scaffold | 4 | *I, M* |
| *E* | Deactivate line | 8 | — |
| *F* | Prefabricate sections | 40 | *B* |
| *G* | Place new pipes | 32 | *F, L* |
| *I* | Fit up pipe and Valves | 8 | *G, K* |
| *J* | Procure valves | 225 | *A* |
| *K* | Place valves | 8 | *J, L* |
| *L* | Remove old pipe and valves | 35 | *C, E* |
| *M* | Insulate | 24 | *G, K* |
| *N* | Pressure test | 6 | *I* |
| *O* | Clean-up and start-up | 4 | *D, N* |

a. Sketch the arrow diagram of this project plan. Hint: at least three dummy arrows are required.
b. Make the forward pass calculations on this network, and indicate the critical path and its length.
c. For obvious reasons, activity *E* "Deactivate line" should be initiated as late as possible. What is the latest allowable time for the initiation of this activity?
d. List the various network paths in decreasing order of criticality.

6. The network plan in exercise 5 is subject to criticism because failure to pass the pressure test could result in several problems. How would you network this project to avoid this criticism, and what is its effect on the expected project duration?

7. Update the network given in Figure 4-6 based on the following activity progress report submitted at the end of the fifth working day.

| *Activity* | *Start Time* | *Finish Time* | *Modifications* |
|---|---|---|---|
| 0-1 | 1 | 3 | — |
| 1-2 | 5 | — | — |
| 0-3 | 0 | 2 | — |
| 3-4 | 3 | — | — |
| 3-7 | 2 | — | — |
| 0-6 | 5 | — | — |
| 5-8 | — | — | Activity duration estimate increased to four days for 5-8. |

a. What is the current status of this project with respect to a scheduled completion time of 15 days?
b. What activities must be expedited to alleviate the situation found in (a)?

8. Referring to Figure 4-7 in the text, trace out the critical paths from end events 109 and 211, using only the forward pass computations shown on the network.
   a. From this information alone, can you say which path is the critical path?
   b. Explain why you need the backward pass computations, or a portion of it, to specify that 210-106-108-211 is the critical path.
9. Redraw the network presented in Figure 4-5 using the node scheme.
   a. Complete only the forward pass computations, and show that they are adequate to identify the critical path, using the same procedure described in the text for Figure 4-5.
   b. Complete the backward pass computations and the path float computations. Check your results against Figure 4-6.
10. Redraw the network presented in Figure 4-7 using the node scheme, and show that the forward and backward pass, and path float calculations lead to the same results as obtained with an arrow diagram.
11. Consider the precedence diagram network shown in Figure 4-19, with activity durations noted inside of each node.

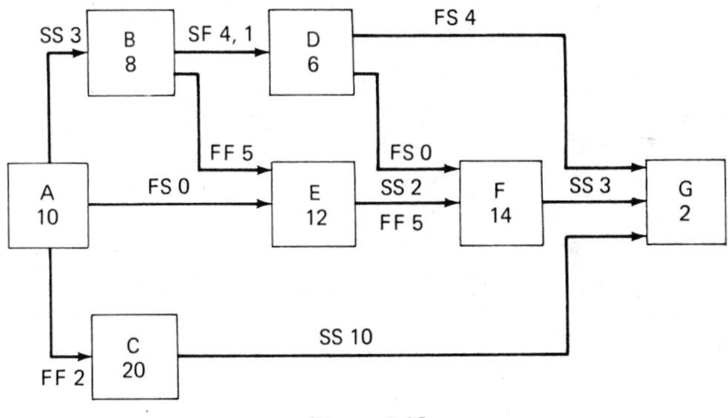


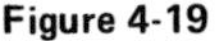
Figure 4-19

   a. Compute the early/late start/finish times for each activity, assuming that no splitting is allowed, the *ES* time for the project is zero, and the *LF* for the project completion is 27 (let $LF = EF = 27$, called the *zero slack convention*). Note that INITIAL TIME = 0 must be utilized to obtain the correct *ES* time for activity *C*, and the TERMINAL TIME = 27 is required to obtain the correct *LF* time for activities *F* and *G*.
   b. Repeat part (a) for the case where splitting is allowed on all activities.
   c. Identify the critical path in parts (a) and (b), note the differences in the computed times, and indicate the type of criticality for each activity.
   d. Is splitting of much significance in this network?
12. In the section called "Arrow versus Precedence Method" in Chapter 3, a project is diagrammed in two forms. The arrow diagram is shown in Figure

3-12 in the form of an early start time bar chart. The equivalent precedence diagram is shown in Figure 3-13.

a. Compute the early/late start/finish times for each activity in Figure 3-13, assuming splitting is allowed. Compare your results with the arrow diagram in Figure 3-12.
b. Repeat part (a) assuming no splitting is allowed, and compare the results with part (a).
c. Draw the precedence diagram for the entire project.
d. Compute the early/late start/finish times for the network in part (c), assuming splitting is allowed.
e. Repeat (d) assuming no-splitting is allowed.
f. Compare your results in parts (d) and (e), and compare them with the bar-chart output from K. C. Crandall's computer program (reference 2), given below in Figure 4-20.

```
                                                  JUNE                             JULY
                                                  1972                             1972

                          MTWTF MTWTF MTWTF MTWTF MTWTF MTWTF MTWTF MT
                          22222 233         11111 12222 22223       11
                          23456 90112 56789 23456 90123 67890 34567 01

                          1   5     10    15    20    25    30    35
 1  BOT REIN FIRST FLR    **XXX -----
 2  MECH ROUGH 1ST FLR      *** ***** **
 3  ELEC ROUGH 1ST FLR       X- XX--- ----- ----- ----- -
 4  TOP MESH FIRST FLR          X--- --XX- ----- ----- ---
 5  POUR FIRST FLR                      XX--- ----- ----- ----- --
 6  BOT REIN SECOND FLR         XXXXX ----- -----
 7  MECH ROUGH 2ND FLR                *** ***** **
 8  ELEC ROUGH 2ND FLR            X- XX--- ----- ----- ----
 9  TOP MESH SECOND FLR                 X ----- --XX- ----- -
10  POUR SECOND FLR                                 XX--- ----- --
11  BOT REIN THIRD FLR                XXXXX ----- ----- -----
12  MECH ROUGH THIRD FLR                          *** ***** **
13  ELEC ROUGH THIRD FLR                X- XX--- ----- ----- ----
14  TOP MESH THIRD FLR                            X ----- --**
15  POUR THIRD FLR                                                **

                 * CRITICAL PATH
                 X NON-CRITICAL SCHEDULE
                 - SLACK (FLOAT)
```

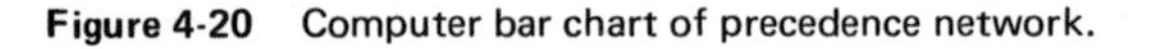

**Figure 4-20** Computer bar chart of precedence network.

**13.** Develop an example to illustrate a precedence diagram activity that is *Bi-Critical* (*BC*); that is, an activity that would cause an *increase* in project duration for either an increase *or* a decrease in its duration.

Hint: Construct an activity that is *reverse critical* as a result of an active predecessor constraint of the *finish* type (*FF* or *SF*) and an active successor constraint of the start time type (*SS* or *SF*). At the same time add constraints that make this same activity *normal* critical.

**14.** a. Complete the forward and backward pass calculations on the precedence diagram given below. Assume the project starts at $T = 0$, and that we have a scheduled project duration of 30 working days (latest allowable completion time). Also, assume that no splitting of activities is allowed.

b. Indicate the critical path by writing down an alternating sequence of activities and precedence constraints, noting the nature of the criticality of each activity.

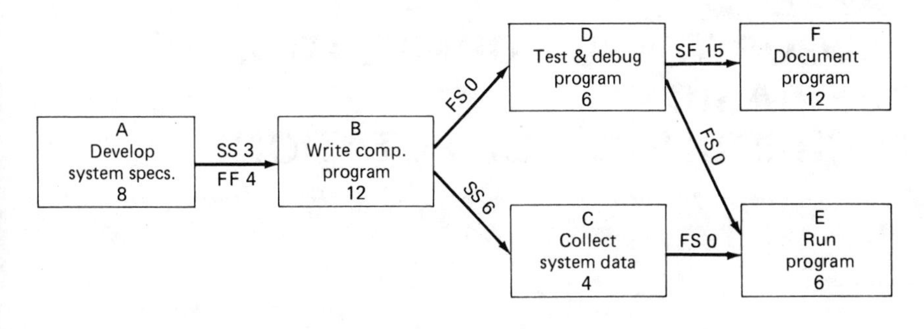

# APPENDIX 4-1 ALGORITHMIC FORMULATION OF BASIC SCHEDULING COMPUTATIONS

The hand-methods of performing the basic arrow diagram network scheduling computations taken up in this chapter placed numerical entries directly on the network. This is by far the most efficient hand method of carrying out these computations. If, however, networks are to be updated frequently, or if involved questions pertaining to resource allocation or time-cost trade-offs are raised, then the basic scheduling computations must be carried out many times on modified input data. For example, in the resource allocation procedures described in Chapter 7, the basic scheduling computations may be repeated hundreds of times in the process of allocating resources to the project activities. In such cases, it is clear that the efficient computer processing of networks is very important. In Chapter 11, the problems associated with the use of computers are considered from a user's point of view. This appendix will deal with the computer processing from the programmer analyst point of view.

Computer logic dictates that the basic scheduling computations must be carried out sequentially in some sort of tabular or matrix form. Thus, to introduce the subject of programming a computer to carry out these computations, tabular methods of hand computation will be described. This will be followed by the presentation of a computer flow diagram to carry out the basic scheduling computations for arrow diagrams. For computer methods to carry out the basic scheduling computations on node diagrams, the reader is referred to Chapter 7 of the second edition of this text, or to the reference by Montalbano[7] given at the end of this chapter. Similarly, for precedence diagrams, see the reference by Crandall.[2]

## TOPOLOGICAL ORDERING OF PROJECT ACTIVITIES

The basic scheduling computational problem can be stated mathematically as follows, which is more brief and more suggestive than the form used earlier in this chapter.

Earliest and Latest Event Times

$E_1 = 0$ by assumption, then

$$E_j = \underset{i}{\text{Max}}\,(E_i + D_{ij}); \qquad 2 \leqq j \leqq n. \tag{1}$$

$E_n$ = (Expected) project duration, and

$L_n = E_n$ or $T_S$, the scheduled project completion time. Then,

$$L_i = \underset{j}{\text{Min}}\,(L_j - D_{ij}); \qquad 1 \leqq i \leqq n - 1 \tag{2}$$

Earliest and Latest Activity Start and Finish Times and Float

$$ES_{ij} = E_i; \qquad \text{all } ij \tag{3}$$

$$EF_{ij} = E_i + D_{ij}; \qquad \text{all } ij \tag{4}$$

$$LF_{ij} = L_j; \qquad \text{all } ij \tag{5}$$

$$LS_{ij} = L_j - D_{ij}; \qquad \text{all } ij \tag{6}$$

$$F_{ij} = L_j - EF_{ij}; \qquad \text{all } ij \tag{7}$$

The above equations represent the basic scheduling computations; the emphasis here is on getting the earliest and latest event times, $E_i$ and $L_i$, since the earliest and latest activity start and finish times follow directly from these in a straightforward fashion. If the computations are to be carried out in a progressive tabular form, then equation (1), for example, requires that the early start times ($E_i$'s) for all activities merging to event $j$ must have been computed prior to the computation of $E_j$. To insure this, the first step in most tabular procedures is to "arrange" the activities in a table or matrix so that the predecessors of any activity will always be found "above" it, or previous to it in the table, and successors will always be found "below," or after it. A table so arranged is referred to as *topologically ordered.* A simple way to accomplish this ordering is to number the network events so that $i < j$ for all activities, and then list the activities according to increasing $i$ or $j$ numbers. Fulkerson[4] has given the following simple procedure to accomplish this end:

1. Number the initial project event (event with no predecessor activities) with 1. (If there is more than one initial project event, they should be numbered consecutively in any order.)
2. Delete all activities from the initial event(s) and search for events in the new network that are now initial events; number these $2, 3, \ldots, k$ in any order.
3. Repeat step 2 until the terminal project event(s) is numbered.

The network originally given in Figure 4-5 is reproduced in Figure 4-21; it has its events numbered in topological order. The information describing this network is also given in topological ordered matrix form. Each entry in the matrix corresponds to an activity in Figure 4-21. For example, in the row labeled initial (predecessor) event 1, we see entries in columns 2, 4, and 7. These are the final (successor) events of the three activities bursting from event 1. The entries of 2, 2, and 1 give the estimated duration times of their respective activities, 1-2, 1-4,

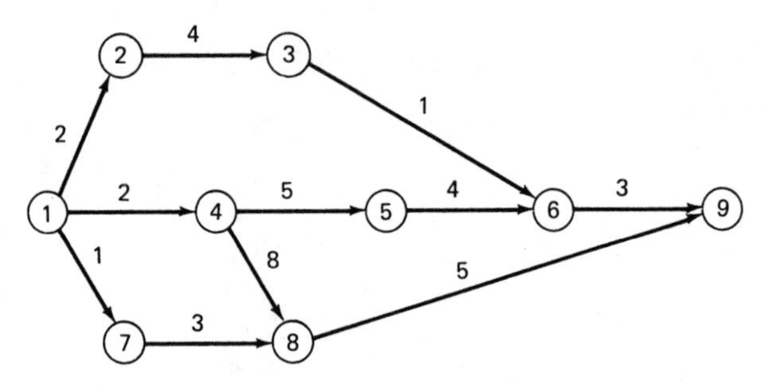


| Initial Event $i$ | Final Event $j$ 1 | 2 | 3 | 4 | 5 | 6 | 7 | 8 | 9 | Earliest Event Time, $E_i$ |
|---|---|---|---|---|---|---|---|---|---|---|
| 1 | — | 2 | | 2 | | | 1 | | | 0 |
| 2 | — | — | 4 | | | | | | | 2 |
| 3 | — | — | — | | | 1 | | | | 6 |
| 4 | — | — | — | — | 5 | | | 8 | | 2 |
| 5 | — | — | — | — | — | 4 | | | | 7 |
| 6 | — | — | — | — | — | — | | | 3 | 11 |
| 7 | — | — | — | — | — | — | — | 3 | | 1 |
| 8 | — | — | — | — | — | — | — | — | 5 | 10 |
| 9 | — | — | — | — | — | — | — | — | — | 15 |
| Latest Event Time, $L_j$ | 0 | 7 | 11 | 2 | 8 | 12 | 7 | 10 | 15 | — |

**Figure 4-21** Arrow diagram and matrix representation of a project network.

and 1-7. Dashes have been placed in the lower left hand portion of this matrix, starting with the diagonal cells, to denote that activities corresponding to these cells are not possible in a topologically ordered matrix. Blanks in the upper right hand portion of this matrix indicate that while the corresponding activity is possible, it is not present in the particular network represented by the matrix.

We begin the computations by using the matrix in Figure 4-21 to carry out the computations of the earliest and latest event times given by equation (1) and (2). We first enter $E_1 = 0$ in the first row of the column giving the $E_i$'s. Next, to find $E_2$, we note the entries in the column headed by $j = 2$. There is only one entry because only one activity in the network precedes event 2. We add this entry, $D_{1,2} = 2$, to the entry in the last column of this same row, $E_1 = 0$, to obtain $E_2 = 2$. The latter is entered in the second row of the $E_i$ column. We proceed sequentially in this manner until we reach column 6, which is the first one to have more than one entry. In this case $E_3 + D_{3,6} = 7$ and $E_5 + D_{5,6} = 11$, so $E_6 = 11$; that is,

$$E_6 = \underset{i=3,5}{\text{Max}} \; (E_3 + D_{3,6} = 7, E_5 + D_{5,6} = 11) = 11.$$

To find the values of $L_j$, we start by arbitrarily letting $L_9 = E_9 = 15$, that is, the zero-float convention described in this chapter. For event 8, we note the entries along row 8. There is only one so that $L_8 = L_9 - D_{8,9} = 15 - 5 = 10$, which is entered in the $L_j$ row for column 8. Again, we proceed sequentially in this manner until we reach row 4, which is the first row that arises with more than one entry. Here we note $D_{4,5} = 5$ and $D_{4,8} = 8$, so that $L_4$ becomes

$$L_4 = \underset{j=5,8}{\text{Min}} \; (L_5 - D_{4,5} = 3, L_8 - D_{4,8} = 2) = 2$$

Finally, we note that $L_1 = E_1 = 0$, which is a check on the accuracy of our computations, since we let $L_9 = E_9 = 15$.

The earliest and latest activity start and finish times and the total activity float can now be computed in a straightforward manner using equations (3) through (7). These results have been given previously in Table 4-2 (with slightly different event numbers) and hence will not be repeated here. The sequential computational procedure illustrated in Figure 4-21 readily admits itself to computer programming.

## COMPUTER PROGRAMMING

Programming the basic scheduling computations for computer processing is an exceedingly interesting exercise, because it permits ingenuity to be exercised to a great degree. Rather than following directly the computational procedure illustrated in Figure 4-21, a more general approach will be taken here which has a minimum of requirements placed on the input network data. In particular, the assumption that $i < j$ for all $i - j$, and the requirement of a unique pair of $i$ and $j$

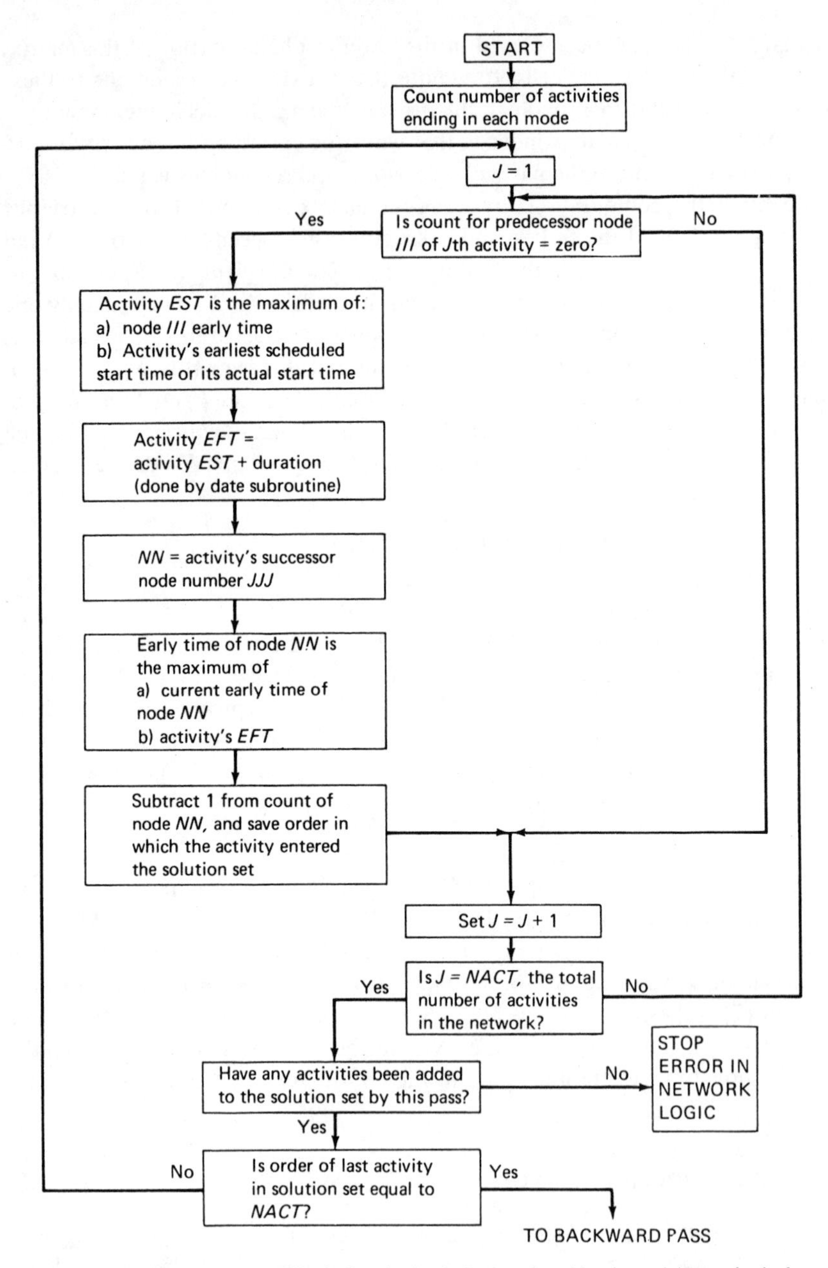


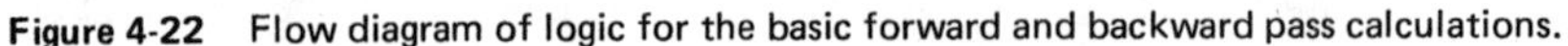

**Figure 4-22** Flow diagram of logic for the basic forward and backward pass calculations.

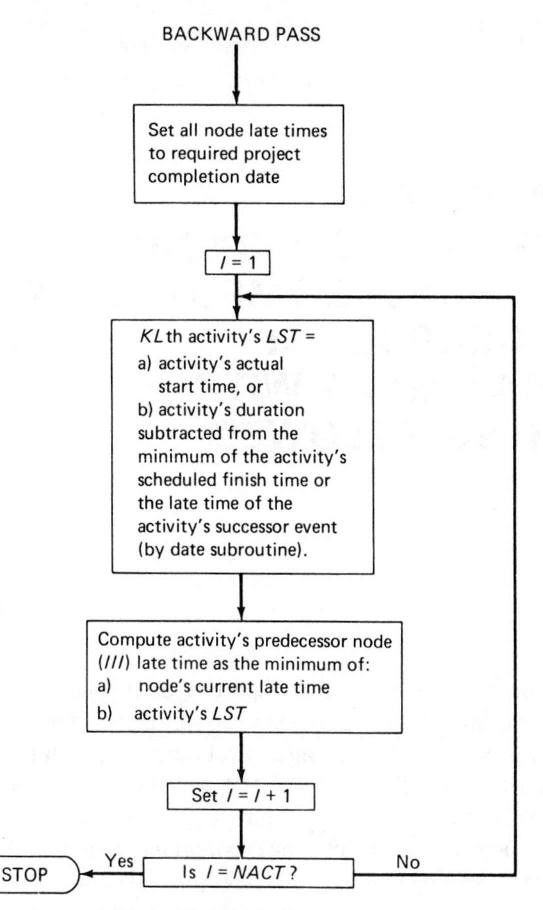


**Figure 4-22** Continued.

numbers for each activity will not be made. The network must, of course, be free of loops and hence no duplication of event numbers is allowed. The ability to detect and to indicate the presence of such a network defect must, of course, be incorporated in the computational procedure to terminate the processing of such networks.

The logic for such a computational procedure is given in Figure 4-22. The presence of a loop in the network is detected by noting whether an activity has been added to the solution set in the forward pass computations, after each pass through the list of network activities. One should also note that the backward pass computations are greatly facilitated by preserving the order of processing the activities in making the forward pass computations, and then processing the activities in the reverse order in the backward pass. The computer code for this logic, written in FORTRAN-IV, is given in the second edition of this text.

# APPENDIX 4-2 ALGORITHM FOR BASIC SCHEDULING COMPUTATIONS FOR PRECEDENCE DIAGRAMMING WITH SPLITTING ALLOWED

This algorithm, which allows activity splitting, is applied in the same manner as described above for the algorithm where no splitting is allowed. That is, it is applied to each project activity in topological sequence, with the same definitions of the terms INITIAL TIME and TERMINAL TIME. There is one major difference, however, and that is the requirement to "evaluate" each activity for computational purposes, to see if splitting is *potentially* required. In this evaluation, an activity need only be considered as having a maximum of two segments, even though it can be shown that an activity can subdivide into many segments. For the precedence diagram network in Figure 4-15, it is shown in Figure 4-16 that activities B, D and F split into 3, 3, and 4 segments, respectively. However, only two segments need be defined in either the forward or backward pass calculations to follow.

If an activity is "split" *during the calculation process*, it is necessary to note the number of days in the first segment as it could impact the early start of followers of the split activity. That is, in those cases where there is a start-to-start constraint and the first segment of the predecessor has fewer days than required by the start-to-start relation, the follower cannot start until after a portion of the second segment is complete. In this case, the start of the follower will be delayed by the amount of the idle time between the split segments of the predecessor.

The issue is then to know easily when an activity is "split." There are two cases that can force splitting; both set the early finish of a follower at a date later than the early start of the follower plus its duration. The first, and by far the most common, case is when a finish-finish factor determines the early finish of follower ($j$). This situation occurs in Figure 4-11e where $FF = 1$ between framing and electrical causes $EF = 11$ for electrical, which is greater than its $ES + D = 2 + 5 = 7$. The result is that electrical is split into two segments with an idle gap of $11 - 7 = 4$ time units. The last segment has a duration of 1 to preserve the continuity of work flow, since $FF = 1$, and the first segment has a duration of $5 - 1 = 4$. The duration of the first segment will be designated by $\alpha_j$, and is computed as follows, for the case where $EF_j$ is set by $EF_i + FF_{ij}$:

$$\alpha_j = D_j - FF_{ij}$$

The second situation where an activity can be split during the forward iteration involves the $SF_{ij}$ factor. When this relationship exists between two activities, the predecessor can cause the follower to have an early finish greater than $ES_j + D_j$. The $EF$ of the follower in this case is:

$$EF_j = ES_i + SF_{ij}.$$

When this condition establishes the early finish of an activity, the length of the first segment must once again be determined to preserve work flow and is clearly:

$$\alpha_j = D_j - SF''_{ij}.$$

The actual value of $\alpha_j$ is important in the evaluation of the early start of followers to activity ($j$), when the relation to such followers involves start-to-start factors. As in all network evaluations, the forward iteration uses local maximums to establish early start and finish values. When the relation between activities includes start-to-start factors, the potential value for the early start of the follower is related to the early start of the predecessor and the number of days of progress required to have been completed on the predecessor. This required progress ($SS_{ij}$) may be greater than the number of time units in the first segment of the predecessor, and if this is true the potential early start of the follower must account for the additional time units still remaining to be accomplished. This is done by the following equation: Potential $ES_j = EF_i - D_i + SS_{ij}$. This evaluation effectively creates a pseudostart for the predecessor, as though there were no split segments, and the potential early start of the follower will be based on the proper number of completed time units on the predecessor. This situation occurs in Figure 4-15 at activity F. Here, the $EF_{\mathrm{F}} = 31$ time is set by the $FF_{\mathrm{EF}} = 7$ factor between activities E and F, i.e., $EF_F = EF_{\mathrm{E}} + FF_{\mathrm{EF}} = 24 + 7 = 31$. This results in an $\alpha_{\mathrm{F}}$ value of 4 for the initial portion of activity F, i.e., $\alpha_{\mathrm{F}} = D_{\mathrm{F}} - FF_{\mathrm{EF}} = 11 - 7 = 4$. But the $SS_{\mathrm{FG}} = 8$ factor between activities F and G calls for the first portion of activity F to have a duration of 8; hence the problem requiring a pseudo early start. If activity F started at its $ES_{\mathrm{F}} = 15$ time, and ran for 4 time units, then it would have to be interrupted for 5 time units before the last portion of F could be started and run for the remaining 7 time units. This interruption is required in order to satisfy the $FF_{EF} = 7$ factor. The 5 time unit delay is accounted for in the pseudo early start calculation for activity F, when evaluating the early start time of activity G. That is, $ES_G = EF_F - D_F + SS_{FG} = 31 - 11 + 8 = 28$, which is 5 time units greater than taking $ES_G = ES_F + SS_{FG} = 15 + 8 = 23$.

The backward iteration evaluates late starts and late finishes and has the same complication as the forward iteration in that activities can also split during these calculations. Basically, an activity splits when its late start is set less (earlier) than the late finish minus its duration, that is, $LS_j < LF_j - D_j$. The situation is completely analogous to the setting of early finish in the forward iteration. Once again the computational procedure must recognize this situation and record the duration in the *second* segment ($\beta_i$) which is thereby created.

The complete computational algorithm for the case where activity splitting is allowed is as follows.

### Forward Pass Calculations—Splitting Allowed

Apply the following 3 steps to each project activity in topological order.

STEP 1. Compute the early start time ($ES_j$) of the activity ($j$) in question. It is the maximum (latest) of the set of start times which includes the INITIAL TIME, and one start time computed from *each* start-time constraint of the form $FS_{ij}$ and $SS_{ij}$, going to activity $j$, from predecessor activities indexed by $i$.

$$ES_j = \underset{\text{all } i}{\text{MAX}} \begin{Bmatrix} \text{INITIAL TIME} & \\ EF_i + FS_{ij} & \text{(for each } FS_{ij} \text{ constraint)} \\ EF_i - D_i + SS_{ij} & \text{(for each } SS_{ij} \text{ constraint with } \alpha_i < SS_{ij}) \\ ES_i + SS_{ij} & \text{(for each } SS_{ij} \text{ constraint with } \alpha_i \geqq SS_{ij}, \text{ or where } \alpha_i \text{ was not required)} \end{Bmatrix}$$

STEP 2. Compute the early finish time ($EF_j$) of the activity ($j$) in question. It is the maximum (latest) of the set of finish times which includes the early start time plus the duration of the activity $j$ in question ($ES_j + D_j$), and a finish time computed from *each* finish-time constraint of the form $FF_{ij}$ or $SF_{ij}$, going to activity $j$, from predecessor activities indexed by $i$.

$$EF_j = \underset{\text{all } i}{\text{MAX}} \begin{Bmatrix} ES_j + D_j & \\ EF_i + FF_{ij} & \text{(for each } FF_{ij} \text{ constraint)} \\ EF_i - D_i + SF_{ij} & \text{(for each } SF_{ij} \text{ constraint with } \alpha_i < SF'_{ij}) \\ ES_i + SF_{ij} & \text{(for each } SF_{ij} \text{ constraint with } \alpha_i \geqq SF'_{ij}\text{, or where } \alpha_i \text{ was not required)} \end{Bmatrix}$$

STEP 3. If $EF_j > ES_j + D_j$, then compute $\alpha$ for activity $j$ as follows:

$$\alpha_j = \begin{Bmatrix} D_j - FF_{ij}; & \text{if } EF_j \text{ was set by an } FF_{ij} \text{ constraint} \\ D_j - SF''_{ij}; & \text{if } EF_j \text{ was set by an } SF_{ij} \text{ constraint} \end{Bmatrix}$$

### Backward Pass Computations—Splitting Allowed

Apply the following 3 steps to each project activity in reverse topological order.

STEP 1. Compute $LF_i$, the late finish time of the activity ($i$) in question. It is the minimum (earliest) of the set of finish times which includes the TERMINAL TIME, and a finish time computed from *each* finish-time constraint of the form $FS_{ij}$ or $FF_{ij}$, going from activity $i$, to successor activities indexed by $j$.

$$LF_i = \underset{\text{all } j}{\text{MIN}} \begin{Bmatrix} \text{TERMINAL TIME} & \\ LS_j - FS_{ij} & \text{(for each } FS_{ij} \text{ constraint)} \\ LS_j + D_j - FF_{ij} & \text{(for each } FF_{ij} \text{ constraint with } \beta_j < FF_{ij}) \\ LF_j - FF_{ij} & \text{(for each } FF_{ij} \text{ constraint with } \beta_j \geqq FF_{ij}\text{, or where } \beta_j \text{ was not required)} \end{Bmatrix}$$

STEP 2. Compute $LS_i$, the late start time of the activity ($i$) in question. It is the minimum (earliest) of the set of start times which includes the latest finish time minus the duration of activity $i$ ($LF_i - D_i$), and a start time computed from *each* start-time constraint of the form $SS_{ij}$ or $SF_{ij}$, going from activity $i$, to successor activities indexed by $j$.

$$LS_i = \underset{\text{all } j}{\text{MIN}} \begin{Bmatrix} LF_i - D_i & \\ LS_j - SS_{ij} & \text{(for each } SS_{ij} \text{ constraint)} \\ LS_j + D_j - SF_{ij} & \text{(for each } SF_{ij} \text{ constraint with } \beta_j < SF''_{ij}) \\ LF_j - SF_{ij} & \text{(for each } SF_{ij} \text{ constraint with } \beta_j \geqq SF''_{ij}, \text{ or where } \beta_j \text{ was not required)} \end{Bmatrix}$$

STEP 3. If $LS_i < LF_i - D_i$, then compute $\beta$ for activity $i$ as follows:

$$\beta_i = \begin{Bmatrix} D_i - SS_{ij}; & \text{if } LS_i \text{ was set by an } SS_{ij} \text{ constraint} \\ D_i - SF'_{ij}; & \text{if } LS_i \text{ was set by an } SF_{ij} \text{ constraint} \end{Bmatrix}$$

## Example Problem

The example previously given in Figure 4-15 shows the results of applying the above algorithm during the forward and backward pass calculations, for the case where splitting of activities is allowed. Some of the calculations are shown below for the forward pass.

*Activity A*

$$ES_A = \{\text{INITIAL TIME} = 0\} = 0$$

$$EF_A = ES_A + D_A = 0 + 12 = 12$$

Computation of $\alpha_A$ is unnecessary since $EF_A = 12$ is not *greater* than $ES_A + D_A = 12$)

*Activity B*

$$ES_B = \underset{A}{\text{MAX}} \begin{Bmatrix} \text{INITIAL TIME} = 0 \\ ES_A + SS_{AB} = 0 + 3 = 3 \end{Bmatrix} = 3$$

$$EF_B = \underset{A}{\text{MAX}} \begin{Bmatrix} ES_B + D_B = 3 + 10 = 13 \\ EF_A + FF_{AB} = 12 + 5 = 17 \end{Bmatrix} = 17$$

(Since $EF_B > ES_B + D_B$, calculate $\alpha_B$)

$$\alpha_B = D_B - FF_{AB} = 10 - 5 = 5$$

*Activity D*

$$ES_D = \{\text{INITIAL TIME} = 0\} = 0$$

$$EF_D = \underset{B}{\text{MAX}} \begin{Bmatrix} ES_D + D_D = 0 + 12 = 12 \\ ES_B + SF_{BD} = 3 + (2 + 10) = 15 \end{Bmatrix} = 15$$

(Since $EF_D > ES_D + D_D$, calculate $\alpha_D$)

$$\alpha_D = D_D - SF''_{BD} = 12 - 10 = 2$$

*Activity C*

$$ES_C = \underset{A}{\text{MAX}} \begin{Bmatrix} \text{INITIAL TIME} = 0 \\ ESA + SS_{AC} = 0 + 3 = 3 \end{Bmatrix} = 3$$

$$EF_C = \underset{D}{\text{MAX}} \begin{Bmatrix} ES_C + D_C = 3 + 18 = 21 \\ EF_D - D_D + SF_{DC} = 15 - 12 + (7 + 14) = 24 \end{Bmatrix} = 24$$

Note: The formula $EF_D - D_D + SF_{DC}$ was required since

$$\alpha_D = 2 < SF'_{DC} = 7.$$

(Since $EF_C > ES_C + D_C$, calculate $\alpha_C$)

$$\alpha_C = D_C - SF''_{DC} = 18 - 14 = 4$$

etc.

A portion of the backward pass calculations are as follows:

*Activity I*

$$LF_I = \{\text{TERMINAL TIME} = 38\} = 38$$

$$LS_I = LF_I - D_I = 38 - 7 = 31$$

(Computation of $\beta_I$ is unnecessary since $LS_I = 31$ is not *less than* $LF_I - D_I = 31$)

*Activity F*

$$LF_F = \{\text{TERMINAL TIME} = 38\} = 38$$

$$LS_F = \underset{G, I}{\text{MIN}} \begin{Bmatrix} LF_F - D_F = 38 - 11 = 27 \\ LS_G - SS_{FG} = 28 - 8 = 20 \\ LS_I - SS_{FI} = 31 - 8 = 23 \end{Bmatrix} = 20$$

(Since $LS_F < LF_F - D_F$, calculate $\beta_F$)

$$\beta_F = D_F - SS_{FG} = 11 - 8 = 3$$

*Activity C*

$$LF_C = \underset{F,H}{\text{MIN}} \left\{ \begin{array}{l} \text{TERMINAL TIME} = 38 \\ LS_F + D_F - FF_{CF} = 20 + 11 - 5 = 26 \\ LF_H - FF_{CH} = 35 - 3 = 32 \end{array} \right\} = 26$$

Note: Middle equation was required since $\beta_F = 3 < FF_{CF} = 5$.

$$LS_C = \underset{F,H}{\text{MIN}} \{LF_C - D_C = 26 - 18 = 8\} = 8$$

(Computation of $\beta_C$ is unnecessary since $LS_C = 8$ is not *less than* $LF_C - D_C = 8$)

etc.

These computational results are shown in Figure 4-15. The two critical paths through this network are shown at the top of the figure.

# 5

# PROJECT COST CONTROL

From the preceding chapters it should be clear that PERT/CPM and precedence diagramming are *time-oriented* methodologies. They permit the development of *time-based* plans and schedules for projects in great detail, and the monitoring of actual versus planned *times*. The basic procedures described in these previous chapters have, in fact, often been termed "time-only" methods because they do not explicitly consider possible constraints on the available resources (e.g., money, men, and equipment) which may be needed for completion of project activities. While specific resource *requirements* (e.g., number of men) are implicitly assumed in deriving the time durations associated with activities in the network diagram, resources *available* are ignored.

Although the developers of CPM included a provision for activity time/cost trade-offs, the scheme was used only as a means of determining the "best" (in terms of minimum total cost) activity times for use in scheduling the project. That scheme, which is decribed in Chapter 8, was not designed for complete project cost control and is not in any sense a cost accounting/control system.

In spite of the time-only orientation of the basic PERT/CPM scheduling methodologies, it was recognized very early on that by adding activity costs to the network diagram a potentially powerful means of improving project cost planning/control could be obtained. Conceptually, this addition was not difficult, since none of the basic network scheduling rules were changed. For example, consider the simple activity-on-node network in Figure 5-1, previously shown as an arrow diagram in Figure 4-6. In addition to time durations for each

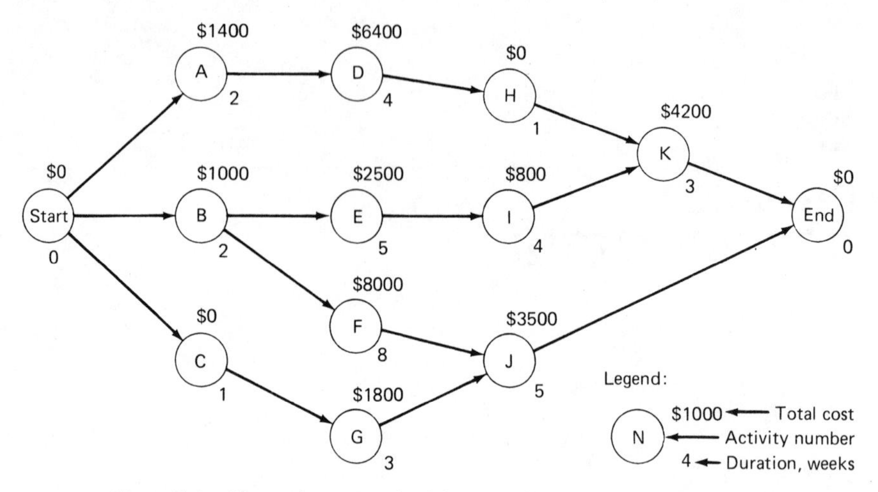


**Figure 5-1** Illustrative network with costs of performing each activity.

activity, the estimated total cost of performing the activity is shown beside each node. The total activity cost is assumed to be spread evenly across activity duration, so that a per period cost can be calculated for each activity. Performing the basic scheduling computations on this network will produce the earliest and latest start and finish times for each activity as shown in Table 5-1, which also summarizes the duration and cost data for each activity.

If all activities are assumed to start at their early start (*ES*) times, and costs are summed cumulatively across time periods, the *ES* cumulative cost curve shown in Figure 5-2 will be obtained. Similarly, by starting activities at their late-start times the *LS* curve shown will be obtained. If the activity costs are the total costs associated with this project the region between these two curves

**Table 5.1. Time and Cost Data for Illustrative Network**

| *Activity* | *Duration, Weeks* | *ES Time* | *LS Time* | *Total Cost, Dollars* | *Cost per Week* |
|---|---|---|---|---|---|
| A | 2 | 0 | 5 | $1400 | $700 |
| B | 2 | 0 | 0 | 1000 | 500 |
| C | 1 | 0 | 6 | 0 | 0 |
| D | 4 | 2 | 7 | 6400 | 1600 |
| E | 5 | 2 | 3 | 2500 | 500 |
| F | 8 | 2 | 2 | 8000 | 1000 |
| G | 3 | 1 | 7 | 1800 | 600 |
| H | 1 | 6 | 11 | 0 | 0 |
| I | 4 | 7 | 8 | 800 | 200 |
| J | 5 | 10 | 10 | 3500 | 700 |
| K | 3 | 11 | 12 | 4200 | 1400 |

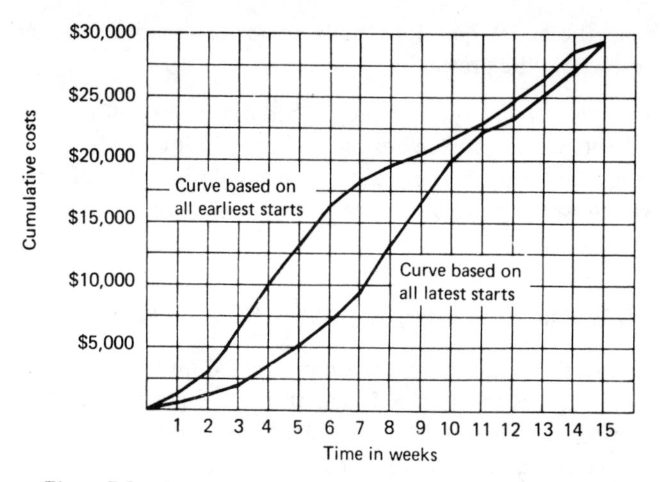


**Figure 5-2** Cumulative cost curves for network in Figure 5-1.

is the area of possible project budgets. A schedule to achieve a particular budget in this region can be determined by starting activities at specific times between their *ES* and *LS* times. For example, in some cases it may be desirable to have a relatively straight line of cumulative costs from project start to finish (i.e., a uniform rate of expenditure). This can be approximately achieved by juggling the scheduled start times of activities within their allowable limits. There are also other, more systematic, procedures for choosing the activity start times which meet such a uniform rate of maximum per-period expenditure constraint. These procedures are covered in Chapter 7, which discusses limited-resource scheduling procedures. These procedures can be applied to the cost planning problem by treating money as a type of resource whose availability by time periods is limited.

As this simple example illustrates, the network and critical path schedule computations can be used as vehicles for producing useful information about project costs. In this case the information produced is about *planned* costs. However there are also other important questions having to do with cost *control* which can be answered with the use of network-based procedures. This chapter discusses some of these procedures, which deal primarily with the following basic questions:

1. What are the actual project costs to date?
2. How do the actual costs to date compare with planned costs to date?
3. What are the project accomplishments to date?
4. How do the actual costs of the accomplishments to date compare with the planned costs of these same accomplishments?
5. By how much may the project be expected to overrun or underrun the total planned cost?

6. How do the above questions apply to various subdivisions and levels of interest within the project?

Some examples of the type of project-based systems and procedures which have been developed to provide answers to these questions will be given in this chapter. First, however, it is helpful to understand some of the important problems associated with their development and implementation as well as their historical background.

## BASIC PROBLEMS IN NETWORK-BASED COST CONTROL

While in theory the concepts of cost control based on the project network are not complex, the design and implementation of a practical cost control system is not readily accomplished. The fundamental problems facing the system designer may be classified as (1) those related to organizational conflicts, and (2) those related to the necessary efficiency of the system. The basic organizational problem is the conflict between the project approach of network cost control and the fundamental approach of cost accounting procedures found most in industry. This conflict is manifest particularly in the design of input and output phases of network control systems. The input to a network-based system requires the development of an *activity accounting* procedure by which actual expenditure data are coded to provide association with activities (or groups of activities) in the project network. The output from the system likewise must be project-oriented to provide *project summary reports*, organized by time period, areas of responsibility, and technical subdivisions of the project.

The efficiency of the system is a problem because network-oriented systems lend themselves to major increases in the amount of detail available to the manager. The level of detail is both the promise and the inherent hazard of such systems, and it is one of the primary tasks of the system designer to achieve the level of detail that provides the greatest return on the investment in the system. A network-based cost control system can easily require routine input data in quantities and frequencies that project personnel find extremely burdensome. Unless the requirements are reduced and the procedures simplified, the system will come to an early death. Similar dangers lie in the design of the data processing and output phases of the system. The use of precedence diagramming, which often reduces the number of network activities by 50 percent or more, offers some promise here.

### Activity Accounting

Basic CPM/PERT systems gained acceptance rapidly because the critical path concept filled a generally recognized need for improved, formal procedures for project planning and scheduling. In the cost control area, however, formal accounting procedures were established long before the beginning of PERT/CPM. Thus, one of the greatest obstacles to the use of network-based cost control, has been the difficulty of developing a network-based system compatible with the established accounting system.

Generally speaking, accounting systems in organizations engaged in project-type endeavors are designed to plan budgets and to report expenditures both by organizational unit and by project. However, the emphasis is on the organizational unit accounts (section, department, division, etc.), inasmuch as the objectives of the accounting system are summaries of expenditures by the functional elements of the organization. Where the accounting codes also permit summaries by project, the project summary represents the lowest level of detailed cost reporting available to the operating management. The purpose of the critical path approach, on the other hand, is to provide more detailed information and control *within* the project.

Cost data, like time data, must be applied to project *activities*. The budgeting and recording of expenses both by cost item and by network activity clearly means that a more elaborate system is required. Not only are there more codes and figures to deal with, but there are many typical questions developed in the application of cost data to a project network:

1. Electronic testing gear is purchased for use in several activities in the project. Should the cost of the gear be assigned entirely to the purchasing activity, when the expenditure actually occurs, or should it be allocated over the activities involving use of the gear?
2. What cost, if any, should be assigned to the curing of concrete? Approval of shop drawings? Negotiation of a subcontract?
3. Should overhead be included or only direct costs? If overhead is included, is it computed the same way for all activities?
4. How should the costs associated with project management be shown, as activities or as overhead?

Such questions arise largely from the fact that basic time-oriented networks and the list of project cost items represent two different sets of data. Although the sets largely overlap, many elements of the project which involve costs have not been shown in the network. This is particularly true of management and other overhead expenses. Certain other activities involve no direct costs but consume time and perhaps should account for a portion of the indirect costs. Various answers to these activity accounting problems have been proposed, and we will give some examples of cost control systems utilizing network activity accounting later. But first it is important to review the history of network-based cost control and to understand the concepts of work breakdown structure (WBS) and work packages.

## HISTORY OF PERT/COST

As noted earlier, the idea of collecting costs on the basis of network diagram activities was recognized very early as a potential means of improving project cost control. A few manual procedures and computer programs to answer the 4 cost questions listed earlier were developed by individual CPM and PERT users in the period 1959-62. In 1962, a major boost to the interest in network-based cost control was provided by agencies of the U.S. Government. The Department of Defense and the National Aeronautics and Space Administration jointly issued a manual entitled *DOD and NASA Guide, PERT/Cost Systems Design*,[4] which emphasized the cost control aspects of "PERT-type systems." Several companies and agencies in the aerospace field had already been working with various PERT Cost procedures and computer programs, but the DOD and NASA Guide served to formalize the interest of the government and thus to inititate active development of the procedures throughout the aerospace industry. By mid-1963, the use of PERT/Cost procedures had become a requirement in certain military research and development projects.

One of the key features of PERT/Cost was the utilization of a "Work Breakdown Structure" (WBS) to show the hierarchy, or levels, of tasks within a project, and the definition of "work packages" at the lower or basic levels of work. The DOD-NASA Guide described these concepts as follows:

> *End Item Subdivisions.* The development of the work breakdown structure begins at the highest level of the program with the identification of project end items (hardware, services, equipment, or facilities). The major end items are then divided into their component parts (e.g., systems, subsystems, components), and the component parts are further divided and subdivided into more detailed units. . . . The subdivision of the work breakdown continues to successively lower levels, reducing the dollar value and complexity of the units at each level, until it reaches the level where the end item subdivisions finally become manageable units for planning and control purposes. The end item subdivisions appearing at this last level in the work breakdown structure are then divided into major work packages (e.g., engineering, manufacturing, testing).

The DOD-NASA Guide suggested that the basic work packages might be formed by the same individual activities, or groups of activities, used in the PERT/CPM network. Figure 5-3 shows, for example, an illustration from the Guide of the level-by-level WBS and use of network activities as work packages.

The specification of network activities as the basic cost control unit was widely misinterpreted as meaning that individual network activities should always comprise work packages. This created numerous problems with respect to the acceptance of PERT/Cost. There was belated recognition of the fact that it was often impractical for contractors to budget and report costs by PERT/CPM

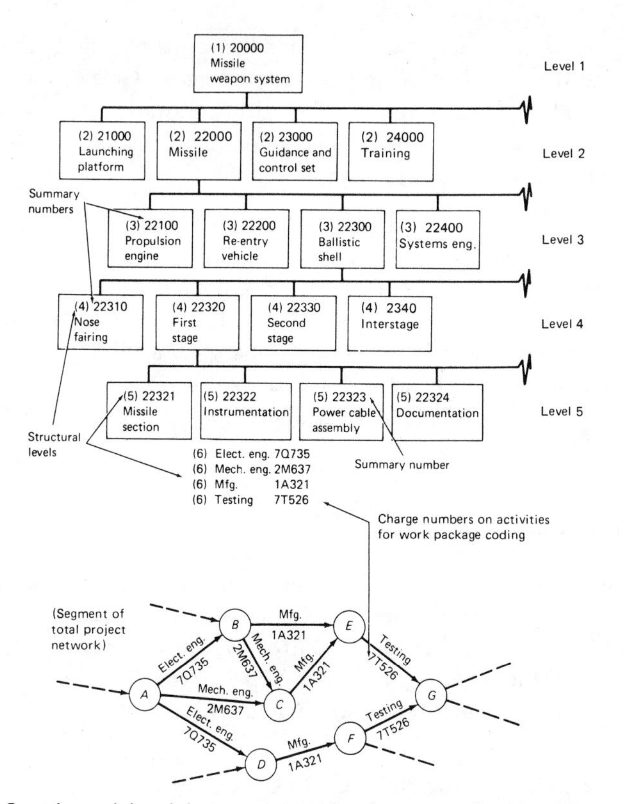


**Figure 5-3** Sample work breakdown structure showing accounting identification codes.[4]

network activities. As J. R. Fox, a member of the original PERT/Cost development team, notes:[5]

> ... the design group intended to make clear that the work breakdown structure, not the network, was the basis for cost planning and control. In retrospect this was far from clear. Confusion arose because many defense contractors as well as other government agencies were developing their own versions of PERT/Cost. Several of these versions called for the use of PERT networks in the budgeting and reporting of costs by network activities, a requirement that resulted in the collection of vast amounts of very detailed cost information.
>
> By 1964 more than ten variations of PERT/Cost existed throughout DOD and NASA, most of which called for costing of PERT network activities and the submission of detailed cost information on a monthly basis. Contractors recognized the impracticality of these systems and created their own PERT/Cost groups to prepare reports for DOD. These groups operated separately

> from management teams responsible for the actual planning, scheduling, budgeting and measurement of program performance. Government auditors decided that PERT/Cost groups were a legitimate overhead expense that could be charged to the government. Many contractors considered PERT/Cost as the basis for negotiating higher overhead rates. . . .

Because of the confusion and resistance fo PERT/Cost throughout the defense industry, the Air Force, in 1964, developed simplified standards by which a contractor's internal cost management system could be measured before the company could qualify for defense work. These specifications contained some essential elements of the original PERT/Cost system, including the WBS concept, but did not include detailed reporting of cost information from PERT/CPM network activities.

## DEVELOPMENT OF C/SCSC

The simplified criteria developed by the Air Force in 1964 were subsequently developed into an improved set of criteria which were issued under DOD instruction 7000.2 "Cost/Schedule Control System Criteria (C/SCSC)." These criteria retained the advantages of some of the essential elements of PERT/Cost, and yet also reflected current industry practice. As Fox notes:[5]

> It became clear from the PERT/Cost pilot tests that cost planning and control could be based on a system that was used on most large commercial development and production programs. This cost information could be based on a WBS that subdivided the program or project according to the manner in which work responsibility was assigned. Project work was traced down through several levels of work definition to the point where short-term work packages could be identified as the basis for planning and controlling manpower. Budgets were then established for each short-term work package. Costs were estimated at every level of the WBS, to arrive at a total cost estimate for the program or project. As a contractor began a development program, actual manhours and costs were assigned to the work packages. As work was completed the contractor could compare estimates of cost for short-term work packages with the actual man-hours and cost required to accomplish the work. Thus the contractor would keep a constant check on whether work was costing more or less than was estimated.

In line with Fox's observation, the *DOD C/SCSC Implementation Guide*[8] continued the WBS philosophy of the earlier DOD-NASA Guide, and commented on work packages as follows:

A work package is simply a lower level task or job assignment. "Work package" is the generic term used to identify discrete tasks which have definable end results. A work package has the following characteristics:

1. It represents the units of work at levels where work is performed.
2. It is clearly distinguished from all other work packages.
3. It is assignable to a single organizational element.
4. It has scheduled start and complete dates.
5. It has a budget of assigned value expressed in terms of dollars, man-hours or other measurable units.
6. Its duration is limited to a relatively short span of time, or it is subdivided by discrete value milestones to facilitate the objective measurement of work performed.
7. It is integrated with detailed engineering, manufacturing or other schedules.

In an attempt to increase the flexibility of its cost accounting requirements the C/SCSC Guide included new provisions for establishment of "cost accounts" and clarified the establishment of organizational responsibilities for elements of the WBS. As the Guide states:

> The lowest level at which functional responsibility for individual WBS elements exists, actual costs are accumulated and performance measurement is conducted, is referred to as the cost account level. While it is usually located immediately above the work package level, cost accounts may be located at higher levels when in consonance with the contractor's method of management.
>
> In addition to its function as a focal point for collecting costs, the cost account in a performance measurement system is also the lowest level in the structure at which comparisons of actual direct costs to budgeted costs are required, although some contractors also collect costs and make comparisons at the work package level.
>
> While the WBS defines and organizes the work to be performed, the contractor's organizational structure reflects the way the contractor has organized the people who will accomplish the work. To assign work responsibility to appropriate organizational elements, the WBS and organizational structure must be interrelated with each other. This interrelationship may be visualized as a matrix with the functional organization element listed on one axis and the applicable WBS elements listed on the other [see Figure 5-4]. Further subdivision of the effort into work packages may be accomplished by assigning work to operating units.

The WBS concept promulgated in C/SCSC, with its associated flexibility in provisions for establishing work packages and cost accounts, represents a prac-

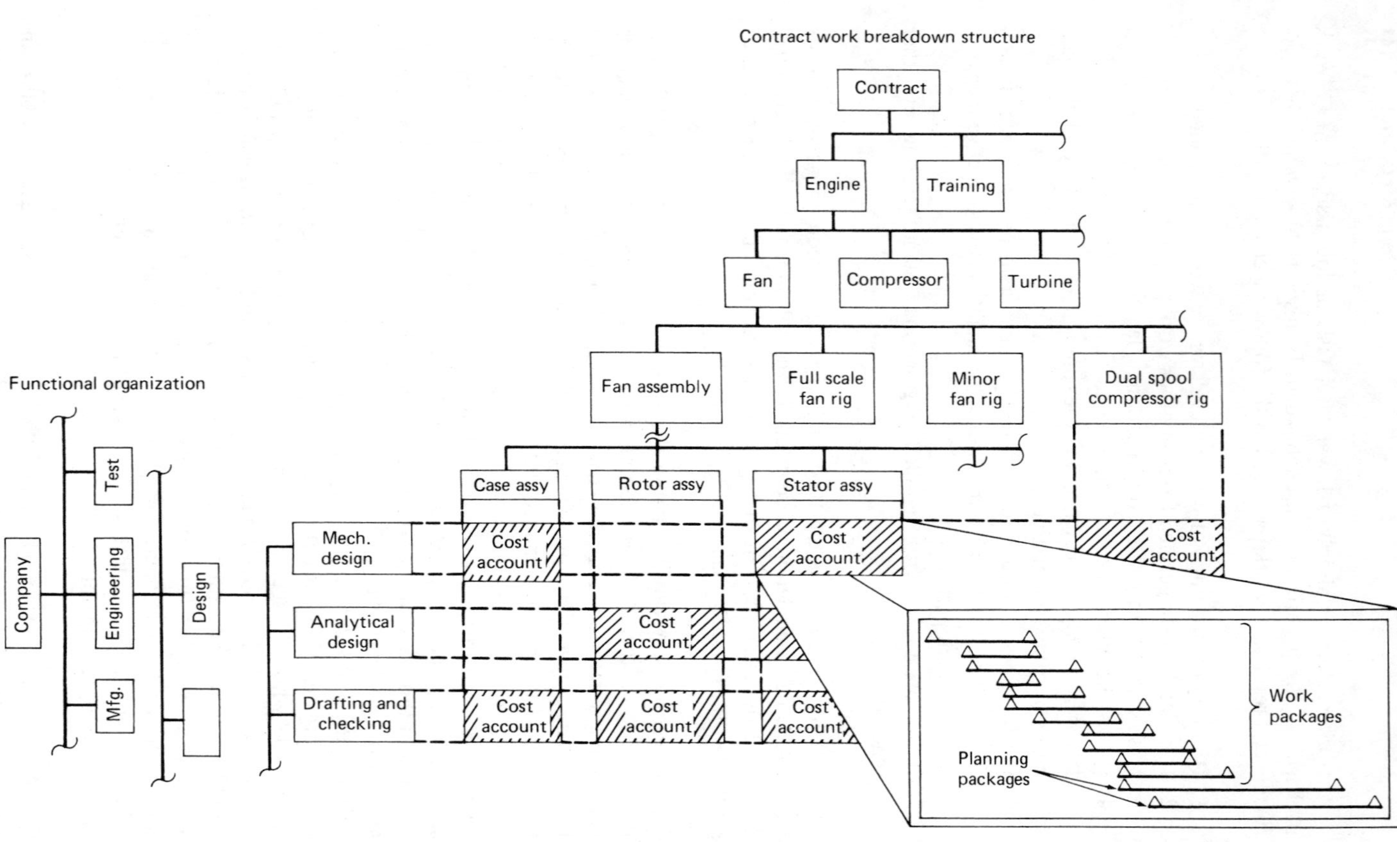


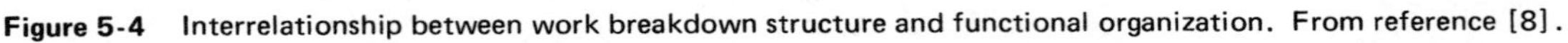
**Figure 5-4** Interrelationship between work breakdown structure and functional organization. From reference [8].

tical compromise in the concept of activity accounting for business organizations, particularly in the aerospace industry. The approach should be applicable in many industries where network-based cost control is desired. In the construction industry, for example, the lowest level work packages may be identical or nearly identical to the bid items in the project. In some cases the work package would be represented on the network by a single activity; in most cases it would be comprised of a group of activities within a division of the project, such as the concrete work for the first floor of a building. The specific work organization and cost accounts established would naturally vary with the type of project, areas of responsibility of the foremen and superintendents, the capabilities of the cost accounting system, and other factors related to the particular company and its projects.

## MONITORING PROJECT COSTS: VALUE OF WORK PERFORMED

As noted earlier, an effective cost control system not only permits the establishment of realistic budgets or plans, but also includes provision for monitoring actual costs. This is done by measuring expenditures against budget and identifying variances so that corrective action may be taken when required. Since schedule progress is also often monitored by measuring actual activity time performance against estimated times, this means that two separate streams of information, time data and cost data, are flowing to the project managers. The collection, manipulation, and analysis of even one of these two streams of data typically present significant problems for management. For example, in the case of direct labor costs, these are typically recorded on weekly time cards which require the distribution of each individual's time over the activities in which he participated that week. If a network-based cost approach like PERT/Cost is used, employee activities must be defined in terms of, or coded to correspond to, specific network activities. Requirements such as this can add significantly to the paperwork burden of project personnel, and can be an important factor in the acceptance or rejection of the system.

It is interesting to note the C/SCSC provisions for data collection and reporting, since they are directed at easing some of the problems stemming from the added requirements of time and cost progress reporting. The C/SCSC objective is an *integrated* time/cost progress monitoring/control system. That is, it provides information about project time and cost performance from a single data collection stream. The particular data collected is based on the WBS structure and utilizes the three key cost performance measures.

The key cost measures, defined in (8), can be summarized as:

*Actual Cost of Work Performed* (ACWP) is the amount reported as actually expended in completing the particular work accomplished within a given time period.

*Budgeted Cost of Work Performed* (BCWP) is the budgeted amount of cost for the work completed in a given time period, including support effort and allocated overhead. (This is sometimes referred to as "earned value of work accomplished".)

*Budgeted Cost of Work Scheduled* (BCWS) is the budgeted amount of cost of the work *scheduled* to be accomplished in a given time period (including support and allocated overhead).

Although C/SCSC requires the collection of these costs by *cost account* level, these costs can be collected at any level of effort (i.e., project task, subtask, work package, etc.). These cost measures are utilized in establishing both cost and schedule (time performance) variances as follows:

$$\text{Cost variance} = \text{BCWP} - \text{ACWP}$$

$$\text{Schedule/Performance variance} = \text{BCWP} - \text{BCWS}$$

The use of both cost and schedule variances calculated in this fashion provides an integrated cost/time reporting system which measures cost performance in relation to the work accomplished, and ensures that both time scheduling and cost budgeting are constructed upon the same data base.

Figure 5-5 shows a popular graphic report from such an integrated time/cost

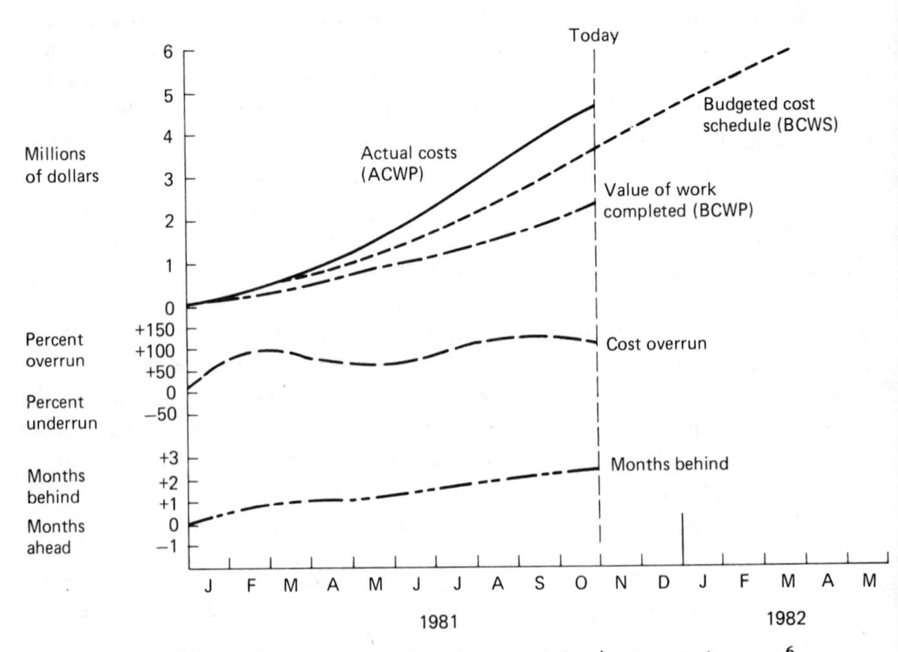


**Figure 5-5** Performance report from integrated time/cost control system.[6]

control system. The Budgeted Cost Schedule curve (BCWS), is essentially a plot of cumulative costs corresponding to a particular schedule of activity start times, as illustrated earlier in this chapter. The report shows that budgeted expenditures up through October 31, 1981 ("today" on the graph) were \$3.6 million. Actual costs expended in the performance of project activities (i.e., ACWP) are plotted in the upper portion of the figure. As can be seen, cumulative actual expenditures up through the current reporting date are \$4.6 million. Also, actual costs have consistently exceeded budget for the past several months.

This much information, i.e., actual costs to date vs. budgeted costs to date, is essentially all that traditional cost control systems report. But this information by itself is not particularly useful and can even be misleading. For example, it could be that the project is actually running ahead of schedule and the higher costs are a reflection of a greater amount of work accomplished than was expected. However, if the project is just on schedule, or behind schedule, then a true cost overrun has occurred. The only way this can be determined for certain is to know what work has been completed, in terms of money originally budgeted for the tasks involved. The "Value of Work Completed" (BCWP) line* displays this information.

The fact that the actual costs curve is above the value of work completed curve indicates that a cost overrun has occurred. The amount of overrun at any point in time is given by the vertical distance between the two curves. Thus at the present time actual costs in this example (\$4.6 million), exceed the budgeted value of work completed, BCWP, (\$2.3 million) by \$2.3 million, which is an overrun of 100 percent. The "Cost Overrun" curve in the lower part of the figure is a plot of the calculated percent over- or underrun at any given time. This percentage is obtained by the calculation:

$$\%\text{ overrun (underrun) } \frac{(\text{ACWP}) - (\text{BCWP})}{(\text{BCWP})}$$

where a negative value indicates an underrun has occurred.

In this example, the Value of Work Completed curve is also below the Budgeted Cost curve, which indicates that the project is behind schedule. The amount of delay at any particular time is taken as the horizontal distance between the Value of Work Completed curve at that point in time, and the Budgeted Cost curve. For example, at the current report date, work budgeted at \$2.3 million has been completed. But that same amount of work was originally scheduled to have been completed about the middle of August, which means that the project is about $2\frac{1}{2}$ months behind schedule. The "Months Behind" Curve at the lower part of the figure is a running tally of the number of months behind. It shows that the project has fallen farther and farther behind schedule.

*The Value of Work Completed is also sometimes referred to as the "earned value" or the "earned value of work performed," as noted earlier. For this reason, this approach is also termed the "earned value approach" to cost control.

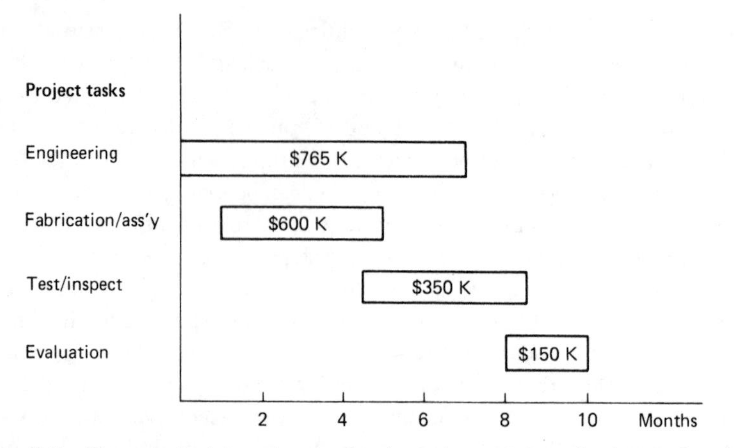


**Figure 5-6** Planned schedule and costs of major tasks, prototype development project.

These cost measurement concepts are also useful for cost control at the task, subtask, or work package level. For example, Figure 5-6 shows the major tasks of a planned 10-month prototype development project. Total project costs are estimated at $1.865 million. Of this, $765,000 is engineering work. Assume that the project has been under way for 3 months and we wish to know how the engineering work is progressing.

A separate schedule of the project engineering work is shown in Figure 5-7, which indicates that 9 major subtasks are involved. The veritcal dashed line in Figure 5-7 indicates the report date; as can be seen, 3 tasks have been completed, 5 have been started and are in progress, and 1 has not yet been started as of the reporting date.

Figure 5-8 shows the type of information which might be collected on the engineering subtasks at the end of the 3rd month, using the cost measures defined earlier (BCWS, BCWP, ACWP). This report shows that each of the 3 completed tasks has experienced a cost overrun, ranging from slight ($500) in the case of Material Tests to substantial ($13,000) in the case of System Design. Given this information, the status of the 5 tasks currently in progress is of considerable interest. Their status is summarized in Figure 5-9, which interprets the significance of the cost measure relationships and projects total engineering costs at completion, based upon progress to date. As can be seen, engineering work is generally behind schedule and over cost. If the present rate of progress continues, total cost for these tasks will be $812,000 instead of the originally estimated $765,000.

The cost measurement concepts illustrated by these examples represent an important step towards the development of effective, implementable cost control procedures in project management. While these procedures are today used primarily on government-related projects, they are equally applicable to private

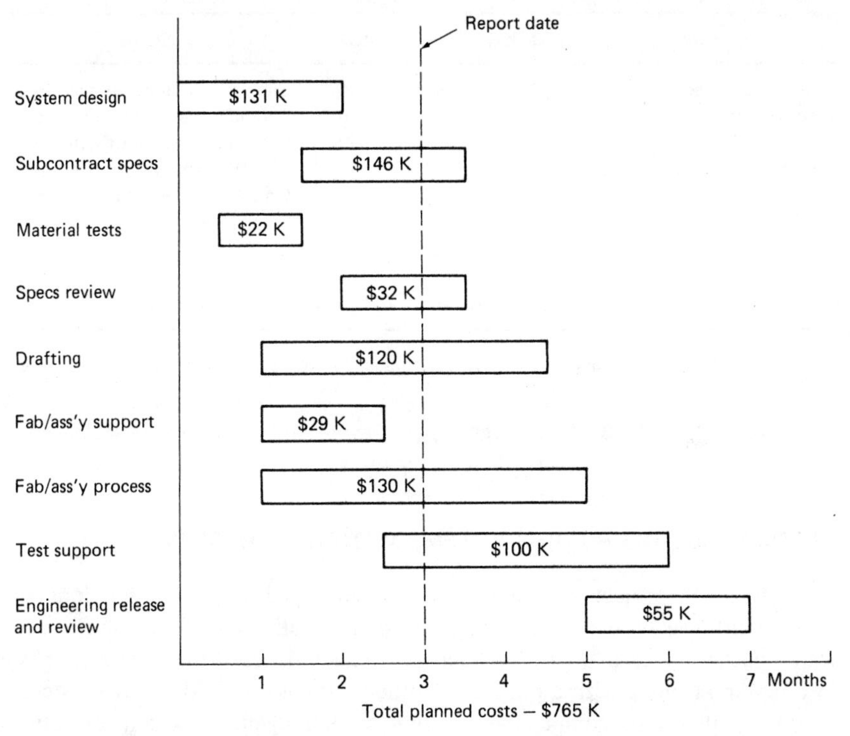


**Figure 5-7** Major engineering subtasks, prototype development project.

**ENGINEERING SUBTASKS STATUS AT END OF MONTH 3**
**(thousands of dollars)**

| *Task* | *Status* | *BCWS* | *BCWP* | *ACWP* |
|---|---|---|---|---|
| System design | Completed | $131.0 | $131.0 | $144.0 |
| Subcontract specs | Started | 132.0 | 82.0 | 84.0 |
| Material tests | Completed | 22.0 | 22.0 | 22.5 |
| Specs review | Started | 22.0 | 15.0 | 15.5 |
| Drafting | Started | 97.0 | 59.0 | 59.0 |
| Fabrication/assembly support | Completed | 29.0 | 29.0 | 36.0 |
| Fabrication/assembly process | Started | 100.0 | 63.0 | 67.0 |
| Test support | Started | 21.0 | 21.0 | 21.0 |
| Engineering release and review | Not Started | | | |
| Total | | $554.0 | $422.0 | $449.0 |

**Figure 5-8** Status report on engineering subtasks at end of month 3.

| Task | BCWS | BCWP | ACWP | Status |
|---|---|---|---|---|
| Subcontract specs | $132 | $82 | $84 | Behind schedule and over cost |
| Specs review | 22 | 15 | 15.5 | Behind schedule and over cost |
| Drafting | 97 | 59 | 59 | Behind schedule but within cost |
| Fabrication/assembly process | 100 | 63 | 67 | Behind schedule and over cost |
| Test support | 21 | 21 | 21 | On schedule and within cost |

$$\text{Estimate to complete: } \frac{449}{422} \times 765 = \$812\text{ K}$$

Summary: Engineering work is behind schedule and a cost overrun is occurring.

**Figure 5-9** Status summary of engineering subtasks in progress and estimate to complete.

sector projects. The evidence suggests that more and more private-sector companies are voluntarily adopting these procedures.

## AN EXAMPLE OPERATING SYTEM: METIER'S ARTEMIS

The ARTEMIS project mangement system developed by the Metier Company is representative of the wide variety of commercially-available computer software "packages" available today. General characteristics of the system are given in Chapter 11 along with a number of other systems. The ARTEMIS system is unusual both for its completeness of time scheduling and cost control options and for its unique software/hardware configuration. It involves a comprehensive collection of software programs fully implemented on a minicomputer. The complete system is designed for easy mobility to distant project sites; success in this regard is suggested by the fact that in 1981 more than 300 separate ARTEMIS installations were reported operating in 22 different countries. The software package offers cost management procedures that incorporate the concept of WBS, Work Package costing, and Earned Value measurement specified in the C/SCSC criteria and illustrated earlier in this chapter.

The demonstration project illustrated here is a waste disposal plant construction project consisting of about 600 activities with a duration of about 24 months. A portion of the project WBS is shown in Figure 5-10; the Organizational Breakdown Structure (OBS) is shown in Figure 5-11. The project has 4 WBS levels and approximately 100 work packages, each containing several network activities. Each work package has a Work Package Code, an OBS code and a description. For each work package the following information is stored:

- baseline start and finish dates
- current planned start and finish dates
- baseline budget cost

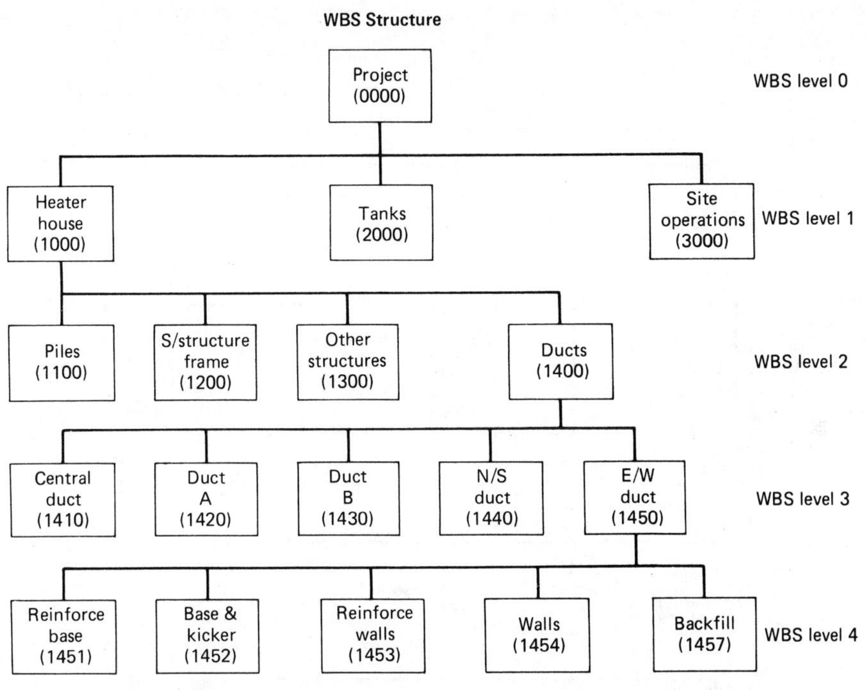


**Figure 5-10** Work breakdown structure for demonstration project.

- ACWP this period
- ACWP to date
- forecast costs to complete
- BCWS this period
- BCWS to date
- BCWP this period
- BCWP to date
- Work Package percentage complete

The baseline data referred to above is calculated as follows: prior to project start the Original Early Start (OES) and Original Early Finish (OEF) are calculated for each network activity. The calculated original schedule start and finish dates are then summarized up to work package level to give the schedule baseline. Also, the original budget costs for each work package are developed and distributed over the original schedule to generate the Project Baseline, which is the BCWS. These cost and schedule baselines are set as reference points at the beginning of the project. This information can be displayed in Project Cost Plan reports organized by either WBS format or OBS format. Figure 5-12(a)

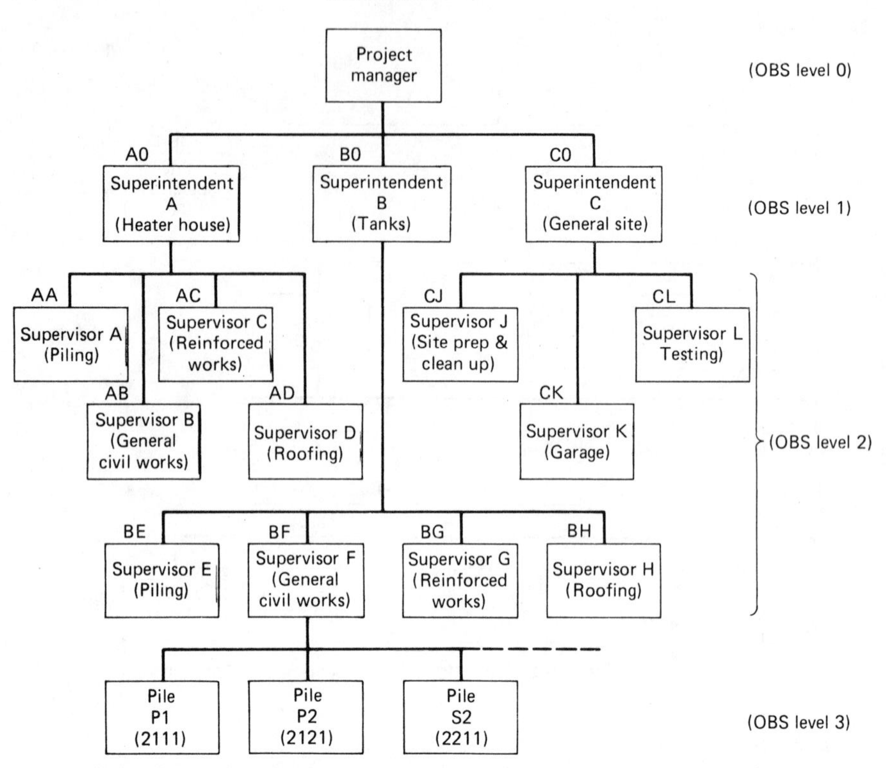


**Figure 5-11** Organizational breakdown structure for demonstration project.

shows a WBS format report for the Heater House portion of the project; Figure 5-12(b) shows an OBS format report for the tank superintendent areas of responsibility.

As actual cost and schedule data are collected in periodic update cycles, new activity schedules are produced and the following calculations and operations occur:

- calculation of new ACWP
- transfer of latest dates from network activities to work packages
- calculation of percent complete of work packages
- calculation of new BCWP

Figure 5-13 is a sample updated activity schedule for a portion of the project, and shows the type of information produced for each activity, including work package and organizational section codes. Figure 5-14 is a Slippage Barchart, which shows graphically the original and current schedule for each activity, reflecting the updated information on project completions, duration changes, and changes in network logic.

Figure 5-15 shows Schedule Performance reports produced after an update, organized by WBS and OBS format. In both reports the actual and forecast schedule performance is compared to the baseline schedule. Predicted schedule overruns for each portion of the project, based on the latest data, are indicated on the right side of each report.

Figure 5-16 is an integrated C/SCSC-type Cost Performance Report produced after an update, organized by WBS and summarized to Level 1. It shows the current and cumulative values of BCWS, BCWP, and ACWP for the elements of the WBS separately and in total, as well as the original and latest estimates of cost at completion. Figure 5-17 shows similar information for the tank superintendent's areas of responsibility of the OBS. Figure 5-18 illustrates the type of detailed cost information available for individual work packages upon request. Schedule information can also be displayed for each work package.

The cost performance index shown at the bottom of the Cost Performance Report (Figure 5-16) is calculated as follows:

$$\text{Cost Performance Index} = \text{CPI} = \frac{\text{BCWP}}{\text{ACWP}}$$

Values of the CPI greater than 1.0 indicate costs below budget. The Schedule Performance Index also shown in Figure 5-16 is calculated as:

$$\text{Schedule Performance Index} = \text{SPI} = \frac{\text{BCWP}}{\text{BCWS}}$$

Values of the SPI greater than 1.0 indicate completions ahead of schedule. The ARTEMIS system utilizes these indices in combination with its four-color graphics capabilities to produce three summary reports for top level management. The first of these, shown as Figure 5-19(a), indicates the movement of the Cost Performance Index over time and shows at a glance whether the CPI is above or below budget or trending in an unfavorable direction. Similarly, the report shown as Figure 5-19(b) indicates the movement of the Schedule Performance Index over time. The third top level report (Figure 5-20) plots CPI against SPI over time to show at a glance whether project performance is currently favorable, unfavorable or marginal, and whether there are any short- or long-term trends in any of these categories.

The ARTEMIS example used here is a good example of a complete network-based cost control system. For example, it utilizies CPM network activities as basic building blocks for the work packages; and it uses periodically-updated information from the CPM schedules to produce time-related cost information

```
                                                                   P R O J E C T   C O S T   P L A N

METIER/ARTEMIS :
DEMONSTRATION OF AN EARNED VALUE SYSTEM
                                                                   CONTRACT NUMBER : 1/08/1932

                                                                   LEVEL 1 SUMMARY : HEATER HOUSE

                                                                   COMPANY         : XYZ COMPANY

                                         APR  JUN  AUG  OCT  DEC  FEB  APR  JUN  AUG  OCT  DEC  FEB  APR  JUN  AUG
WBS        WBS ELEMENT          BUDGET   79   79   79   79   79   80   80   80   80   80   80   81   81   81   81
CODE       DESCRIPTION            $      I----I----I----I----I----I----I----I----I----I----I----I----I----I----I----I

1100  PILES                     11200    I    I BBB I    I    I    I    I    I    I    I    I    I    I    I    I
1200  S/STRUCTURE FRAME         38400    I    I     I    I    I    I    I    I    IBBBBBBBBBBBBBBB I    I    I    I    I    I
1300  OTHER STRUCTURES          19592    I    I     I    I    I    I    I    BBBBBBBBBB    I    I    I    I    I    I    I
1400  DUCTS                    109772    I    I  BBBBBBBBBBBBBBBBBBBBBBBBBBBBBBBBBBB  I    I    I    I    I    I    I    I    I    I

SUMMARY
-------
1000  HEATER HOUSE             178964    I    I BBBBBBBBBBBBBBBBBBBBBBBBBBBBBBBBBBBBBBBBBBBBBBBBBBBBBBBBBBBBBBBBBBBBBBBBBBBBB I    I    I    I    I    I
```

(a) WBS Level 1 Summary.

```
                                                                              PROJECT  COST  PLAN
METIER/ARTEMIS
DEMONSTRATION OF AN EARNED VALUE SYSTEM
                                                                              CONTRACT NUMBER    : 1/08/1932
                                                                              COMPANY            : XYZ COMPANY
                                                                              PROJECT MANAGER    :  J.CARTER

                                                                              SUPERINTENDENT     : TANKS - D. HARVEY

                                        APR    JUN    AUG    OCT    DEC    FEB    APR    JUN    AUG    OCT    DEC    FEB    APR    JUN    AUG
                               BUDGET   79     79     79     79     79     80     80     80     80     80     80     81     81     81     81
          SUPERVISOR              $     I------I------I------I------I------I------I------I------I------I------I------I------I------I------I------I

PILING - J.P.LANE               79976   I      I   BBBBBBBBBBBBBBBBBBBBBB     I      I      I      I      I      I      I      I      I      I      I
GENERAL CIVIL WORKS - J.MAHER  133288   I      I      I      I BBBBBBBBBBBBBBBBBBBBBBBBBBBBBBBBBBBBBBBBBBBBBBBB I      I      I      I      I      I      I
REINFORCED WORKS - E.MARTIN    160040   I      I      IBBBBBBBBBBBBBBBBBBBBBBBBBBBBBBBBBBBBBBBBBBBBBB I      I      I      I      I      I      I      I
ROOFING - T.C.ALNAN             91296   I      I      I      I      I      I   BBBBBBBBBBBBBBBBBBBBBBBBBBBBBBBBB I      I      I      I      I      I

OVERALL WORK SPAN
-----------------
                               464600   I      I   BBBBBBBBBBBBBBBBBBBBBBBBBBBBBBBBBBBBBBBBBBBBBBBBBBBBBBBBBBBBBBBBBBBBBB I      I      I      I      I      I
```

(b) OBS format (tank superintendent's areas).

**Figure 5-12** Project cost plan reports.

ACTIVITY SCHEDULE

PAGE 1

WASTE DISPOSAL PLANT

RUN DATE: 1-JUL-79

| PREC ACT. | SUCC ACT. | DESCRIPTION | PROJECT SECTION | WORK PACKAGE | DURATION | EARLIEST START | EARLIEST FINISH | LATEST START | LATEST FINISH | TOTAL FLOAT |
|---|---|---|---|---|---|---|---|---|---|---|
| 86 | 88 | COMPLETE REINF BASE TANK S1 | S1 | 2212 | 6 | 8-OCT-79 | 15-OCT-79 | 23-NOV-79 | 30-NOV-79 | 34 |
| 18 | 22 | COMPLETE REINF BASE TANK P1 | P1 | 2112 | 14 | 8-OCT-79 | 25-OCT-79 | 8-OCT-79 | 25-OCT-79 | 0 |
| 18 | 20 | FORM RING BEAM TANK P1 | P1 | 2113 | 8 | 8-OCT-79 | 17-OCT-79 | 29-OCT-79 | 7-NOV-79 | 15 |
| 128 | 130 | STRIKE WALLS (TANK P2) | P2 | 1415 | 8 | 8-OCT-79 | 17-OCT-79 | 19-NOV-79 | 28-NOV-79 | 30 |
| 208 | 210 | STRIKE WALLS DUCT B | | 1435 | 2 | 8-OCT-79 | 9-OCT-79 | 18-FEB-80 | 19-FEB-80 | 95 |
| 256 | 258 | FORM DUCT "A" (TANK P2) | P2 | 1426 | 8 | 8-OCT-79 | 17-OCT-79 | 4-DEC-79 | 13-DEC-79 | 41 |
| 210 | 212 | BACKFILL DUCT B | | 1437 | 8 | 10-OCT-79 | 19-OCT-79 | 20-FEB-80 | 29-FEB-80 | 95 |
| 130 | 132 | FORM (TANK P2) ROOF SOFFIT | P2 | 1416 | 8 | 18-OCT-79 | 29-OCT-79 | 29-NOV-79 | 10-DEC-79 | 30 |
| 130 | 134 | COMMENCE FIX REINF TO ROOF CEN | | 1416 | 8 | 18-OCT-79 | 29-OCT-79 | 29-NOV-79 | 10-DEC-79 | 30 |
| 154 | 156 | CONC RING BEAM TANK P1(1) | P1 | 2113 | 2 | 18-OCT-79 | 19-OCT-79 | 8-NOV-79 | 9-NOV-79 | 15 |
| 260 | 262 | COMPLETE FIX REINF TO ROOF DUC | | 1426 | 2 | 18-OCT-79 | 19-OCT-79 | 14-DEC-79 | 17-DEC-79 | 41 |
| 156 | 158 | STRIKE RB TANK P1(1) AND FORM | P1 | 2113 | 3 | 22-OCT-79 | 24-OCT-79 | 12-NOV-79 | 14-NOV-79 | 15 |
| 262 | 264 | CONC ROOF DUCT A | | 1426 | 2 | 22-OCT-79 | 23-OCT-79 | 18-DEC-79 | 19-DEC-79 | 41 |
| 264 | 370 | DELAY | | | 1 | 24-OCT-79 | 24-OCT-79 | 20-DEC-79 | 20-DEC-79 | 41 |
| 264 | 266 | DELAY | | | 1 | 24-OCT-79 | 24-OCT-79 | 21-DEC-79 | 21-DEC-79 | 42 |
| 158 | 160 | CONC RING BEAM TANK P1(2) | P1 | 2113 | 2 | 25-OCT-79 | 26-OCT-79 | 15-NOV-79 | 16-NOV-79 | 15 |
| 370 | 372 | STRIKE FORM SOFFIT DUCT A | | 1426 | 2 | 25-OCT-79 | 26-OCT-79 | 21-DEC-79 | 24-DEC-79 | 41 |
| 266 | 268 | STRIKE FORM ROOF DUCT "A" | | 1426 | 2 | 25-OCT-79 | 26-OCT-79 | 24-DEC-79 | 25-DEC-79 | 42 |
| 22 | 24 | FIX FORM TO CTR BASE TANK P1 | P1 | 2114 | 10 | 26-OCT-79 | 8-NOV-79 | 26-OCT-79 | 8-NOV-79 | 0 |
| 160 | 162 | STRIKE FORM RB TANK P1(2) | P1 | 2113 | 2 | 29-OCT-79 | 30-OCT-79 | 19-NOV-79 | 20-NOV-79 | 15 |
| 372 | 268 | DELAY | | | 1 | 29-OCT-79 | 29-OCT-79 | 25-DEC-79 | 25-DEC-79 | 41 |
| 134 | 136 | COMPLETE FIX REINF TO ROOF CEN | | 1416 | 2 | 30-OCT-79 | 31-OCT-79 | 11-DEC-79 | 12-DEC-79 | 30 |
| 268 | 270 | BACKFILL DUCT A | | 1427 | 8 | 30-OCT-79 | 8-NOV-79 | 26-DEC-79 | 4-JAN-80 | 41 |
| 162 | 164 | FORM RING BEAM TANK S1(1) | S1 | 2213 | 8 | 31-OCT-79 | 9-NOV-79 | 27-NOV-79 | 6-DEC-79 | 19 |
| 136 | 138 | CONC ROOF CENT DUCT | | 1416 | 2 | 1-NOV-79 | 2-NOV-79 | 13-DEC-79 | 14-DEC-79 | 30 |
| 138 | 358 | DELAY | | | 1 | 5-NOV-79 | 5-NOV-79 | 17-DEC-79 | 17-DEC-79 | 30 |
| 138 | 140 | DELAY | | | 1 | 5-NOV-79 | 5-NOV-79 | 18-DEC-79 | 18-DEC-79 | 31 |
| 358 | 360 | STRIKE FORM CENT DUCT | | 1416 | 2 | 6-NOV-79 | 7-NOV-79 | 18-DEC-79 | 19-DEC-79 | 30 |
| 140 | 142 | STRIKE FORM CTR DUCT ROOF | | 1416 | 2 | 6-NOV-79 | 7-NOV-79 | 19-DEC-79 | 20-DEC-79 | 31 |
| 360 | 142 | DELAY | | | 1 | 8-NOV-79 | 8-NOV-79 | 20-DEC-79 | 20-DEC-79 | 30 |
| 24 | 26 | CONC CTR BASE TANK P1 | P1 | 2114 | 4 | 9-NOV-79 | 14-NOV-79 | 9-NOV-79 | 14-NOV-79 | 0 |
| 142 | 144 | BACKFILL CENT DUCT | | 1417 | 3 | 9-NOV-79 | 13-NOV-79 | 21-DEC-79 | 25-DEC-79 | 30 |
| 270 | 374 | EXC TRIM & BLIND TANK P2 | P2 | 2122 | 12 | 9-NOV-79 | 26-NOV-79 | 7-JAN-80 | 22-JAN-80 | 41 |
| 164 | 166 | CONC RING BEAM TANK S1(1) | S1 | 2213 | 2 | 12-NOV-79 | 13-NOV-79 | 7-DEC-79 | 10-DEC-79 | 19 |
| 144 | 362 | COMMENCE EXC TRIM AND BLIND HH | HH | 1451 | 2 | 14-NOV-79 | 15-NOV-79 | 26-DEC-79 | 27-DEC-79 | 30 |
| 166 | 168 | STRIKE RB TANK S1(1) AND FORM | S1 | 2213 | 3 | 14-NOV-79 | 16-NOV-79 | 11-DEC-79 | 13-DEC-79 | 19 |
| 26 | 28 | STRIKE FORM CTR BASE TANK P1 | P1 | 2114 | 4 | 15-NOV-79 | 20-NOV-79 | 15-NOV-79 | 20-NOV-79 | 0 |
| 362 | 364 | COMPLETE EXC TRIM AND BLIND HH | HH | 1451 | 2 | 16-NOV-79 | 19-NOV-79 | 31-DEC-79 | 1-JAN-80 | 31 |
| 362 | 366 | COMMENCE REINF BASES E & W DUC | | 1451 | 3 | 16-NOV-79 | 20-NOV-79 | 28-DEC-79 | 1-JAN-80 | 30 |
| 168 | 170 | CONC RING BEAM TANK S1(2) | S1 | 2213 | 2 | 19-NOV-79 | 20-NOV-79 | 14-DEC-79 | 17-DEC-79 | 19 |
| 580 | 582 | COMMENCE EXC TRIM & BLIND HH D | HH | 1441 | 8 | 20-NOV-79 | 29-NOV-79 | 27-FEB-80 | 7-MAR-80 | 71 |
| 88 | 90 | FIX FORM CTR BASE TANK S1 | S1 | 2214 | 4 | 21-NOV-79 | 26-NOV-79 | 3-DEC-79 | 6-DEC-79 | 8 |
| 28 | 30 | FIX SCREED & CONC INCL BASE TA | P1 | 2115 | 4 | 21-NOV-79 | 26-NOV-79 | 21-NOV-79 | 26-NOV-79 | 0 |
| 366 | 368 | COMPLETE REINF BASES E & W DUC | | 1451 | 3 | 21-NOV-79 | 23-NOV-79 | 9-JAN-80 | 11-JAN-80 | 35 |

**Figure 5-13** Activity schedule (portion of project).

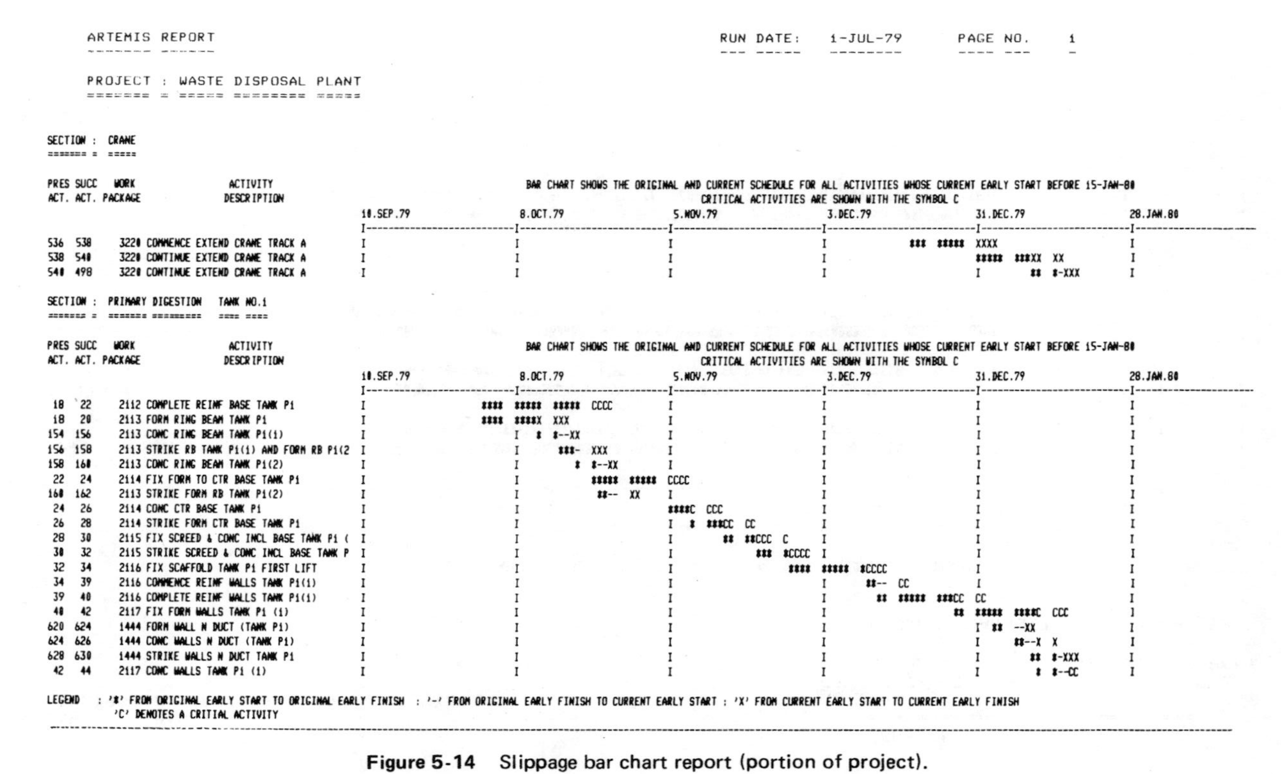

```
ARTEMIS REPORT                                                    RUN DATE:   1-JUL-79      PAGE NO.   1

PROJECT : WASTE DISPOSAL PLANT

SECTION :  CRANE

PRES SUCC   WORK           ACTIVITY                 BAR CHART SHOWS THE ORIGINAL AND CURRENT SCHEDULE FOR ALL ACTIVITIES WHOSE CURRENT EARLY START BEFORE 15-JAN-80
ACT. ACT. PACKAGE         DESCRIPTION                                     CRITICAL ACTIVITIES ARE SHOWN WITH THE SYMBOL C
                                           10.SEP.79          8.OCT.79          5.NOV.79          3.DEC.79          31.DEC.79          28.JAN.80
                                           I------------------I-----------------I-----------------I-----------------I------------------I-----------------
536  538    3220 COMMENCE EXTEND CRANE TRACK A   I            I                 I                 I         ### #####  XXXX             I
538  540    3220 CONTINUE EXTEND CRANE TRACK A   I            I                 I                 I                  ##### ###XX  XX    I
540  498    3220 CONTINUE EXTEND CRANE TRACK A   I            I                 I                 I                  I      ## #-XXX    I

SECTION :  PRIMARY DIGESTION   TANK NO.1

PRES SUCC   WORK           ACTIVITY                 BAR CHART SHOWS THE ORIGINAL AND CURRENT SCHEDULE FOR ALL ACTIVITIES WHOSE CURRENT EARLY START BEFORE 15-JAN-80
ACT. ACT. PACKAGE         DESCRIPTION                                     CRITICAL ACTIVITIES ARE SHOWN WITH THE SYMBOL C
                                           10.SEP.79          8.OCT.79          5.NOV.79          3.DEC.79          31.DEC.79          28.JAN.80
                                           I------------------I-----------------I-----------------I-----------------I------------------I-----------------
 18   22    2112 COMPLETE REINF BASE TANK P1       I     #### ##### ##### CCCC  I                 I                  I                 I
 18   20    2113 FORM RING BEAM TANK P1            I     #### ####X  XXX        I                 I                  I                 I
154  156    2113 CONC RING BEAM TANK P1(1)         I            I  #  #--XX     I                 I                  I                 I
156  158    2113 STRIKE RB TANK P1(1) AND FORM RB P1(2  I       I     ###-  XXX   I                 I                  I                 I
158  160    2113 CONC RING BEAM TANK P1(2)         I            I       #  #--XX  I                 I                  I                 I
 22   24    2114 FIX FORM TO CTR BASE TANK P1      I            I         ##### ##### CCCC        I                  I                 I
160  162    2113 STRIKE FORM RB TANK P1(2)         I            I          ##--  XX I                 I                  I                 I
 24   26    2114 CONC CTR BASE TANK P1             I            I                 ####C  CCC      I                  I                 I
 26   28    2114 STRIKE FORM CTR BASE TANK P1      I            I                 I  #  ###CC  CC I                  I                 I
 28   30    2115 FIX SCREED & CONC INCL BASE TANK P1 (  I       I                 I      ##  ##CCC  C                  I                 I
 30   32    2115 STRIKE SCREED & CONC INCL BASE TANK P  I       I                 I          ###  #CCCC I             I                 I
 32   34    2116 FIX SCAFFOLD TANK P1 FIRST LIFT   I            I                 I             #### ##### #CCCC      I                 I
 34   39    2116 COMMENCE REINF WALLS TANK P1(1)   I            I                 I                 I  ##--  CC      I                 I
 39   40    2116 COMPLETE REINF WALLS TANK P1(1)   I            I                 I                 I   ## ##### ###CC  CC               I
 40   42    2117 FIX FORM WALLS TANK P1 (1)        I            I                 I                 I          ## ##### ####C  CCC     I
620  624    1444 FORM WALL N DUCT (TANK P1)        I            I                 I                 I                  I  ##  --XX     I
624  626    1444 CONC WALLS N DUCT (TANK P1)       I            I                 I                 I                  I    ##--X  X   I
628  630    1444 STRIKE WALLS N DUCT TANK P1       I            I                 I                 I                  I      ## #-XXX I
 42   44    2117 CONC WALLS TANK P1 (1)            I            I                 I                 I                  I       # #--CC I

LEGEND  : '#' FROM ORIGINAL EARLY START TO ORIGINAL EARLY FINISH : '-' FROM ORIGINAL EARLY FINISH TO CURRENT EARLY START : 'X' FROM CURRENT EARLY START TO CURRENT EARLY FINISH
          'C' DENOTES A CRITIAL ACTIVITY
```

**Figure 5-14** Slippage bar chart report (portion of project).

```
METIER/ARTEMIS :                                                          SCHEDULE PERFORMANCE REPORT
DEMONSTRATION OF AN EARNED VALUE SYSTEM                                   ===========================

                                                                          CONTRACT NUMBER     : 1/08/1932
                                                                          COMPANY             : XYZ COMPANY

                                                                          PROJECT MANAGER     :  J.CARTER

                              ACHIEVEMENTS : BASELINE SCHEDULE (B) : ACTUAL (A) : FORECAST (F) : TIMENOW (*)        FORECAST
                                                                                                                    OVERRUN
SUPERINTENDENT          KEY APR   JUN   AUG   OCT   DEC   FEB   APR   JUN   AUG   OCT   DEC   FEB   APR   JUN      (DAYS)
                            79    79    79    79    79    80    80    80    80    80    80    81    81    81
                            I-----I-----I-----I-----I-----I-----I-----I-----I-----I-----I-----I-----I-----I--I

HEATER HOUSE - R.BROWN   B  I     I   BBBBBBBBBBBB*BBBBBBBBBBBBBBBBBBBBBBBBBBBBBBBBBBBBBBBBBBBBBBBBBBBB I     I     I     I  I        0
                        AF  I     I    AAAAAAAAAAA*FFFFFFFFFFFFFFFFFFFFFFFFFFFFFFFFFFFFFFFFFFFFFFFFFFFFFFFI     I     I     I  I        6

TANKS - D. HARVEY        B  I     I   BBBBBBBBBBBB*BBBBBBBBBBBBBBBBBBBBBBBBBBBBBBBBBBBBBBBBBBBBBBBB     I     I     I     I  I        0
                        AF  I     I    AAAAAAAAAAA*FFFFFFFFFFFFFFFFFFFFFFFFFFFFFFFFFFFFFFFFFFFFFFFFFF     I     I     I     I  I        6

GENERAL SITE - I.ROBERTS B  I BBBBBBBBBBBBBBBBBBBBBB*BBBBBBBBBBBBBBBBBBBBBBBBBBBBBBBBBBBBBBBBBBBBBBBBBBBBBBBB I     I     I  I        0
                        AF  I  AAAAAAAAAAAAAAAAAAAAA*FFFFFFFFFFFFFFFFFFFFFFFFFFFFFFFFFFFFFFFFFFFFFFFFFFFFFFFF I     I     I  I        4

OVERALL WORK SPAN
-----------------

                         B  I BBBBBBBBBBBBBBBBBBBBBB*BBBBBBBBBBBBBBBBBBBBBBBBBBBBBBBBBBBBBBBBBBBBBBBBBBBBBBBB I     I     I  I        0
                        AF  I  AAAAAAAAAAAAAAAAAAAAA*FFFFFFFFFFFFFFFFFFFFFFFFFFFFFFFFFFFFFFFFFFFFFFFFFFFFFFFF I     I     I  I        4


PRODUCED BY METIER/ARTEMIS
```

(a) OBS format (primary superintendent's areas).

```
                                                                        SCHEDULE PERFORMANCE REPORT
METIER/ARTEMIS :                                                        ===========================
DEMONSTRATION OF AN EARNED VALUE SYSTEM
                                                                        CONTRACT NUMBER  : 1/08/1932

                                                                        LEVEL 1 SUMMARY  : SITE OPERATIONS

                                                                        COMPANY          : XYZ COMPANY

                                  ACHIEVEMENTS : BASELINE SCHEDULE (B) : ACTUAL (A) : FORECAST (F) : TIMENOW (*)        FORECAST
WBS          WBS ELEMENT                                                                                                OVERRUN
CODE         DESCRIPTION     KEY APR   JUN   AUG   OCT   DEC   FEB   APR   JUN   AUG   OCT   DEC   FEB   APR   JUN      (DAYS)
                                 79    79    79    79    79    80    80    80    80    80    80    81    81    81
                                 I-----I-----I-----I-----I-----I-----I-----I-----I-----I-----I-----I-----I-----I--I

3100  SITE PREPARATION        B  I BBBBBBBBBBB I   *I    I     I     I     I     I     I     I     I     I     I  I      0
                             AF  I  AAAAAAAAA  I   *I    I     I     I     I     I     I     I     I     I     I  I      2

3200  CRANAGE                 B  I     I     I BBBBBB*BBBBBBBBBBBBBBBBBBBBBBBBBBBBBBBBBBBBBBBBBBBBBBBBBBBBBBI     I     I     I  I      0
                             AF  I     I     I AAAAA*FFFFFFFFFFFFFFFFFFFFFFFFFFFFFFFFFFFFFFFFFFFFFFFFFFFFFFFI     I     I     I  I      4

3400  SITE CLEAN-UP           B  I     I     I    *I     I     I     I     I     I BBBBBBBBBBBBBBBBBBBBBB I     I     I  I      0
                             AF  I     I     I    *I     I     I     I     I     I FFFFFFFFFFFFFFFFFFFFFF I     I     I  I      4

SUMMARY
-------
3000  SITE OPERATIONS         B  I BBBBBBBBBBBBBBBBBBBBBB*BBBBBBBBBBBBBBBBBBBBBBBBBBBBBBBBBBBBBBBBBBBBBBBBBBBBBBBBBBBBBBBBBB I     I     I  I      0
                             AF  I  AAAAAAAAAAAAAAAAAAAA*FFFFFFFFFFFFFFFFFFFFFFFFFFFFFFFFFFFFFFFFFFFFFFFFFFFFFFFFFFFFFFFFFFF I     I     I  I      4
```

(b) WBS format.

**Figure 5-15** Schedule performance reports.

COST PERFORMANCE REPORT
WORK BREAKDOWN STRUCTURE

REPORT DATE : 1-JUL-79

METIER/ARTEMIS :
DEMONSTRATION OF AN EARNED VALUE SYSTEM

PROJECT NAME : WASTE DISPOSAL PLANT
CONTRACT TYPE/NUMBER : FP. 1/08/1932

TOTAL PROJECT SUMMARY

COMPANY : XYZ COMPANY
CONTRACTOR : BUCKSHEE ENG.
LOCATION : CINCINATTI

REPORT PERIOD : OCT-1979

| ITEM | CURRENT PERIOD | | | | | CUMULATIVE TO DATE | | | | | AT COMPLETION | | |
|---|---|---|---|---|---|---|---|---|---|---|---|---|---|
| | BUDGETED COST | | ACTUAL | VARIANCE | | BUDGETED COST | | ACTUAL | VARIANCE | | | | |
| | WORK SCHEDULED | WORK PERFORM'D | COST WORK PERF'D | SCHEDULE | COST | WORK SCHEDULED | WORK PERFORM'D | COST WORK PERF'D | SCHEDULE | COST | BUDGETED | LATEST REVISED ESTIMATE | VARIANCE |
| HEATER HOUSE | 13473 | 11889 | 15800 | -1584 | -3911 | 64996 | 54643 | 56461 | -10353 | -1818 | 178964 | 189227 | -10263 |
| TANKS | 22996 | 22146 | 18324 | -850 | 3822 | 103476 | 102854 | 105446 | -622 | -2592 | 488120 | 499139 | -11019 |
| SITE OPERATIONS | 2040 | 2880 | 288 | 840 | 2592 | 57694 | 56605 | 56820 | -1089 | -215 | 107848 | 109296 | -1448 |
| TOTAL PROJECT SUMMARY | 38509 | 36914 | 34412 | -1595 | 2502 | 226166 | 214102 | 218728 | -12064 | -4626 | 774932 | 797662 | -22730 |

CURRENT TOTAL BUDGET = $774932

ESTIMATED FINAL TOTAL COST = $797662

COST PERFORMANCE INDEX =0.98
SCHED PERFORMANCE INDEX =0.95

PRODUCED BY METIER/ARTEMIS

**Figure 5-16** Cost performance report organized by WBS (Level 1 summary).

COST PERFORMANCE REPORT
FUNCTIONAL STRUCTURE

REPORT DATE : 1-JUL-79

METIER/ARTEMIS :
DEMONSTRATION OF AN EARNED VALUE SYSTEM

PROJECT NAME : WASTE DISPOSAL PLANT
CONTRACT TYPE/NUMBER : FP. 1/08/1932
SUPERINTENDENT : TANKS - D. HARVEY

SUPERVISOR : REINFORCED WORKS - E.MARTIN

COMPANY : XYZ COMPANY
CONTRACTOR : BUCKSHEE ENG.
LOCATION : CINCINATTI

REPORT PERIOD : OCT-1979

| ITEM | CURRENT PERIOD | | | | | CUMULATIVE TO DATE | | | | | AT COMPLETION | | |
|---|---|---|---|---|---|---|---|---|---|---|---|---|---|
| | BUDGETED COST: | | ACTUAL | VARIANCE | | BUDGETED COST | | ACTUAL | VARIANCE | | | | |
| | WORK SCHEDULED | WORK PERFORM'D | COST WORK PERF'D | SCHEDULE | COST | WORK SCHEDULED | WORK PERFORM'D | COST WORK PERF'D | SCHEDULE | COST | BUDGETED | LATEST REVISED ESTIMATE | VARIANCE |
| REINFORCE BASE (P1) | 5040 | 0 | 960 | -5040 | -960 | 6720 | 961 | 1075 | -5759 | -114 | 6720 | 7526 | -806 |
| RING BEAM (P1) | 5906 | 0 | 880 | -5906 | -880 | 5906 | 810 | 898 | -5096 | -88 | 7296 | 8172 | -876 |
| REINFORCE WALLS (P1) | 0 | 0 | 0 | 0 | 0 | 0 | 0 | 0 | 0 | 0 | 13200 | 13200 | 0 |
| REINFORCE BASE (P2) | 0 | 0 | 0 | 0 | 0 | 0 | 0 | 0 | 0 | 0 | 11904 | 11904 | 0 |
| RING BEAM (P2) | 0 | 0 | 0 | 0 | 0 | 0 | 0 | 0 | 0 | 0 | 12136 | 12136 | 0 |
| REINFORCE WALLS (P2) | 0 | 0 | 0 | 0 | 0 | 0 | 0 | 0 | 0 | 0 | 16080 | 17040 | -960 |
| REINFORCE BASE (S1) | 6720 | 6720 | 6720 | 0 | 0 | 6720 | 6720 | 7526 | 0 | -806 | 6720 | 7526 | -806 |
| RING BEAM (S1) | 304 | 0 | 0 | -304 | 0 | 304 | 0 | 0 | -304 | 0 | 7296 | 7296 | 0 |
| REINFORCE WALLS (S1) | 0 | 0 | 0 | 0 | 0 | 0 | 0 | 0 | 0 | 0 | 17040 | 17040 | 0 |
| REINFORCE BASE (S2) | 0 | 0 | 0 | 0 | 0 | 0 | 0 | 0 | 0 | 0 | 11904 | 12576 | -672 |
| RING BEAM (S2) | 0 | 0 | 0 | 0 | 0 | 0 | 0 | 0 | 0 | 0 | 7296 | 7296 | 0 |
| REINFORCE BASE (C1) | 1926 | 796 | 440 | -1130 | 356 | 8560 | 8560 | 9058 | 0 | -498 | 8560 | 9058 | -498 |
| RING BEAM (C1) | 0 | 0 | 0 | 0 | 0 | 0 | 0 | 0 | 0 | 0 | 0 | 0 | 0 |
| REINFORCE WALLS (C1) | 350 | 1439 | 1440 | 1089 | -1 | 402 | 1439 | 1613 | 1036 | -174 | 4320 | 4838 | -518 |
| REINFORCE BASE (C2) | 569 | 3786 | 3816 | 3217 | -30 | 1281 | 7976 | 8337 | 6695 | -361 | 9168 | 10268 | -1100 |
| RING BEAM (C2) | 0 | 0 | 0 | 0 | 0 | 0 | 0 | 0 | 0 | 0 | 0 | 0 | 0 |
| REINFORCE WALLS (C2) | 224 | 0 | 240 | -224 | -240 | 224 | 246 | 269 | 22 | -23 | 4320 | 4838 | -518 |
| TOTAL - SUPERVISOR | 21039 | 12741 | 14496 | -8298 | -1755 | 30118 | 26712 | 28776 | -3406 | -2064 | 160040 | 167755 | -7715 |

CURRENT BUDGET = $ 160040 ESTIMATED FINAL COST = $ 167755

**Figure 5-17** Cost performance report organized by OBS (for reinforced works supervisor, tanks area).

```
WORK PACKAGE / COST CODE REPORT - WP       2121

METIER/ARTEMIS :
DEMONSTRATION OF AN EARNED VALUE SYSTEM
                                                     REPORT PERIOD : OCT-1979
                                                     CONTRACT NUMBER - 1/08/1932

                                                     WORK PACKAGE #     2121    DESCRIPTION : PILES (P2)
                                                     SUMMARY WBS CODE    212    DESCRIPTION : PRIMARY TANK NO.2

WORK PACKAGE SUMMARY
====================

DATES :
-----

BUDGET DATES  :   START = 12-JUL-79   FINISH = 16-NOV-79
CURRENT DATES :   START = 17-JUL-79   FINISH = 29-NOV-79

COST INFORMATION
---- -----------

CURRENT BUDGET : $10976.00        ACTUAL COST TO DATE : $8691.20        REVISED ESTIMATE  : $12293.12

          PERCENTAGE COMPLETION TO DATE :    76.90

COST CODE DETAIL TO DATE
==== ==== ====== == ====
```

| COST CODE | BUDGET MANHOURS | BUDGET $ | ACTUALS TO DATE MANHOURS | ACTUALS TO DATE $ | FORECAST MANHOURS | FORECAST $ |
|---|---|---|---|---|---|---|
| A103 | 192 | 2016 | 0 | 0 | 197 | 2258 |
| A105 | 640 | 8960 | 697 | 8691 | 659 | 10035 |

```
-------------------------------------------------------------------------------
REPORT SHOULD BE RETURNED TO SUPERVISOR : PILING - J.P.LANE


PRODUCED BY METIER/ARTEMIS
```

**Figure 5-18** Sample of detailed cost information for individual work package.

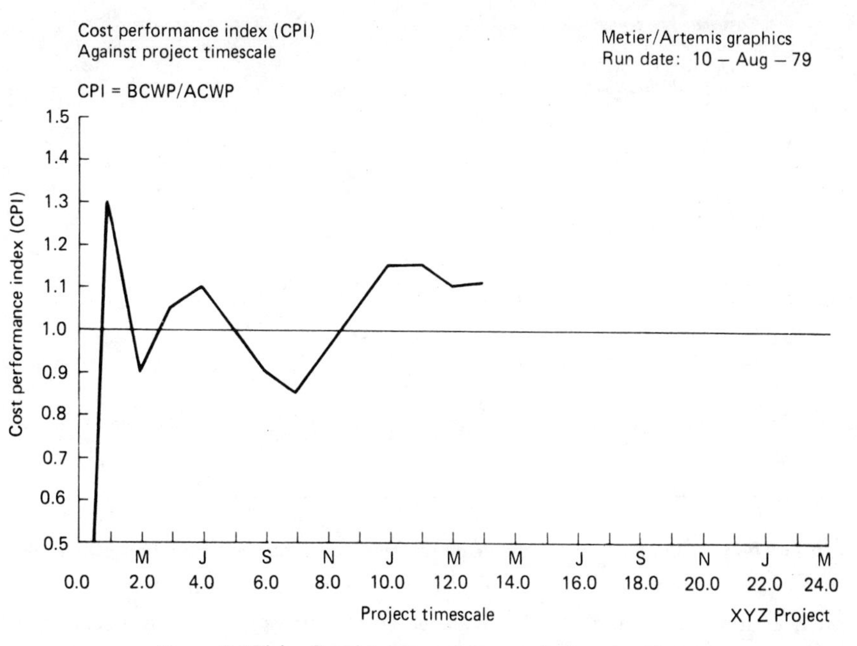


**Figure 5-19(a)** Project cost performance index over time.

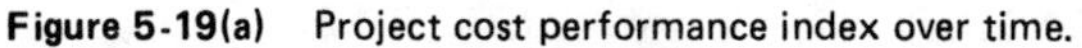

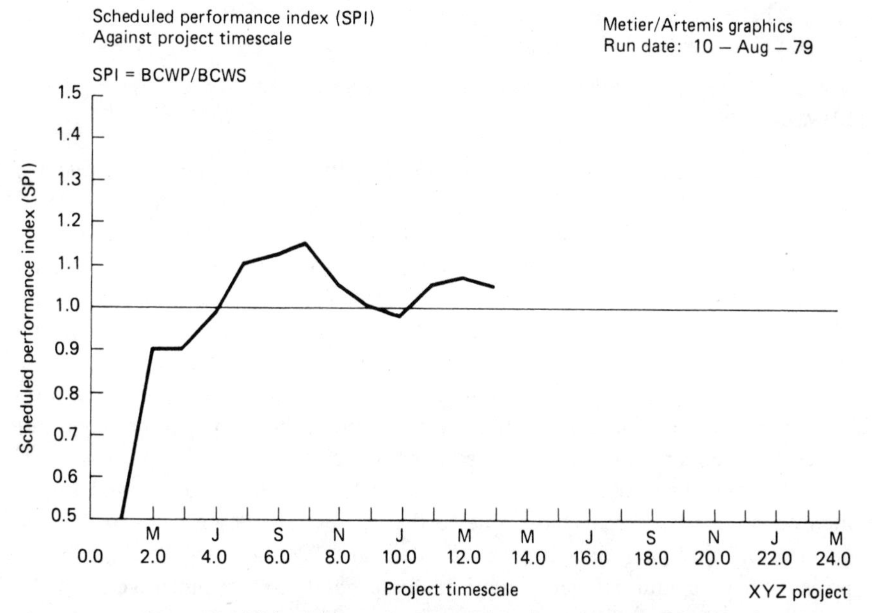


**Figure 5-19(b)** Project schedule performance index over time.

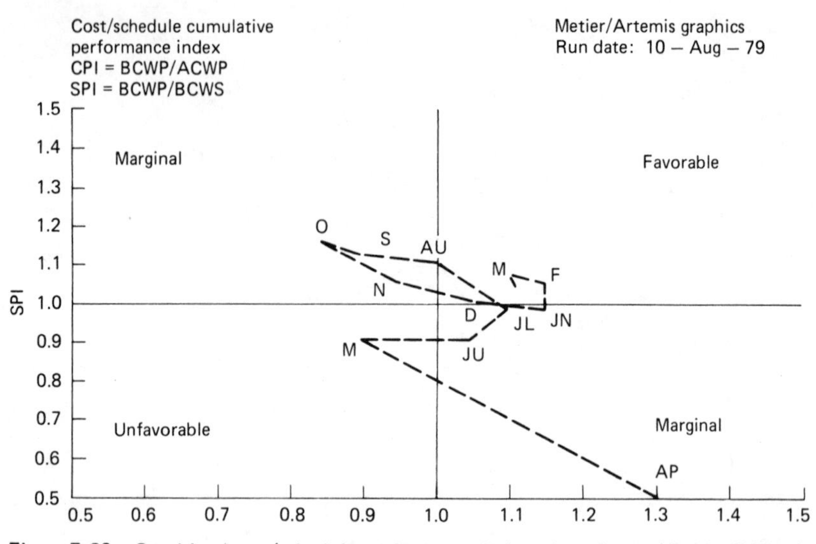


**Figure 5-20** Combined cost/schedule performance index plotted monthly, April-March.

such as the distribution of work package costs over time. It is not necessary to have such close interlinkages between the CPM scheduling and the cost monitoring portions of the overall system, as pointed out earlier in this chapter. However, as these interlinkages are reduced in number and frequency the potential benefits of improved accuracy and reliability of the information produced will correspondingly decline.

## SUMMARY

The concept of network-based cost control offers to project management a basically sound and powerful way of relating expenditures to the work done, a way of reporting what management paid for what it got. Unlike other critical path techniques, however, network cost control cannot be applied easily to only selected projects, or only during certain phases of a project. The design and implementation of such a control system involves modifications in established payroll and cost accounting procedures of the organization. To obtain an adequate return on this investment, the system should be consistently applied to the projects undertaken by the organization. Furthermore, the full use of network cost control concepts requires data processing equipment.

In the aerospace industry, these concepts have been developed and implemented on a fairly broad basis, due to the interest of the government and the general availability of data processing talent and facilities. In the other major project-oriented industry, construction, there has been less incentive from the customers and relatively less use of computers. The special field of petrochemical facilities construction is an exception to this general rule; developments there have been at an accelerated pace over the past few years and show no signs of abating. With the increasing economies of data processing equipment and services, and continuing pressures of inflation and high interest rates, there can be little doubt that network-based cost control systems will be developed and utilized even more widely in the next few years.

## REFERENCES


1. Control Data Corporation, *3400/3600 Computer Systems, PERT Reference Manual*, Palo Alto, California, 1964.
2. Archibald, R. D., *Experience Using the PERT Cost System as a Management Tool*, presented at the Institute of Aerospace Sciences, Los Angeles, June 21, 1962.
3. Beutel, M. L., "Computer Estimates Costs, Saves Time, Money," *Engineering News-Record*, February 28, 1963.
4. *DOD and NASA Guide, PERT Cost Systems Design*, by the Office of the Secretary of Defense and the National Aeronautics and Space Administration, U.S. Government Printing Office, Washington, D.C., June 1962.
5. Fox, J. Ronald, *Arming America*, Division of Research, Graduate School of Business Administration, Harvard University, 1974.
6. McDonnell Douglas Automation Company, "Automated Cost Control Tools and The Management Scheduling and Control System (MSCS)," St. Louis, 1974; from original work by J. D. Wiest.
7. *Network Analysis System*, Regulation No. 1-1-11, Department of the Army, Office of the Chief of Engineers, March 15, 1963.
8. U.S. Air Force Systems Command, *Cost/Schedule Control Systems Criteria Joint Implementation Guide*, Oct. 1976.
9. Metier Management Systems Company, *Demonstration Of An Earned Value System, ARTEMIS*, 1980.


## EXERCISES

1. For the network in Figure 5-1, assume that it is decided to plan that all activities will begin at the ES times, except for the following activities, which are scheduled to begin at the times listed below:

| *Activity* | *Scheduled Start* |
|---|---|
| A | 3 |
| D | 5 |
| H | 9 |
| C | 4 |
| G | 5 |

Compute the planned cumulative costs for the project and plot the results on Figure 5-2.

2. Using the methods given in Chapter 7, schedule the activities in Figure 5-1 such that the project duration is minimized, subject to the constraint that the planned weekly expenditures never exceed $2,500.
3. The ARTEMIS Cost Performance Report shown in Figure 5-16 provides data on the 3 major elements of the Waste Disposal Plant project (Heater, House, etc.). Using the information provided in this report, summarize the performance status of each of the 3 major elements in the current period, over the total period to date, and what is indicated as likely at completion.
4. The text lists several problems inherent in the design of network-based cost control systems having to do with activity accounting procedures. How do these problems appear to be handled in the ARTEMIS system?

# 6

# COMMENTS ON PRACTICAL APPLICATIONS

One might say that critical path methods are applicable to the management of a project from the cradle to the grave. At each of the stages in the life of a project, there are a number of areas related to the practical applications of critical path methods that are not covered in the other chapters but are deserving of attention. This chapter will treat several of these topics, from the project proposal to the project updating and control phase.

Special attention is given to techniques in the construction industry, in which many firms employ network methods. Concluding the chapter is an introduction to the organizational aspects of project management.

## PREPARING PROJECT PROPOSALS

The numerous potential applications of critical path methods in the preparation of project proposals are fairly obvious and straightforward. These applications are based on the various project planning and scheduling techniques discussed in this text, except perhaps that these techniques are applied with less detail than would normally be used on a project that is definitely to be executed. If the project is to be executed, the proposal network serves as a framework for developing a detailed plan and schedule to be used in carrying out the project.

In cases where the project completion time is specified in the contract, critical path methods are useful in determining a project plan which will meet the time specification. It may turn out that this plan requires the use of certain time saving features which add unexpected costs, or risks, to those which would normally be required to complete the project tasks. For example, if performance time is extremely critical, "crash" time performance of critical path activities may be required. In addition, it may be necessary to perform certain activities concurrently, which would normally be performed in series. The added risks of the expedited plan over the normal plan must, of course, be considered in preparing the proposal cost estimate.

If the proposal requires consideration of alternate completion times and costs, then the time-cost trade-off procedures described in Chapter 8 may be appropriate in arriving at project costs. Similarly, if the project under consideration will be competing for a fixed set of resources, then the procedures discussed in Chapter 7 may be applicable. It may turn out that the proposed project schedule dovetails nicely with the phasing out of current projects, or it may put an extreme burden on certain critical resources. This type of analysis may greatly influence the profit margin that management places on the project, or influence the decision of whether of not to submit a bid on the proposed project.

A number of firms in various industries are now using critical path methods routinely, in varying degrees of detail, in preparing project proposals. One large metal-working firm requires that all internal proposals for capital expenditures over $25,000 be accompanied by a CPM network showing the project plan and schedule. Some advertising firms submit to prospective clients networks of proposed promotional campaigns, showing how the activities involved in product distribution, space advertising, television and radio commercials, surveys, and other facets of the campaign are to be coordinated. The implementation of a computerized management system has proved to be a type of project that is difficult to schedule and control; now it is not unusual for project networks to be submitted to management, along with flow charts, proposed report formats, and other elements of proposed projects, and for the networks to be updated weekly or biweekly to maintain status control over such implementation projects.

## CONTRACTUAL REQUIREMENTS

As mentioned in Chapter 1 and elsewhere, the use of network methods has been boosted greatly by contractual requirements that network methods be used to report the plans and progress of projects. This has been particularly true in the two largest project-oriented industries, aerospace and construction. In the aerospace industry, where a major customer is the U. S. Government and its various military and space agencies, the accommodation to network requirements was widespread and relatively quick. Contractors in the aerospace industry are accustomed to extensive Government reporting requirements: the firms are relatively large and equipped with computers and capable technical staffs; and the contracts are often on a cost-plus basis. For these reasons, contractors have been able to meet network requirements promptly. Some projects produced networks of several thousand activities in a few months.

In the construction industry, however, the circumstances have been quite different. The Government has played a major role here also, through the Corps of Engineers, the Bureau of Yards and Docks, NASA, and other agencies responsible for large construction programs. But construction companies tend to be relatively small, low-overhead organizations without computer equipment or data processing personnel. Also, the contracts are usually on a fixed-price basis. Under these circumstances, which exist throughout the commercial as well as Government segments of the industry, requirements for critical path methods encounter a variety of difficulties.

The first problem in this application is to prepare specifications for the network reporting system desired. Since the contractors are often not familiar with network methods, are not administratively geared to handle increases in technical reporting, and often must pay for outside assistance, the firms will take very conservative approaches to critical path method specifications. The authors have seen specifications consisting of only one or two sentences, saying only that "the contractor shall report progress on the project monthly by means of a Critical Path Method chart and schedule." Requirements that are this brief, of course, invite the contractor to submit a five-activity network or anything that he thinks might meet the minimum technical requirements of a network "chart and schedule."

At the other extreme, one construction office of a Government agency prepared and repeatedly used standard CPM specifications that covered eight pages and contained almost 1000 words. This specification attempted to detail exactly how all the initial updating computations were to be made and how the computer reports were to be formatted (including the column headings for such items as an "actual early date" and other unique descriptions). These specifications were so confusing that only a few specialized computer service bureaus in the region could satisfy them, and the contractors were dependent upon the use of these service bureaus.

For those who wish to prepare contractual specifications for network reporting, the considerations listed below are suggested as guidelines for reasonable and adequate requirements.

1. *Definition of Network method.* An available document or textbook should be referenced as comprising a definition of the basic technique to be employed. The terms used throughout the specification should be consistent with the terms used in the reference.

2. *Level of Detail.* The best way to specify the desired level of network detail is to specify a range of the number of activities to be included. The range selected, however, should be based on practical experience with the type of project involved.
3. *Graphic Format.* If a particular network format is desired, such as the activity-on-arrow or the activity-on-node format, this must be specified. As mentioned above, the best way to do this is to refer to an available text and adopt the terminology used in that text to write this specification. One may also elect to have progress reports submitted in the network format rather than in computer reports. This may also be specified.
4. *Computer Report.* If computer reports are to be required, the data desired on the reports should be spelled out, item for item. The desired sequences (sorts) of the reports should also be specified, taking care not to require more reports than will actually be used. (See Chapter 11 for more complete discussion of computer methods.) Normally two output sorts, one by I-J or activity number and one by total float, should be adequate.
5. *Updating.* As described in Chapter 11 some updating procedures give misleading or erroneous results. Therefore, it may be necessary to reference or to fully describe the updating calculation required. The frequency of updating reports, number of copies desired, whether revised networks are required, and similar details should be mentioned. In this connection, it should be noted that a requirement to maintain a network on a time scale may increase the updating cost.
6. *Cost Reporting.* If the network is also to be used for reporting the cost of work completed, the specific means of allocating all costs to activities and of determining the percent completion on activities must be fully detailed.

## IMPLEMENTATION

Generally speaking, the implementation of critical path methods involves the six steps outlined in Chapter 1 and summarized as follows: (1) planning, (2) time and resource estimation, (3) scheduling, (4) time-cost trade-offs, (5) resource allocation, and (6) updating and controlling the project.

However, it should be emphasized that the full application of all the techniques presented in this text is not required in order to accomplish these steps and obtain worthwhile improvements in the plan and conduct of projects. *Nor is the size of the project a critical factor in the practical economic implementation of critical path methods.* Networks sketched on the backs of envelopes have proved to be useful ways of quickly analyzing and communicating the plan of a small project, such as the preparation of a technical report. The use of additional techniques, such as the PERT statistical approach and time-cost trade-offs may also be applied in concept, if not in full detail, to relatively small projects. Thus, the implementation of critical path concepts can be considered a matter of routine management practice, rather than an investment justified only for large or complex projects.

The implementation of network planning and control techniques should take place after a preliminary study has been made to determine how the project tasks are to be broken down and assigned to key personnel within the organization or to subcontractors. In large projects, particularly in the aerospace industries, this preliminary study is quite important; it first requires that the overall mission and performance goals of the system be refined to the satisfaction of the systems engineer. Then, the functional analysis of the system can begin, which will lead to the design requirements for the proposed system configuration. The establishment of the base-line design requirements is a major milestone in the systems engineering portion of a program definition study. It is at this time that the formal application of network methods can be made most effectively.

### Utilizing Personnel Effectively

The preparation of the project network is a job for the key management of the project, the person or team of persons who know the most about the objectives, technology, and resources of the project. To conserve the valuable time of these personnel, the networking effort should be a concerted and concentrated one, not a secondary activity that becomes drawn out and perhaps never completed. There is a period of time that management must devote to planning, and the network should be used as a vehicle for, an aid to, and a documentation of this valuable effort. Technicians can relieve much of the load on the project management by transcribing sketches into legible networks, by making computations, and by making preliminary analyses of the schedule. For this reason, many organizations have trained young men and women as part-time or full-time critical path analysts.

### Working with Subcontractors

When subcontractors play a major role in a project, they should play a major role in the critical path planning, scheduling, and control. In some cases this may require group meetings to develop the network and time estimates. However, in many instances it is not practical to call in a group of people unfamiliar with network theory and expect them all to contribute effectively to the early draft of a network. In these circumstances it may be better for the prime contractor to develop the rough draft as far as possible, then call in the subcontractors to comment and add time estimates to their particular areas. It is often possible for a person to read and effectively criticize a network, even though he may not have the training or experience to develop a network originally. One may also elect to hold short courses in network preparation, then ask each subcontractor to develop a subnetwork of his portion of the project. The feasibility of this approach depends not only on the size of the project and the scope of interest of the subcontractors, but also on how complex are the interrelationships among the various areas of responsibility.

Incidentally, one of the important side effects of critical path applications is the fact that it brings the subcontractors and the prime contractor together to meet and discuss the project. The group generally discovers technical problems and begins to work in advance toward solutions of these as well as in cooperation on the planning and scheduling of the project.

## PROJECT CONTROL

Once a project is underway, the critical path network and schedule should serve as a guide to the accomplishment of every activity in proper sequence and on schedule. It is in the fundamental nature of projects, however, that activities will seldom start or finish exactly as scheduled. Therefore, updating the plan and schedule is an important link in the critical path concept. There are no rigorous or standard updating procedures in general use, except where computer programs are involved. The procedures offered here are suggestions regarding the general approach, with some specific recommendations for particular circumstances.

There are four functions performed in the process of updating the schedule alone (without regard to cost revision). These are:

1. denoting actual progress on the network,
2. revising the network logic or time estimates of uncompleted activities, if appropriate,
3. recomputing the basic schedule of earliest and latest allowable times, and
4. revising the scheduled activity start times and denoting new critical paths.

Item 2 simply requires the erasure of previous notation and replacement with the new. The other changes may be categorized as Progress Notation and Revised Scheduling.

### Progress Notation

A field supervisor or other person close to the actual progress of a project should be assigned the responsibility of making progress notations directly on a copy of the network. This is the most reliable way to maintain accurate records for input to the network updating procedure. Requiring the responsible first-line supervisor to make these routine daily or weekly notations also helps insure that the network serves its purpose as a detailed schedule of work.

Exactly how the notations are made is not particularly important, as long as they are clearly understood and complete. To be complete it is necessary only that both the actual start and the actual finish of each activity be recorded. Percent completion notes are not normally required, except where cost control data are involved.

On the working copy of the new network used to make these notations, it is often useful to use colored pencils for the notation, using a different color for each updating period. Then when converting the notes to computer input forms, the person doing the encoding knows to pick up only the green dates, or whatever was the color of the most recent period.

If the node scheme is used, the actual start and finish dates may be marked on the left and right sides of the nodes, respectively. If the network is computed manually, the node symbol may be designed to allow spaces for the actual dates to be recorded.

## Revised Scheduling

Hand computation of a revised and updated network requires no special arithmetic. One merely begins the forward pass computation at the last uncompleted activity on each path. However, it is important that the first dates computed for each path be consistent with the effective date of the computation. That is, no *ES* or *EF* date can be earlier than the effective or "cut-off" date of the progress report. If the first uncompleted activity has not started (or finished), then its *ES* (or *EF*) date must be recorded as at least equal to the effective date, and the forward pass is carried on from there.

Hand computation does, however, require some attention to the mechanics of erasure, lest this humble function become a major problem. Erasure of all remaining schedule times is required, of course, when a complete recomputation is to be made. One technique that reduces the erasure problem is to make the scheduling computations on a reproducible print of the network, instead of on the original tracing. Erasures may be made on the reproducible print without damage to the arrows and nodes of the network.

Recomputation may be necessary every time a network is updated and/or revised. If the actual progress and expected future progress are very close to the network schedule, there is no need to make a completely revised computation. The critical path network protects itself, in a sense, from the need for many recomputations, even when significant delays or accelerations occur, for the network reveals most of the effects of schedule changes without recomputations. One need only to refer to the float figures, for example, to see whether a specific delay will affect other portions of the project. With this information plainly visible, it is not always necessary to erase and recompute. Recomputation of the backward pass is not necessary at all if no changes are made in the network logic or in the estimated activity duration times. If the scheduling computations are made by computer, then the revision procedures specified by the computer program must be used.

### Time-Scaled Networks

A "time-scaled" network is one in which the arrows and nodes are located by a time scale along the horizontal border. With the time scale, the length of each activity arrow (or its shadow projection on the horizontal scale) represents the activity's estimated duration. The location of the arrow represents its scheduled start and finish times. Similarly, the location of each node represents its scheduled occurrence time. The arrows and nodes may be located by their earliest times, latest times, or at selected times between these. Slack is customarily indicated by dashed lines.

The obvious advantage of time scaling is the visual clarity it provides for the analysis of concurrent activities. Time-scaled networks have revealed problems of concurrency (such as the intended use of a test facility by two groups at the same time) that were overlooked in the analysis of the schedule data in the tabulated form of a computer output. Time-scaled and condensed networks are also advantageous in presentations to top levels of management, since the schedule is communicated more quickly and more emphatically by the graphical technique.

The disadvantage of time-scaling is not so obvious but is nonetheless important. Using manual methods, the maintenance of time-scaled networks through updating and revision periods is expensive. A change in one time estimate or a delay in the actual progress on one path can change the location of a number of succeeding arrows and nodes, necessitating the redrawing of a large portion of the network. Under contracts that require frequency network revisions, the use of time-scaled networks could keep a technician or draftsman busy almost full-time, and the networks could be delayed days or weeks in reaching the contracting agency.

Many attempts have been made to simplify network revision, so that time-scaled networks could be quickly and economically updated. Listed below are some of the materials and techniques that have been tried.

1. *Gummed Labels and Tape.* Transparent materials with adhesive backing can be placed and moved about on tracing paper or mylar sheets. Only high quality materials should be used, lest they blister, peel, or make poor reproductions.

2. *Computer-Drawn Networks.* As explained in Chapter 11, some software packages for network processing include features for computer-drawn networks.
3. *Steel Panels with Magnets for Events, and Tape or Chalk Lines for Arrows.* Networks constructed in this way have worked for some applications but cannot be easily reproduced, even by photography, for distribution.
4. *Scheduling Boards with Moveable Bars.* Slotted aluminum boards with moveable bars in the slots to represent activities have been popular in certain applications. The bars are labeled and color-coded.

Time-scaling can be particularly worthwhile where it is done as an aid to planning and not intended for use after the project is underway. An interesting example is shown in Figure 6-1, which is a network for a complex surgical procedure.[1] The only purpose of this network application is to help develop an efficient, well-coordinated plan, especially in the critical period between the 30th and 45th minutes. Once the plan is developed, the network has no further use.

Another example is shown in Figure 6-2. Here three key resources in the project are summarized by time period, illustrating the utility of time-scaling networks where there are resource allocation problems. As in the previous example, this effort is usually worthwhile only in the planning stage of the project. In some cases, though, the resource problems are important enough to the time

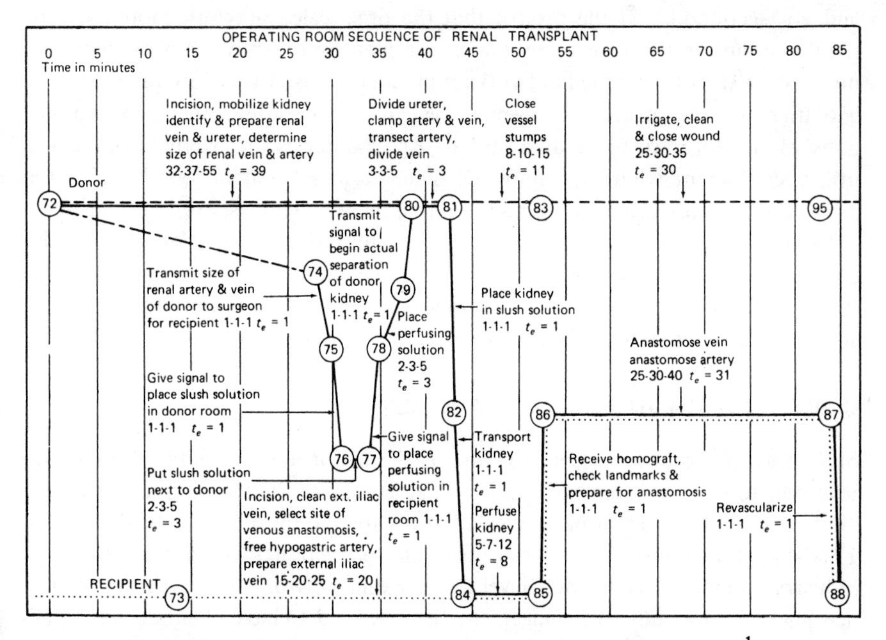


**Figure 6-1** Time-scaled network of a kidney transplant procedure.[1]

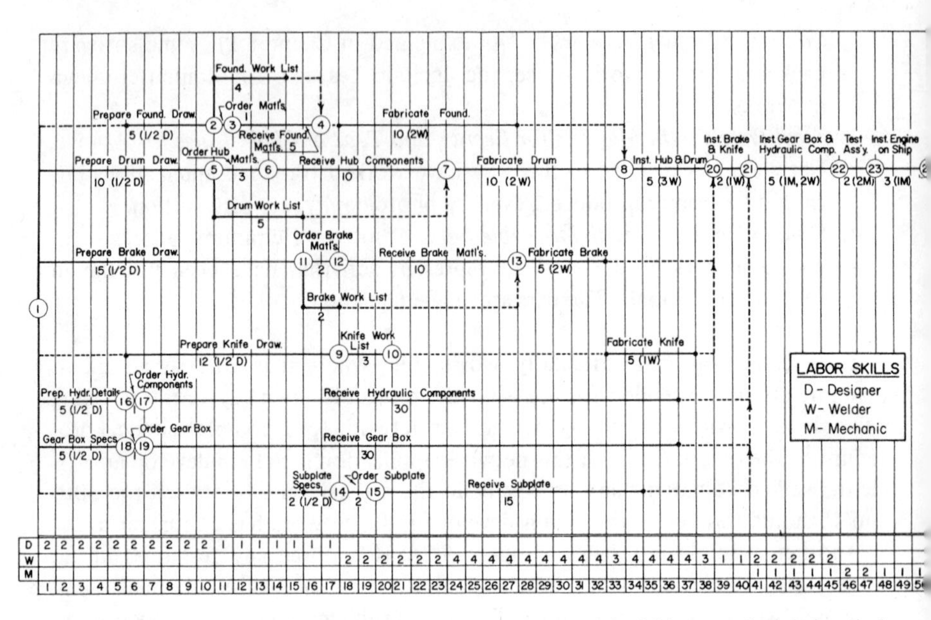


**Figure 6-2** Time-scaled network with resource summary. Activities are scheduled to limit designer requirements to two and welders to four.

and cost economics of the project that the time-scaled network should also be used to monitor the project progress. As mentioned above, however, project monitoring with time-scaled networks is practical, only if there are relatively few activities and the project management is centralized enough that copies of the network do not have to be distributed. An example is the overhaul of electric utility equipment where the network is set up in a "control room" where the project supervisors gather to discuss changes, revise schedules, etc.

Computer software has advanced to the point that time-scaled charting by computer is much more practical than manual methods in many applications. This trend can be expected to continue. At this writing, however, it is still advisable to be cautious about the expense of maintaining time-scaled networks.

## CONSTRUCTION INDUSTRY PRACTICES

Since the construction industry is one of the major users of network methods, it is well to consider some of the practical "tricks of the trade" in construction applications. In an informal survey of firms for the third edition of this text in 1980–81, the authors found that network practitioners in the construction industry varied widely in their level and style of network use. The larger engineering and construction management firms tended to have planning and scheduling departments with trained personnel and computer systems (or frequent access to an outside computer service). The smaller firms often employed scheduling consultants on a project basis, sometimes on their own initiative but more often because the project owner or architect required network-oriented plans and progress reports. Smaller scale homebuilders, paving contractors, subcontractors, and other types of firms may not use the techniques at all or may have had only limited exposure via a general contractor.

The practices of some of the scheduling consultants was of interest because of their full-time, professional attention to network methodology and their variety of applications. Two examples of such firms are illuminating.

*DDR International* is a consulting firm based in Atlanta, Georgia that specializes in construction planning and scheduling for general contractors, owners, and architects.[2] DDR uses computer processing of networks only when required by the client, which is seldom. Rather, they employ time-scaled networks to convey both the sequence and the schedule aspects of project plans. Networks are prepared on tracing paper with translucent mylar labels for activities. The labels have adhesive backing that permits moving them around on the chart at will. The activities are placed on the time scale according to "scheduled" starts rather than *ES* or *LS* times. Updated progress is denoted directly on the network, including notation revealing the ahead or behind schedule status of each path. The activities are not rescheduled, however (meaning a revision in the network), unless a significant change is made in the project plan or schedule.

A major purpose of the DDR technique is to keep the schedule from becoming too complicated for client personnel, including managers, field supervisors, subcontractors, and architects. Computerized analysis and their tabular reports are considered objectionable for this reason. The most essential facts of the plan and schedule can best be communicated, according to the consultant, by time-scaled networks based on agreed-upon "scheduled" start times for each activity. Forward and backward passes are made manually to identify the critical path, but the computations are not otherwise documented.

For an example of this kind of network presentation, see Figure 6-3. Note that the format appears to be activity-on-arrow, since each activity is represented by a time-scaled bar. However, the bars are interconnected by dependency lines following the activity-on-node or precedence method. The network is actually a precedence diagram, and lead/lag factors are sometimes employed. DDR personnel feel that precedence diagramming of this type is "two or three times more effective" than the arrow format.

DDR consultants occasionally assist their clients in resource leveling, where one, two, or three crews of given type may need to be scheduled. This is always done manually, usually as the network is prepared originally. They do not offer network-integrated cost control systems, but they do produce cash requirements schedules and earned value reports, such as those shown in Chapter 5.

*Hanscomb Associates, Inc.* is a large international engineering and construc-

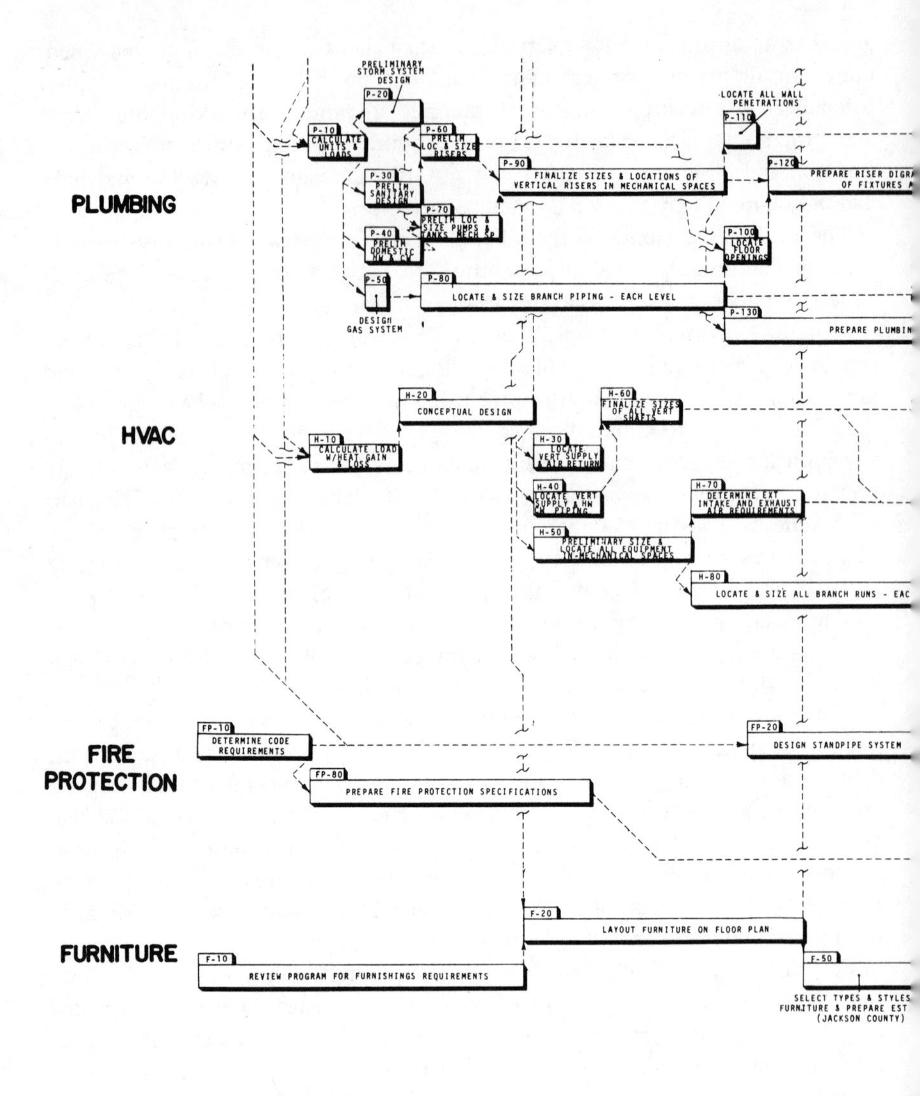

**Figure 6-3** Portion of a consultant's network illustrating time-scaled activity-on-node format. Courtesy of DDR International, Inc., Atlanta.

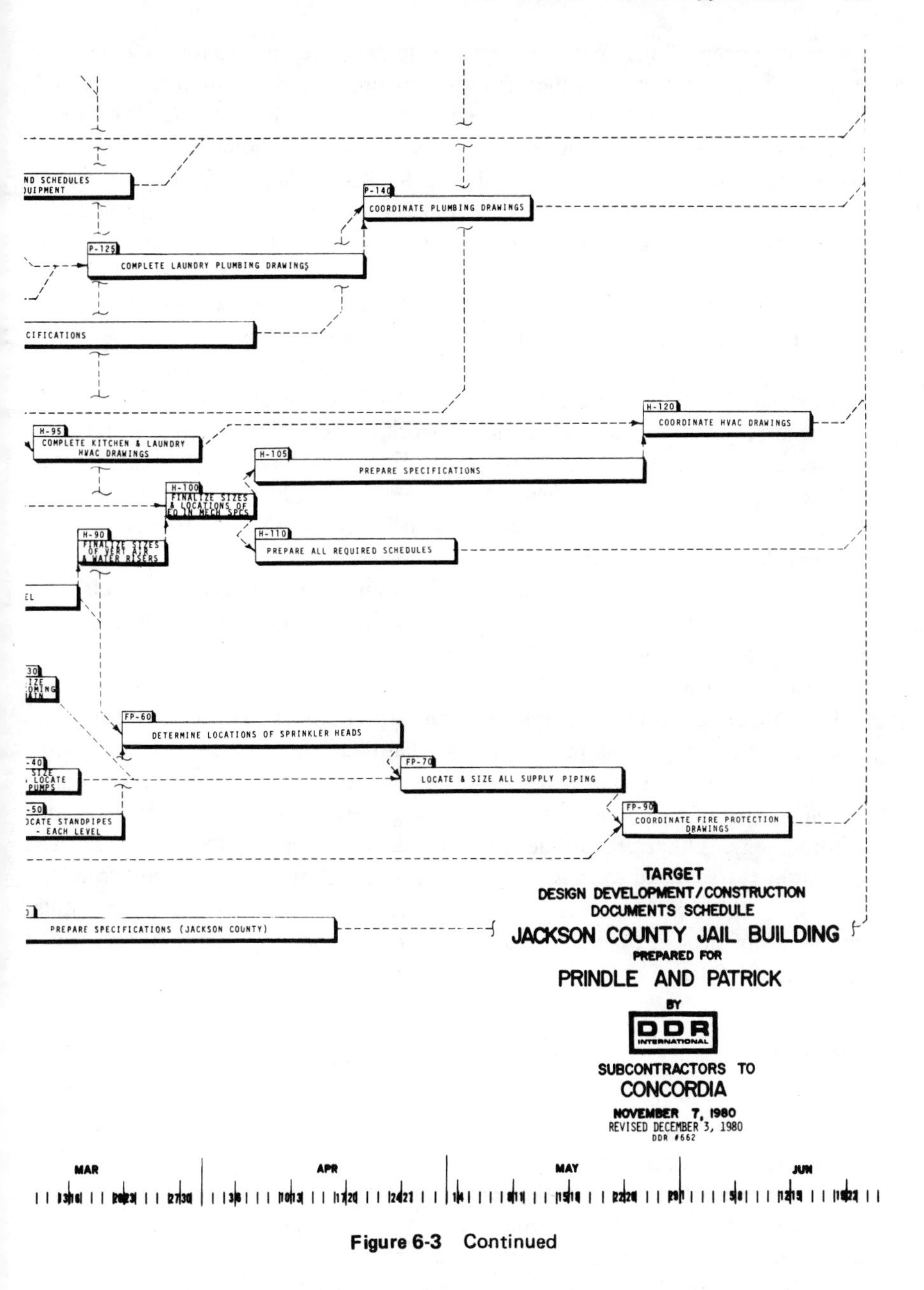


Figure 6-3 Continued

tion management firm. They provide a wide range of services for the construction industry, only one of which is project planning and scheduling. Several of the offices of the firm maintain specialists in network methods. Each office is independent, however, in the methodology employed. Some favor computer methods and some favor manual methods. At this writing, the Chicago office of the firm preferred computer methods, using one of the most advanced and flexible software packages. The Atlanta office never used computers, favoring time-scaled network presentations instead. As with some other consultants, the specialists at the Atlanta office of Hanscomb explained that an overriding consideration in their practice was the communication of clear and essential facts to their client. To avoid confusion or excessive detail they do not even make forward and backward passes; the critical path and near critical path is revealed by the time-scaling process when the network is prepared. For an example of a Hanscomb (Atlanta Office) network, see Figure 6-4. The network is based upon "scheduled" start times for each activity, not necessarily *ES* or *LS* times. The flags denote significant start dates or key delivery dates. The solid portion of the bars denotes completed work.

According to a senior project manager at Hanscomb, general contractors "do not want and will not pay for computer outputs." They also do not allocate resources, employ time-cost trade-off methods, or apply a network-integrated cost accounting system.[3]

In both of the above examples the consultants stress the necessity for clear communication with the project management and field personnel, unencumbered by details or sophisticated calculations that may confuse the most important facts of the schedule. They also stress economy and efficiency in methodology. Other practitioners, however, employ some of the most advanced computer systems and exercise a greater degree of control over the construction process. The variation depends upon the types and sizes of projects, the skills and styles of project managers, the preferences of the consultants (if involved), and other factors.

## HUMAN AND ORGANIZATIONAL ISSUES

While this text deals primarily with network techniques as they apply to planning, scheduling, and controlling projects, it is important to recognize that successful project management encompasses a broader body of knowledge. Indeed, it appears that the development of network methods as a distinctly different approach to planning and controlling projects has helped to focus management theorists and educators on other differences in projects as compared with other forms of endeavor. The term "project management" (synonymous with "program management" in government parlance) has now become a recognized subdivision of management theory. Libraries often list a number of titles under

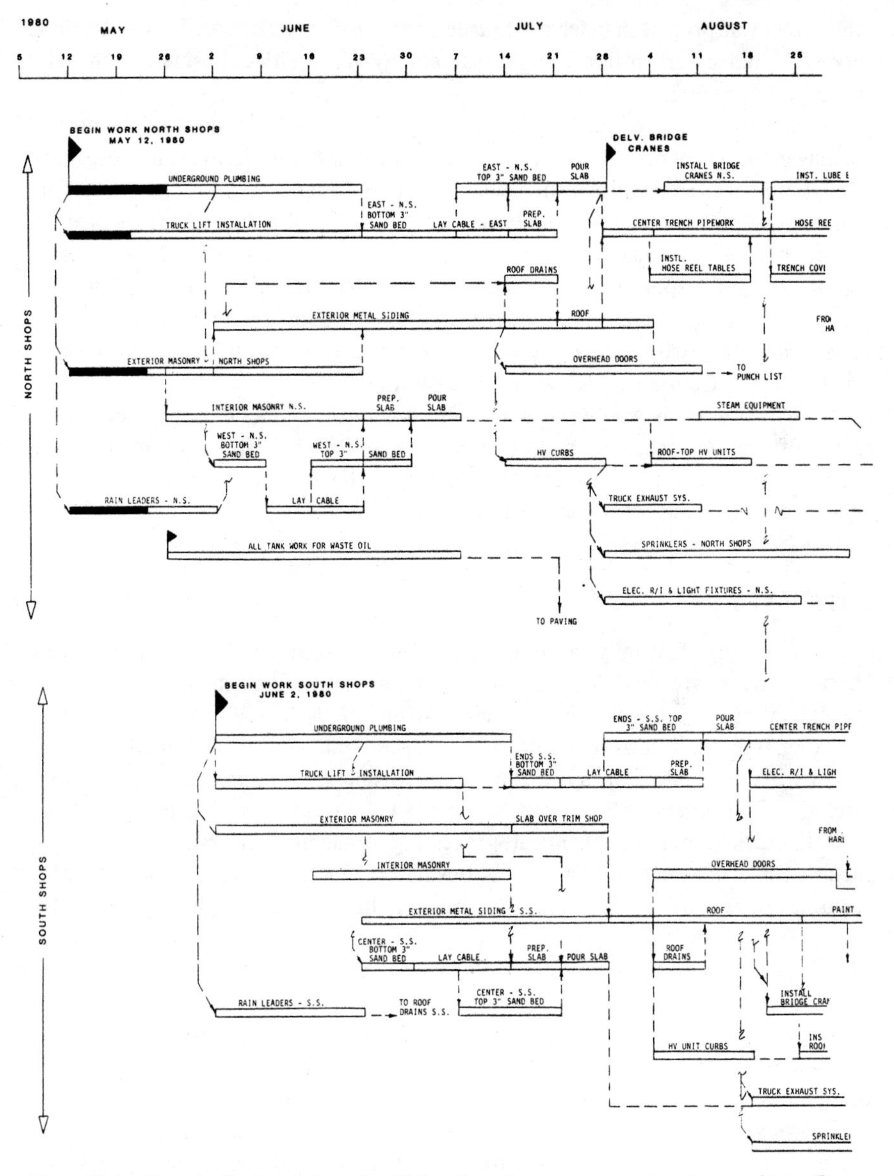


**Figure 6-4** Sample "network-bar chart" for shopping center construction project. Courtesy of Hanscomb Associates, Inc.

the subject of project management, usually in the 658.4 series of Dewey Decimal codes. Courses are offered in the subject by universities, business firms, and government agencies.

In addition to the quantitative and graphical methods that aid in project management, the broader subject now encompasses the human and organizational aspects that distinguish project-oriented endeavors from the more continuous or process-oriented endeavors. The successful management of projects is recognized as requiring different supervisory skills, different definitions of authority and responsibility, different criteria for the selection and evaluation of personnel, different operating policies, and other aspects of management. The reader who wishes to study these topics in greater depth is referred to other texts, such as Galbraith,[4] Kerzner,[5] and Martin.[6]

Nevertheless, it is well worthwhile here to identify and comment briefly upon some of the key issues in the human and organizational aspects of project management, in order to introduce the reader to the broader subject and to help place network methods in perspective.

### Type of Business

One of the first factors governing the role of project management in any situation is the degree of project orientation characterizing the business of the organization. Here we use "business" to mean the overall line of endeavor, whether it is run by private enterprise, governmental, educational institutions, or others. At one end of the spectrum are the businesses that produce a continuous stream of products, goods, and services. Of course these businesses include manufacturing, mining, chemical processing, financing, insurance, hospitality, health care, and many others. These businesses tend to be characterized by high volumes of repetitive operations, which lend themselves to high degrees of efficiency in the use of labor, materials, and facilities. For want of a better term, let us categorize these businesses as the "process" type.

At the other end of this spectrum are the project-oriented businesses. Major examples are the construction industry and research and development. As described in Chapter 1, these endeavors are characterized by limited repetition, uniqueness, and defined time limits for goals or results.

Between the two ends of this spectrum there are many businesses that consist of both process and project features simultaneously, or which shift back and forth from a process mode to a project mode. Examples include job shop manufacturing, where product changes occur frequently; advertising and sales promotion, where the goals and the methods undergo frequent revision; and any business that is frequently expanding, adding facilities, changing products or services, or developing new concepts while also producing a volume of products or services.

The organizations that are formed to operate in the project-oriented business environment are clearly the ones that must be skilled in project management concepts. It is their way of life. Not only do they use network methods for planning and scheduling, but they also employ project cost accounting schemes. The management structure tends to emphasize the project manager's role, which is a strong but temporary position of authority and responsibility.

At the same time, most project-oriented organizations of any size also maintain "functional" units which remain more-or-less permanent, such as engineering, marketing, finance, and distribution. The relationship between the permanent functional units of the organization and the project teams is one that presents particular management problems. Usually the functional units play supportive roles while the project teams are primal; yet the project teams are frequently disbanded or reorganized as projects end and others begin. The functional personnel are often expected to give up or loan key people to project teams at times, and to accept and employ project personnel who may be temporarily unassigned. If not handled well, these requirements can cause friction and/or inefficiencies in the use of personnel and facilities. As pointed out by Martin[5] even the organizations that are highly oriented toward the project mode, such as construction firms or hardware development organizations, must recognize that each project places a unique set of requirements on the management structure, because each project differs in many ways from those that came before and will come after.

In the business that involves a mixture of process and project modes, there is likewise a mixed need for the traditional management heirarchy along with project management schemes. This environment presents the most complex and unstable needs for organizational structure, and thus the most frequent problems.

In these cases, the general management must be aware of the potential problems and be skilled in establishing policies and practices that minimize the built-in conflicts between the goals of the mainstream process functions and the project goals. These skills include a sound working knowledge of project methodology and organizational structures.

## Organizational Structures

There are at least three well-defined alternatives for structuring organizations that involve a significant degree of project-mode operations. They are:

1. Traditional or pure organization *as applied to a project.*
2. Line-Staff organization.
3. Matrix organization.

The traditional or pure functional structure is pictured in Figure 6-5 in a typical form. This is the classical structure for process businesses, but it may

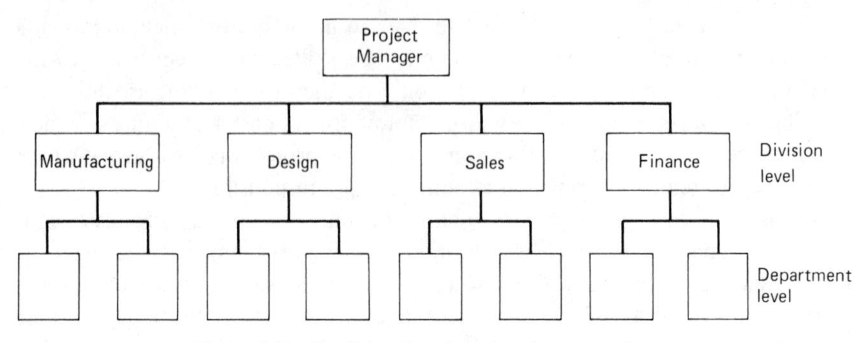


**Figure 6-5** Traditional or functional organization.

be applied, at least theoretically, to a project within a firm. (Martin calls this application "completely projectized management.") The simple structure provides some clear advantages over other organizational schemes, including (but not limited to) the following:

1. Strong management control over the project.
2. Fast response to changing conditions.
3. Personnel dedication to the project, clear lines of authority and personnel management.
4. Strong control over project schedules and costs; very suitable for profit center designation.

At the same time, application of the traditional organizational structure for process activities to a project-oriented business brings some definite disadvantages. Among them are:

1. High cost of maintaining duplicate functional personnel and facilities for each project.
2. Difficulty of maintaining technical quality when the functional units are not united.
3. Difficulty of balancing workloads when projects proceed through life cycles, phasing in and out, resulting in unreimbursed costs.

For these reasons the traditional structure is not likely to be suitable for project management unless the project is very large and expected to live for several years, and the sharing of resources with other projects is unnecessary or undesirable.

The line-staff structure is much more common, especially where both process and project activities exist together. As pictured in Figure 6-6 (adapted from Kerzner[5]), the project leader or manager is in a staff relationship to the line managers. In this position the project manager temporarily employs the re-

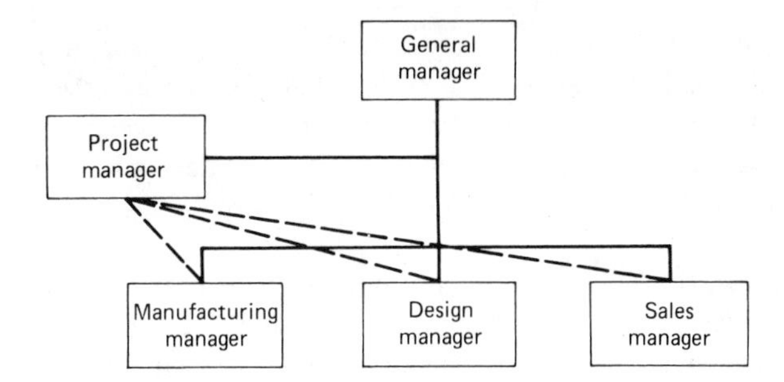


**Figure 6-6** Typical line-staff organization. Solid lines represent formal authority; dashed lines represent informal authority or liaison.

sources of the mainline departments to carry out the project. (See also Youker[7] and Stuckenbruck.[8])

The key to the effectiveness of the line-staff structure lies in the balance of authority between the project and department managers. In some cases the project manager is only a coordinator whose prime responsibilities are to attempt to "influence" the department managers to keep up their schedules on the project and to keep the general manager informed of the status. As explained by Kerzner and Galbraith, project managers in this role must depend on their personal competence and their political or interpersonal skills in order to be effective. (To illustrate this relationship graphically, one might revise Figure 6-6 to place the project manager's box below the level of the functional managers.) More about this coordination role will be mentioned in the next section.

In some cases, more clearly defined authority may be assigned to the project manager. Personnel, equipment, and facilities may be temporarily but fully reassigned from the departments to the project. Martin calls this the "partially projectized management" scheme. In these situations the personnel have a "home" to return to when the subject is done, and their merit reviews and career plans continue to be managed by the functional department heads.

Obviously the line-staff approach has some strong advantages:

1. Resources are effectively shared, thus minimizing losses due to idle time or duplicate effort.
2. Important technical skills continue to be developed and maintained in the line departments.
3. The project manager can concentrate on the product of the project without being distracted by support functions.

Significant disadvantages also exist. Among them are:

1. A high, sometimes unrealistic degree of cooperation is expected of line and staff managers, who have different objectives.

2. The general manager's skills are crucial in setting up the proper degrees of authority and in adjudicating conflicts that arise through the project.
3. Projects conducted under this structure tend to overrun schedules and cost estimates, due to the lack of clear priorities and full authority for execution.

Nevertheless, the line-staff structure is by far the most common arrangement in process businesses and in operations that are mixed between process and project modes. Overall it is cost-effective and flexible with changing conditions. Project teams are easily disbanded when the project is finished, and new ones can be formed fairly readily.

Project teams should be easily disbanded when the project is finished (at least theoretically), and new ones can be formed as needed. (In practice, however, project teams often resist dissolution; their tenacity for life can become a problem. See Benningson.[9])

Suppose one were to more fully develop the line-staff approach to the point that there was a strong project manager position for each of a number of projects, and the project managers were endowed with authority that approximately equaled that of the functional managers. Essentially, this organizational form would be what has come to be known as the matrix. Graphically it looks like Figure 6-7 in its typical form.

In concept, the matrix attempts to achieve the best of both the functional and the line-staff structures. All the technical and specialist personnel have a "home" in their functional departments, where they receive project assignments,

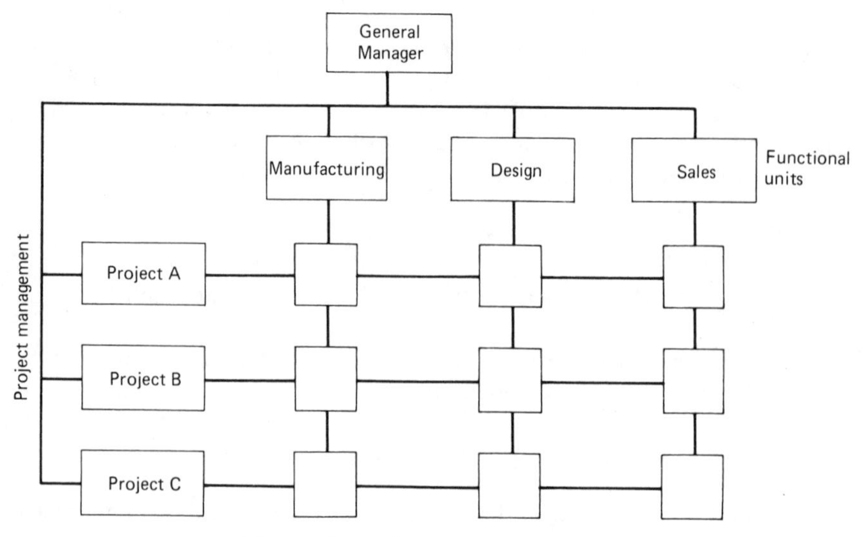


**Figure 6-7** Typical matrix organization.

technical training and guidance, and merit reviews. At any one time, a technician is also assigned to a project and receives specific task direction from the project management.

Thus the technical staff person has two bosses, which would appear to be a confusing weakness in the structure. However, practitioners of matrix management point out that the multiple boss situation is common; it begins from the time one is born with two parents, and continues through school days and even into traditional corporate and government organizations. Even the production line supervisor in a single-product plant is aware of the separate authorities of his department manager, the quality control manager, the personnel director, the union steward, and perhaps others. Thus the two-boss arrangement in a matrix organization is not unique nor necessarily confusing. (See Cleland.[10])

The project manager in a matrix enjoys the advantage of more clearly defined authority and responsibility. He or she should not have to cajole or influence functional heads in order to carry out the project tasks. The matrix does require that the project and functional leaders have a mutually cooperative relationship, but the matrix structure itself can enhance that relationship by its equal recognition of both sides of the organization, thus providing a unifying set of goals for the organization as a whole.

Other advantages of the matrix are:

1. Quick response to changing project conditions.
2. Personnel recognize project work as the mainstream activity and thus develop project loyalty, rather than regarding the project as a distraction.
3. The project manager has full control over the project resources, including personnel and costs.
4. The functional departments perform essentially as support for the project teams.

Naturally, there are also disadvantages to the matrix:

1. With its emphasis on project execution, the functional departments tend to be weaker in certain respects, such as development of the state-of-the-art in technical fields, or the development of new markets.
2. The structure tends to require more attention by general management and administrative personnel in order to maintain the proper balance of power, cost control, and performance control. Vertical and horizontal communications must be open and freely exercised.
3. Multiple, separate projects are vulnerable to duplication of effort or disparate development of techniques and markets.

Clearly, the matrix structure is most appropriate for operations that are predominantly project oriented, such as engineering and construction firms, research and development organizations, and the like. However, the matrix has also found wide application in some process businesses for a variety of purposes. For a more complete discussion see Cleland.[10]

The reader is reminded that this review of the organizational aspects of project management is offered as a brief introduction to the subject. More comprehensive treatment is provided in other texts.

## COORDINATING INTERDEPARTMENTAL ACTIVITIES

As indicated above, a project manager in a line-staff or matrix organization will spend a greater proportion of his time "coordinating" activities than a manager in a traditional or functional organization. For this reason, this task deserves some consideration here. L. A. Benningson[9] has developed a network based analytical tool called TREND (Transformed Relationships Evolved from Network Data) to deal with this problem of coordinating project activities. This technique involves the concepts of "interdependence," "uncertainty," and "prestige."

The sequential *interdependencies* among project activities are conveniently displayed by the project network. If activity A precedes B, then B's work depends on the work of A. In this case, the integrating mechanisms are plans and schedules. Now consider the concept of "prestige" and "authority" associated with the groups conducting activities A and B; then collaboration will be more difficult.[11] While prestige can be inferred from the position of each group in the organization chart of the firm, relative authority may be established in terms of interdependencies, where the party that is dependent is considered to have low relative authority. So, if activity A precedes B, then B would have low relative authority. Conflict would be predicted if group B depended heavily on group A, but group B had much higher prestige in the organization than did group A. In this situation, the project manager could anticipate the particular need for his services in a coordinating role.

Finally, the concept of "uncertainty" is important. Lawrence and Lorsch[12] state that in typically healthy organizations, individuals and groups that must cope with high degrees of uncertainty in their jobs (such as basic research or marketing) are different in important ways from their counterparts who do work involving low degrees of uncertainty (such as manufacturing or accounting). The differences in these two groups can be measured in terms of their time horizons, degree of reliance on formal authority, and degree of value placed on task achievements versus interpersonal relations. A surrogate measure of uncertainty might be the range of activity duration time estimates (pessimistic time minus optimistic time)* used in the PERT system described in Chapter 9. It is con-

*The coefficient of variation (ratio of the standard deviation to the mean) might be useful here. For example, a coefficient of 10% or more might be classified as a high degree of uncertainty.

jectured that greater differences in uncertainty among groups A and B implies greater differences in organizational orientation and greater difficulty in achieving effective collaboration between the groups. Such groups literally speak different organizational languages, and, if they are to relate effectively, require a third party (project manager) who speaks both languages to translate and integrate for them. Coordination of two groups both having high uncertainty can be more direct, possibly handled by committees, while two groups, both having low uncertainty usually present no special coordination problems.

Thus, it is possible for a project manager to use the project network, along with the principles of behavioral science, to pinpoint possible areas which require special attention to coordinate related project activities.

## REFERENCES


1. Long, J. M., "Applying PERT/CPM to Complex Medical Procedures," *Proceedings of Seminar on Scientific Program Management*, Department of Industrial Engineering, Texas A & M University, June 1967.
2. Davis, Gordon, President, DDR International, Atlanta, Georgia, personal interview, November 1981.
3. Jones, Patricia Cukor, Senior Project Manager, Hanscomb Associates, Inc., Atlanta, Georgia, personal interview, December 1981.
4. Galbraith, Jay, *Designing Complex Organizations*, Addison-Wesley Publishing Co., Reading, Mass., 1973.
5. Kerzner, Harold, *Project Management, A Systems Approach to Planning, Scheduling, and Controlling*, Van Nostrand Reinhold Co., New York, 1979.
6. Martin, Charles C., *Project Management: How to Make it Work*, AMACOM, New York, 1976.
7. Youker, Robert, "Organization Alternatives for Project Managers," *Management Review*, November 1977.
8. Stuckenbruck, L. C., *The Implementation of Project Management*, Addison-Wesley, Reading, Mass., 1981.
9. Benningson, Lawrence, "The Strategy of Running Temporary Projects," *Innovation*, September 1971, pp. 31–41.
10. Cleland, David I., "The Cultural Ambience of the Matrix Organization," *Management Review*, November 1981.
11. Seiler, J. A., "Diagnosing Interdepartmental Conflict," *Harvard Business Review*, September–October 1963.
12. Lawrence, P. R., and J. W. Lorsch, "New Management Job: The Integrator," *Harvard Business Review*, November–December 1967, pp. 142–151.

# II

# ADVANCED TOPICS

# 7
# RESOURCE CONSTRAINTS IN PROJECT SCHEDULING

Resource allocation is probably receiving more attention today than any other aspect of PERT/CPM. One reason for this is the worldwide inflation spiral since the oil crisis of 1973-74, which has pushed the cost of resources required for project execution higher and higher (thus making them relatively more important). Another reason is that soaring interest rates have significantly increased the cost of borrowing money, which directly affects many large plant construction and capital investment projects. Slippage of such loan-financed projects can have a devastating impact on project budgets and cost expectations.

Another reason lies in the continued advances in computer technology over the past two decades which have made more cost-feasible the application of sophisticated solution procedures that are often necessary when resource limitations enter the project scheduling picture.

The basic PERT/CPM procedures which produce a detailed project schedule are limited in the sense that resource availabilities are not considered in the scheduling process. Those basic procedures implicitly assume that available resources are unlimited and that only technological (i.e., precedence) requirements constrain job start/finish times. One consequence of this is that the schedules produced may not be realistic when resource constraints *are* considered. For this reason the basic time-only PERT/CPM forward-backward pass procedure has been called by some seasoned users, "a feasible procedure for producing nonfeasible schedules."

It is important to note in discussing the subject of resource constraints here that time/cost trade-off procedures are not included. Those procedures typically assume unlimited resources, are basically different from the procedures discussed in this chapter, and are covered in the Appendix to Chapter 8. Of course, alternative project durations and costs could be covered in the context of simulation of alternative critical resource levels and their associated costs. Such an approach is possible with the procedures discussed here, but it requires extensive computational resources.

Resource constraints, while increasingly important, also unfortunately complicate and alter some of the basic notions of PERT/CPM. For example, the longest sequence of activities through the project when resource availabilities are constrained may not be the same critical path determined by the basic time-only PERT/CPM approach. Another difference is that with the basic time-only procedures there is one unique Early Start time (*ES*) schedule, while under resource constraints many different *ES* schedules may exist. To understand these differences it is necessary to look at how limited resources affect schedule slack (float).

## HOW LIMITED RESOURCES AFFECT SCHEDULE SLACK

Figure 7-1 shows a simple 11-activity network (activity-on-node format) with activity times indicated beside each node. Figure 7-2(a) shows the all-*ES* bar chart schedule for this network. As can be seen, the project duration is 18 weeks, the critical path is the activity (job) sequence A-C-I-J-K, and jobs B, D, E, F, G, and H all have positive slack.

Now assume that jobs C and G each require the use of a special piece of equipment, such as a hoist crane. But only one crane is available. And assume that jobs E and F each require a special bulldozer, but only one is available.

The direct result of these resource constraints is that neither jobs C and

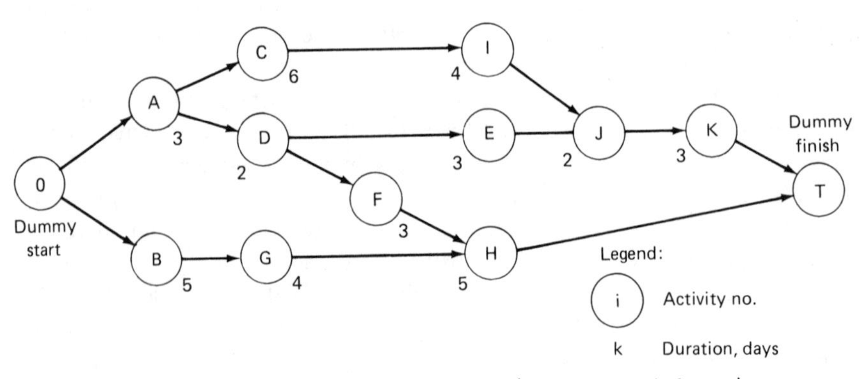


**Figure 7-1** Illustrative 11-activity network (activity on node format).

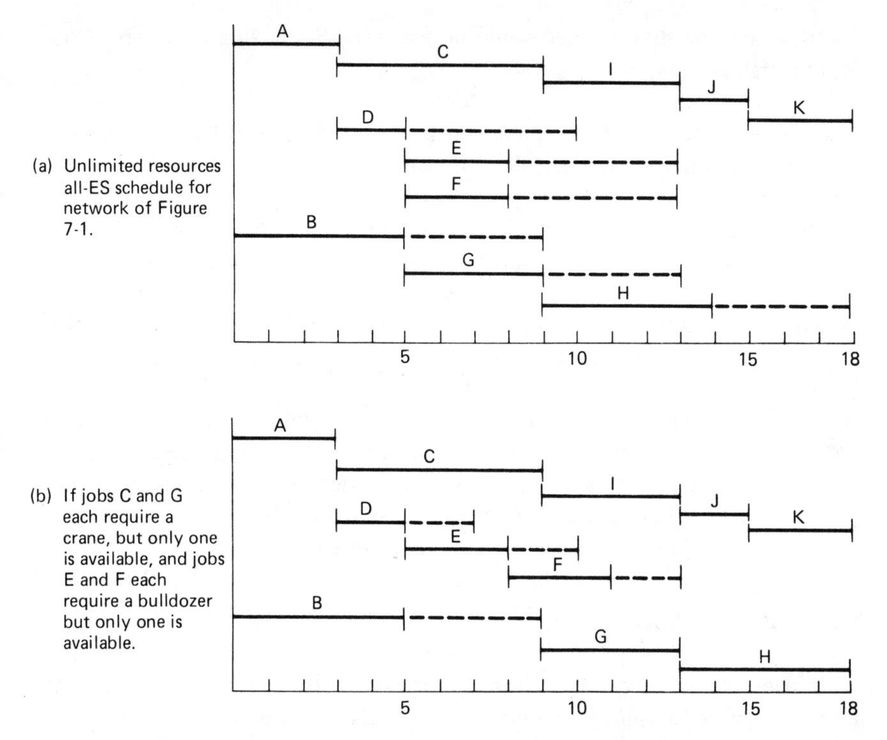


**Figure 7-2** Bar chart schedules for network of Figure 7-1.

G nor jobs E and F can be performed simultaneously as indicated by the *ES* time-only schedule. One or the other of the jobs in each pair must be given priority and each pair must be sequenced so there is no overlap, as shown in Figure 7-2(b).

Examination of Figure 7-2(b) shows that, when resources for activities C/G and E/F, respectively, are constrained, the following is apparent:

- Activities G and H become critical, with slack reduced to zero.
- Activities D, E, and F have their slack reduced significantly.
- With activity E given priority over F as shown (an arbitrary choice), the slack of jobs D and E become dependent upon job F.
- No job can start earlier than shown, given the precedence relations and resource constraints, so this represents an early start schedule. However, this schedule is not unique (as is the case with unlimited resources), since job F could have been sequenced before job E in resolving the resource conflict. In that case, the resulting schedule, though only slightly different

from the one shown here, would be another ES schedule for the resource-constrained case.

As this example illustrates, slack can be affected in significant ways when resources are limited. In general, the following is true:

- Resource constraints reduce the total amount of schedule slack.
- Slack depends both upon activity precedence relationships *and* resource limitations.
- The early and late start schedules are typically not unique. This means that slack values are not unique. These values depend upon the scheduling rules used for resolving resource conflicts.
- The critical path in a resource-constrained schedule may not be the same continuous chain(s) of activities as occurring in the unlimited resources schedule. A continuous chain of zero-slack activities *may* exist, but since activity start times are constrained by resource availabilities as well as precedence relations this chain may contain different activities.

## MULTIPROJECT SCHEDULING

The impact of resource constraints illustrated by the single-project example above is magnified in scheduling multiple projects, i.e., situations where several separate, nominally independent projects are linked together through their dependence upon a pool of common resources. When the pool is tightly constrained relative to total concurrent requirements, activity sequencing choices similar to that illustrated above must be made. However in this situation the ramifications of the sequencing decision may extend beyond the boundaries of the project containing the particular activities involved.

Figure 7-3 illustrates a hypothetical 3-project scheduling situation involving only 3 resource types. To further simplify the example imagine that activities requiring a resource use only one unit of any one of the three types, and only one unit of any type is available (e.g., as in a machine-limited workshop). All activities are shown on their respective resource bar charts at their early start times.

It is not difficult to visualize the dominolike series of events that might occur (depending upon activity float, and project finish times) as a result of delaying activities to resolve particular resource conflicts. For example, delaying activity 1-3 of project 1 (to resolve the conflict with activity 1-3 of project 2) might cause the following:

a. delays in successor activities 3-4, 3-5, and 4-5 of project 1;
b. as a result of (a), the creation of additional resource conflicts among activities requiring resource types 2 and 3 (which must be resolved);

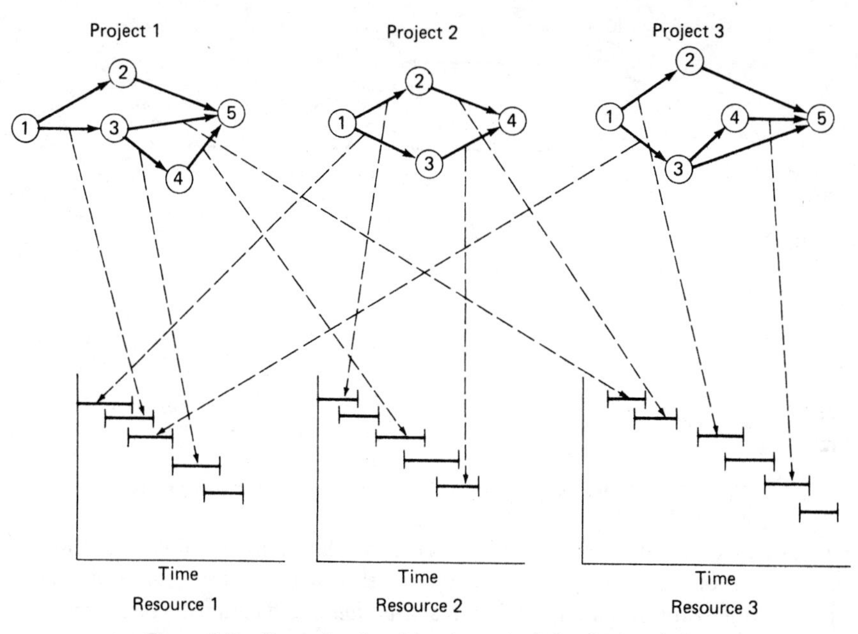


**Figure 7-3** Example of multiproject scheduling interactions.

c. as a result of (b), additional delays in projects 2 and 3, and possibly even project 1 again.

In practice, the schedules for such multiproject problems, involving tens of projects, dozens of resource types, and thousands of activities, are capable of development only with the aid of large-scale computers. Some examples of such computer systems are given in Chapter 11; the point here is that it is the aspect of resource constraints which elevates the multiproject problem from a relatively simple exercise, when the usual time-only CPM procedures are involved, to a problem of truly formidable proportions that can require sophisticated computational routines and powerful computers for solution.

## RESOURCE LOADING

One of the main advantages of the network model for project planning is the ease with which information about resource requirements over the duration of the project can be generated. The only condition for obtaining this information is that the resource requirements associated with each project activity shown on the network be identified separately.

For example, Figure 7-4 shows the same network as Figure 7-1 with manpower requirements of two different types indicated above each activity. By utilizing these resource requirements in conjunction with both an early-start

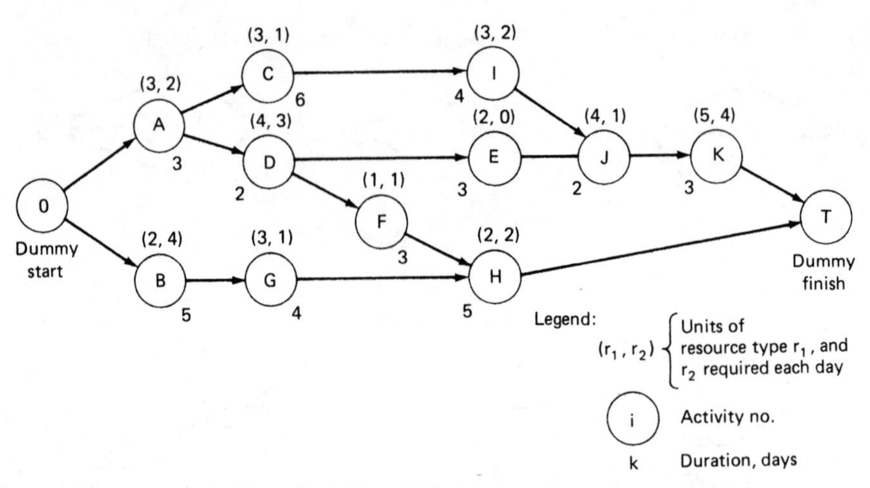


**Figure 7-4** Network of Figure 7-1 with resource requirements added.

schedule (such as shown in Figure 7-2(a)) and a late-start schedule (not shown) the profiles of resource usage over time as shown in Figure 7-5 are obtained. These profiles are commonly called *resource loading diagrams*. Such diagrams are extremely important in project management; they highlight the period-by-

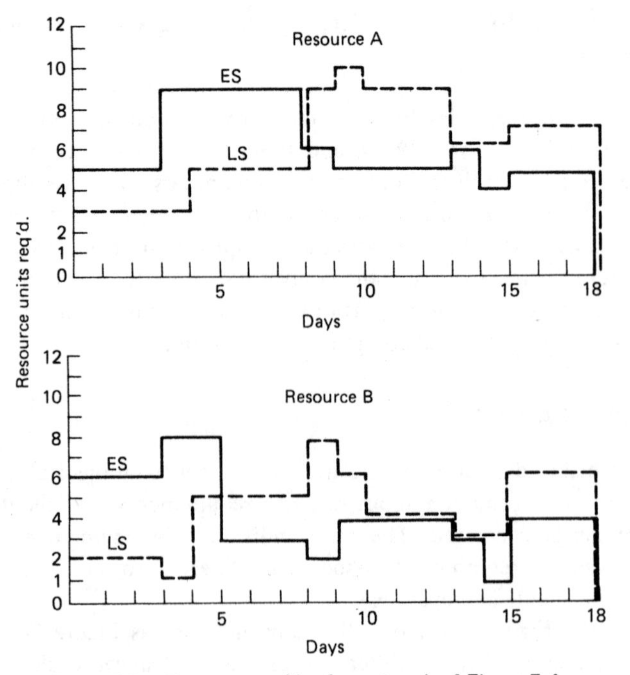


**Figure 7-5** Resource profiles for network of Figure 7-4.

period resource implications of a particular project schedule and provide the basis for improved scheduling decisions.

Resource Loading diagrams are probably more widely used than any other resource analysis technique in project management. The technique is simple, provides a useful visual picture, and does not always require a computer (except for large-sized networks). In many cases the resource loading data are more important as rough planning guides rather than as precise goals to be obtained, because of schedule uncertainty. One major pharmaceutical firm, for example, obtains resource loading information from all CPM schedules for research and development projects, to provide resource feasibility checks on the amounts of high-skilled medical and technical research personnel required. Such checks are particularly useful in the case of multiple on-going projects, which each require varying amounts of resources that must be shared by all projects.

### Cumulative Resource Requirements

Figure 7-6(a) shows the period-by-period total requirements of units of resource A of network Figure 7-4, for both the early-start (*ES*) and late-start (*LS*) schedules. The period totals were used in constructing the resource loading diagram shown in Figure 7-5. Also shown in Figure 7-6(a) are the cumulative requirements for each period. These latter data can be usefully displayed in the form of *cumulative resource requirement curves* as shown in Figure 7-6(b).

The cumulative requirements curves can be used to develop project scheduling information both before and during project implementation. For example, as time progresses after the project has started, the cumulative resources required should lie within the closed area bounded by the *ES* and *LS* cumulative curves. If actual cumulative resources fall under the *LS* curve the project is either behind schedule or the resource requirements were overestimated. Conversely, if they exceed the *ES* curve the project is either ahead of schedule or the resource requirements were underestimated.

### Use in Resource Planning

Another important use of the cumulative curves is in preliminary resource allocation planning. The magnitude of total cumulative requirements and the slope of the average requirements line drawn to this end point can be used to develop rough indicators of resource constraint "criticality" and of the likelihood of project delay beyond the initial forward-pass determined critical path duration. For example, line I in Figure 7-6(b) indicates the average daily requirements for resource A as:

$$\text{avg. daily requirement} = \frac{111 \text{ total units}}{18 \text{ days}} = 6.2 \text{ units/day}$$

| Period | ES Schedule Total Units | ES Schedule Cum. Units | LS Schedule Total Units | LS Schedule Cum. Units |
|---|---|---|---|---|
| 1 | 5 | 5 | 3 | 3 |
| 2 | 5 | 10 | 3 | 6 |
| 3 | 5 | 15 | 3 | 9 |
| 4 | 9 | 24 | 3 | 12 |
| 5 | 9 | 33 | 5 | 17 |
| 6 | 9 | 42 | 5 | 22 |
| 7 | 9 | 51 | 5 | 27 |
| 8 | 9 | 60 | 5 | 32 |
| 9 | 6 | 66 | 9 | 41 |
| 10 | 5 | 76 | 10 | 51 |
| 11 | 5 | 76 | 9 | 60 |
| 12 | 5 | 81 | 9 | 69 |
| 13 | 5 | 86 | 9 | 78 |
| 14 | 6 | 92 | 6 | 84 |
| 15 | 4 | 96 | 6 | 90 |
| 16 | 5 | 101 | 7 | 97 |
| 17 | 5 | 106 | 7 | 104 |
| 18 | 5 | 111 | 7 | 111 |

**Figure 7-6(a)** Early and late start schedule requirements of resource A, network of Figure 7-4.

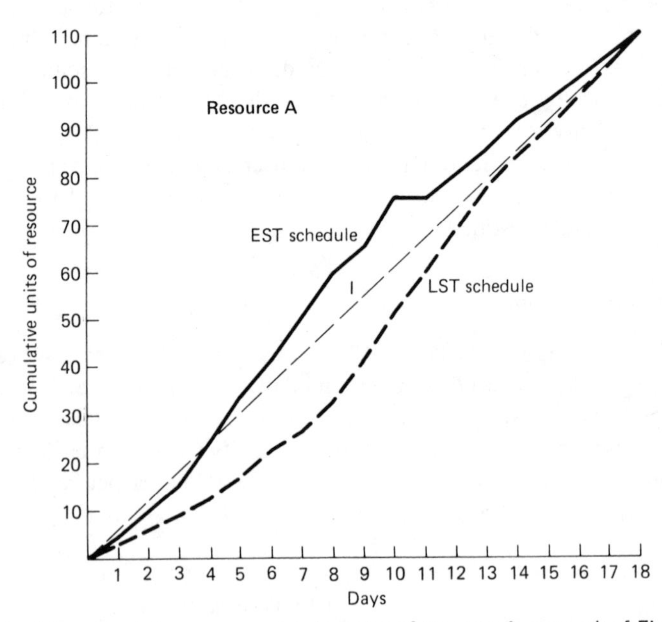


**Figure 7-6(b)** Cumulative resource requirements of resource A, network of Figure 7-4.

Suppose resource A is available at a maximum level of 7 units/day. In this case a total of 126 units *could be* expended over the 18-day critical path duration, which is considerably more than the 111 units required over that period. Thus, there is unlikely to be a project delay beyond the 18-day duration because of constraints on Resource A. This conclusion can also be drawn from the ratio of resources required to available, which is a rough measure of resource "tightness," or criticality. That is,

$$\text{criticality index} = \frac{\text{avg. daily units req'd.}}{\text{max. am't. avail. daily}} = \frac{6.2}{7.0} = 0.88$$

The situation above can be contrasted to the case where only 6 units of resource A are available daily. In this case

$$\text{criticality index} = \frac{6.2}{6.0} = 1.03,$$

In 18 days a total of only 108 units of resource will have been expended (or "in place"), leaving some work unfinished and thus requiring an extension of the project beyond 18 days.

In general, higher values of the resource criticality index calculated here are associated with the most critical (i.e., most tightly constrained) resources. Very little research has been conducted in this area, but what has been done[7] suggests that, while exceptions exist, values significantly below 1.0 typically are associated with nonconstraining resources, while values around and above 1.0 indicate that project delays beyond the original critical path duration will be encountered.

Figure 7-7 shows another type of situation which can exist when using cumulative resource requirement curves. This shows *ES* and *LS* cumulative requirements curves for one type resource (cubic yards of concrete poured) for a hypothetical project. From the position of the *ES* and *LS* curves relative to the project start date, it can be seen that concrete is not required during the very beginning or end periods of project duration. It is assumed here that total project duration has been estimated from the forward pass computations to be 200 working days, with concrete required only between days 65 and 185 (*ES* schedule). A total of 1,000,000 cu. yds. is assumed to be needed and a maximum of 6,000 cu. yds./day is available.

Line I shown in Figure 7-7, constructed as in the previous example from project start to the point of total cumulative resource requirements at project completion, gives the average daily requirement for concrete as:

$$\text{avg. daily requirement over project duration} = \frac{1{,}000{,}000 \text{ cu. yds.}}{200 \text{ days}}$$

$$= 5000 \text{ cu. yds./day}$$

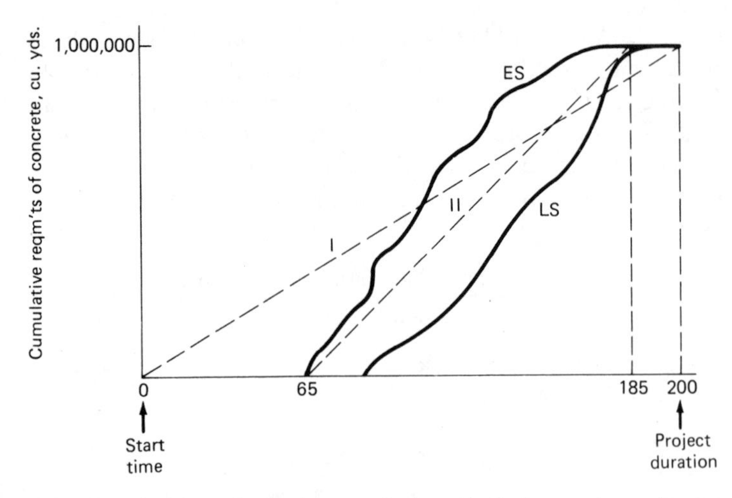


**Figure 7-7** Cumulative requirements curves for hypothetical concrete pouring example.

The use of this average for estimating resource criticality, as in the previous example, will lead to a conclusion that concrete is not a constraining resource. But this is misleading; this average is clearly a lower level of usage than is really required, because the usage is projected over the entire project duration instead of the shorter period of actual need. A better measure of the average requirement is given by Line II, which is drawn from the start of the *ES* curve to the total cumulative requirements level at the end of the *ES* curve. The average requirement given by the slope of this line is calculated as:

$$\text{avg. daily requirement over actual use period} = \frac{1{,}000{,}000 \text{ cu. yds.}}{(185 - 65) \text{ days}}$$

$$= 8333 \text{ cu. yds./day.}$$

If this average is used as before for estimating resource criticality a different conclusion is drawn: The criticality index is greater than 1.0, only 720,000 cu. yds. can be poured over the indicated period of 120 days, and project delays beyond day number 200 will occur. The project can be successfully completed within 200 days only if concrete pouring can start sufficiently earlier than day 65 (e.g., by changing network logic, etc.) to accomplish installation of the 1,000,000 cu. yds. by day 200, and other resources do not become more critical than concrete.

Other situations similar to Figure 7-7 include cases of intermittent or irregular requirement of the resource. For example, Figure 7-8 illustrates a case where a relatively small amount of some resource is required shortly after project start, followed by a period of time of zero requirements, with significant requirements in succeeding periods until shortly before project end (*ES* schedule).

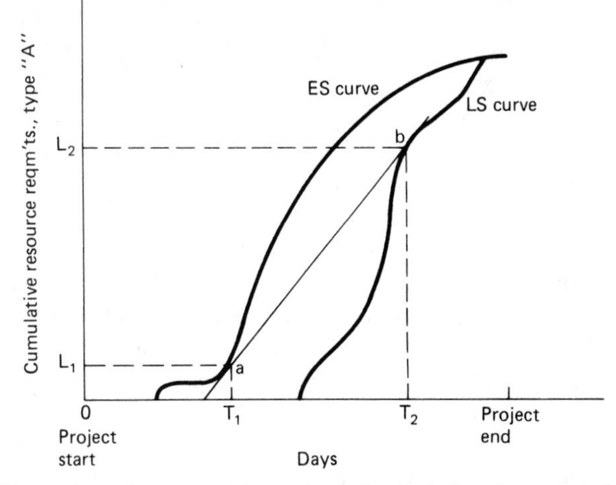


**Figure 7-8** Illustration of resource lower bound calculation from cumulative resource requirement curves.

In cases such as this an estimate of the minimum level requirements of the resource is obtained by construction of the minimum-slope line which can be drawn within the *ES*, *LS* "envelope" of cumulative requirements. Drawn tangent to the envelope in Figure 7-8 at points (a) and (b), it gives the following lower-bound estimate:

$$\text{lower bound estimate of avg. daily requirement} = \frac{L_2 - L_1}{T_2 - T_1} \text{ resource units/day}$$

The estimate of the average calculated in this fashion can then be used in calculating the index of resource criticality, as shown earlier.

The use of the cumulative resource requirements curves as described generally in this section provides an approach for estimating the impact of resource constraints, either in advance of or in the absence of more sophisticated procedures. The procedure could also be used, if desired, as a basis for setting activity priorities in project scheduling. To do this, all resources required by the project could first be ranked by criticality; this listing could then be used to rank the activities, putting those that require the most critical resource(s) at the top of the list. Then the activities could be scheduled in the order listed.

All good computer programs for CPM analysis today will produce resource loading information (*ES*, *LS* daily profiles and cumulative requirements) in the form of tables or figures even though they may lack the more sophisticated capability of automatic schedule adjustment to meet desired resource loading requirements. The latter capability falls into the categories of resource leveling or scheduling under fixed resource constraints, which are discussed in the next section.

## OVERVIEW OF BASIC SCHEDULING PROCEDURES

Scheduling procedures for dealing with resource constraints can be roughly divided into 2 major groups, according to the problem addressed: (1) resource leveling, and (2) fixed-limits resource scheduling.

1. *Resource Leveling.* This type of problem occurs when sufficient total resources are available, and the project must be completed by a specified due date, but it is desirable or necessary to reduce the amount of variability in the pattern of resource usage over the project duration. This type of situation occurs frequently, for example, in the construction industry, where the cost of hiring and laying off of personnel or physical resources can be substantial. Thus, the objective is to level, as much as possible, the demand for each specific resource during the life of the project. Project duration is not allowed to increase in this case.

2. *Fixed Resource Limits Scheduling.* This category of problem, which is much more common, arises when there are definite limitations on the amount of resources available to carry out the project (or projects) under consideration. The scheduling objective in this case is to meet project due dates *insofar as possible*, subject to the fixed limits on resource availability. Thus, project duration may increase beyond the initial duration determined by the usual "time-only" CPM calculations. The scheduling objective is equivalent to minimizing the duration of the project (or projects) being scheduled, subject to stated constraints on available resources. This category of problem can be further subdivided according to whether the fixed limits on resource availabilities are *constant* at some level or *allowed to vary* over activity or project duration. A further useful subdivision is possible according to whether approximate, rule-of-thumb procedures, or mathematically exact procedures are employed. Davis[6] presents a general classification scheme employing these and other ideas.

The basic general approach followed in both resource leveling and fixed resource limits scheduling is similar: Set activity priorities according to some criterion and then schedule activities in the order determined, as soon as their predecessors are completed and adequate resources are available. Because of this common general approach, it is useful to understand somewhat better the nature of the criteria employed before proceeding with a discussion of the different problems.

## USE OF HEURISTICS IN SCHEDULING

The task of scheduling a set of project activities such that both precedence relationships and constraints on resources are satisfied is not an easy one, even for projects of only modest size. The difficulty is increased if simultaneously some objective such as minimum project duration or minimum total cost is sought.

The limited resource project scheduling problem falls into a category of mathematical problems known as *combinatorial* problems. This is because, for any given problem, a very large number of possible combinations of activity start times exist, with each combination representing a different project schedule. The number of combinations is extremely large, even for fairly small problems of 20 or 30 activities and increases rapidly with an increase in the number of activities. In fact, the number is typically so large for realistic-sized problems as to prohibit enumeration of all alternatives, even with the aid of a computer. Analytical methods such as mathematical programming have not proven very successful on these combinatorial problems. Instead, various heuristic-based procedures have been developed.

Simply stated, a heuristic is a rule of thumb—a simple, easy to use guide or aid used in problem-solving situations, to reduce the amount of effort required in coming up with a solution. Heuristics may not always produce the *best* solution in every case, but their usefulness in finding *good* solutions with a minimum of effort is well-known, based on experience and research studies.

Relatively simple heuristics such as "shortest job first," or "minimum slack first" are effective aids in establishing activity priorities on many types of limited-resource scheduling problems. An example application of such simple heuristic rules will be illustrated here on both categories of problem listed above. More complicated heuristic procedures—typically consisting of combinations of heuristics and modifying rules—will be discussed later.

## PARALLEL vs. SERIAL METHODS

There are two general methods of applying heuristics in project resource allocation problems. A *serial scheduling* procedure is one in which all activities of the project are ranked in order of priority as a single group, using some heuristic, and then scheduled one at a time (i.e., serially). Activities that cannot be started at their early start time are progressively delayed until sufficient resources are available.

In *parallel scheduling*, all activities starting in a given time period are ranked as a group in order of priority and resources allocated according to this priority as long as available. When an activity cannot be scheduled in a given time period for lack of resources, it is delayed until the next time period. At each successive time period a new rank-ordering of all eligible activities is made and the process continued until all activities have been scheduled.

Even though the parallel method requires more computer time to reorder the eligible activities at each time period, it appears to be the more widely used of the two methods, being employed in a number of commercially-available computer programs for project scheduling. The only published research comparing the effectiveness of the two approaches in limited resource scheduling indicated that the serial approach could produce shorter-duration schedules for some categories of networks, but that there were also disadvantages with this procedure in terms of the special scheduling conditions (such as activity splitting) which could be handled.[10]

## RESOURCE LEVELING

As noted earlier, in many project scheduling situations it is the *pattern* of resource usage, such as frequent changes in the amount of a particular manpower category required, or peaks of resource demand at undesirable times, that must be improved. The concentration of highest levels of demand for resource type R1 in the earlier schedule periods, as seen in Figure 7-5, might be such an example. Resource-leveling techniques are useful in situations such as this; they provide a means of distributing resource usage over time to minimize the period-by-period variations in manpower, equipment, or money expended.

The essential idea of resource leveling centers about the rescheduling of activities within the limits of available slack (float) to achieve better distribution of resource usage. The slack available in each activity is determined from the basic scheduling computations, without consideration of resource requirements or availabilities. Then, during the rescheduling, or "juggling" of activities to smooth resources, project duration is not allowed to increase.

While the essential ideas of resource-leveling are easy to grasp, implementation in practice can pose difficulties. For example, what activity schedule will minimize resource costs for the network of Figure 7-4 given that project duration is not to exceed 18 days? Questions such as this are not always easy to answer just by looking at resource profiles such as shown in Figure 7-5. Instead, some systematic procedure for arriving at the best (i.e., most level resource utilization) schedule is needed which provides an easily calculated measure of effectiveness.

The first edition of this text described a systematic procedure developed by Burgess[2] for leveling resources. This method utilized a simple measure of effectiveness given by the sum of the squares of the resource requirements for each "day" (period) in the project schedule. It is easy to show that, while the sum of the daily resource requirements over the project duration is constant for all complete schedules, the sum of the squares of the daily requirements decreases as the peaks and valleys are leveled. Further, this sum reaches a minimum for a schedule that is level (or as level as can be obtained) for the project in question.

The Burgess procedure which utilizes the above measure of effectiveness is simple and can be utilized under a variety of problem assumptions (e.g., activity splitting). The basic method, which is representative of a general class of resouce-leveling procedures, is explained in the following section.

**Burgess Leveling Procedure**

*Step 1.* List the project activities in order of precedence by arranging the arrow head or node numbers in ascending order. When using activity-on-arrow format and two or more activities have the same head number, list them so that the arrow tail numbers are also in ascending order. (This assumes that the network events are numbered so that activity tail numbers are always less than the head numbers.) The procedure given in Appendix 4-1 includes a routine to accomplish this. Add to this listing the duration, early start, and slack values for each activity.

*Step 2.* Starting with the last activity (the one at the bottom of the diagram), schedule it period by period to give the lowest sum of squares of resource requirements for each time unit. If more than one schedule gives the same total sum of squares, then schedule the activity as late as possible to get as much slack as possible in all preceding activities.

*Step 3.* Holding the last activity fixed, repeat Step 2 on the next to the last activity in the network, taking advantage of any slack that may have been made available to it by the rescheduling in Step 2.

*Step 4.* Continue Step 3 until the first activity in the list has been considered; this completes the first rescheduling cycle.

*Step 5.* Carry out additional rescheduling cycles by repeating Steps 2 through 4 until no further reduction in the total sum of squares of resource requirements is possible, noting that only movement of an activity to the right (schedule later) is permissible under this scheme.

*Step 6.* If this resource is particularly critical, repeat Steps 1 through 5 on a different ordering of the activities, which, of course, must still list the activities in order of precedence.

*Step 7.* Choose the best schedule of those obtained in Steps 5 and 6.

*Step 8.* Make final adjustments to the schedule chosen in Step 7, taking into account factors not considered in the basic scheduling procedure.

If Steps 2–4 of the Burgess procedure are applied to the sample network of Figure 7-4, the sequence of activity shifts, total units of each resource type required and calculated sum of squares at each major step are as shown in Figure 7-9, along with the final activity positions. Figures 7-10(a) and (b) show, respectively, the sum of squares and resource profiles before and after application of the Burgess procedure. As can be seen, application of the approach produces both a lower sum of the squares and generally smoother profiles for both resources, although the peak level of resource A has actually increased slightly compared to its initial level. Resource A can be held to a peak period requirement of 8 units through Step 8 of the procedure ("final adjustments to the schedule"; here by changing slightly the positions of activities E and G). How-

| Activity | Time 1 | 2 | 3 | 4 | 5 | 6 | 7 | 8 | 9 | 10 | 11 | 12 | 13 | 14 | 15 | 16 | 17 | 18 |
|---|---|---|---|---|---|---|---|---|---|---|---|---|---|---|---|---|---|---|
| A* | 3A<br>2B | 3A<br>2B | 3A<br>2B | | | | | | | | | | | | | | | |
| B | 2A<br>2B | 2A<br>4B | 2A<br>4B | 2A<br>4B | 2A<br>4B | | | | | | | | | | | | | |
| C* | | | | 3A<br>1B | 3A<br>1B | 3A<br>1B | 3A<br>1B | 3A<br>1B | 3A<br>1B | | | | | | | | | |
| D | | | | | | 4A<br>3B | 4A<br>3B | | | | | | | | | | | |
| I* | | | | | | | | | | 3A<br>2B | 3A<br>2B | 3A<br>2B | 3A<br>2B | | | | | |
| E | | | | | | | | | | | 2A<br>0B | 2A<br>0B | 2A<br>0B | | | | | |
| G | | | | | | | 3A<br>1B | 3A<br>1B | 3A<br>1B | 3A<br>1B | | | | | | | | |
| F | | | | | | | | 1A<br>1B | 1A<br>1B | 1A<br>1B | | | | | | | | |
| J* | | | | | | | | | | | | | | 4A<br>1B | 4A<br>1B | | | |
| H | | | | | | | | | | 2A<br>2B | 2A<br>2B | 2A<br>2B | 2A<br>2B | 2A<br>2B | | | | |
| K* | | | | | | | | | | | | | | | | 5A<br>4B | 5A<br>4B | 5A<br>4B |
| Level of Resource A Assigned | 5 | 5 | 5 | 9<br>5 | 9<br>5 | 9<br>8<br>5<br>3<br>7 | 9<br>8<br>6<br>10 | 9<br>7 | 6<br>7<br>9<br>7 | 5<br>3<br>4<br>9<br>9<br>7 | 5<br>7 | 5<br>7 | 5<br>7 | 6 | 9<br>6 | 5 | 5 | 5 |
| Level of Resource B Assigned | 6 | 6 | 6 | 8<br>5 | 8<br>5 | 3<br>7<br>1<br>4 | 3<br>7<br>5 | 3 | 2<br>3 | 4<br>2<br>3<br>4 | 4 | 4 | 4 | 3 | 1<br>3 | 4 | 4 | 4 |

*Critical activity

**Sequence of Major Moves**

(*Each activity shifted in 1-period increments for total shown*)

(1) Delay H one period
$\Sigma R^2 = 1132$

(2) Delay F two periods
$\Sigma R^2 = 1118$

(3) Delay G one period
$\Sigma R^2 = 1116$

(4) Delay E five periods
$\Sigma R^2 = 1112$

(5) Delay D two periods
$\Sigma R^2 = 1062$

| | Sequence (1) | (2) | (3) | (4) | (5) |
|---|---|---|---|---|---|
| $\Sigma A^2$ = | 747 | 733 | 727 | 723 | 715 |
| $\Sigma B^2$ = | 385 | 385 | 389 | 389 | 397 |
| Total | 1132 | 1118 | 1116 | 1112 | 1062 |

| Resource | *All Activities at ES Time* | | *All Activities at LS Time* | | *After Leveling via Burgess* | |
|---|---|---|---|---|---|---|
| | *Max. Res. Level* | *Sum of Squares* | *Max. Res. Level* | *Sum of Squares* | *Max. Res. Level* | *Sum of Squares* |
| A | 9 | 743 | 10 | 779 | 10 | 715 |
| B | 8 | 389 | 8 | 387 | 6 | 347 |
| | | 1132 | | 1166 | | 1062 |

**Figure 7-10(a)** Comparison of Burgess leveling criterion for network of Figure 7-4.

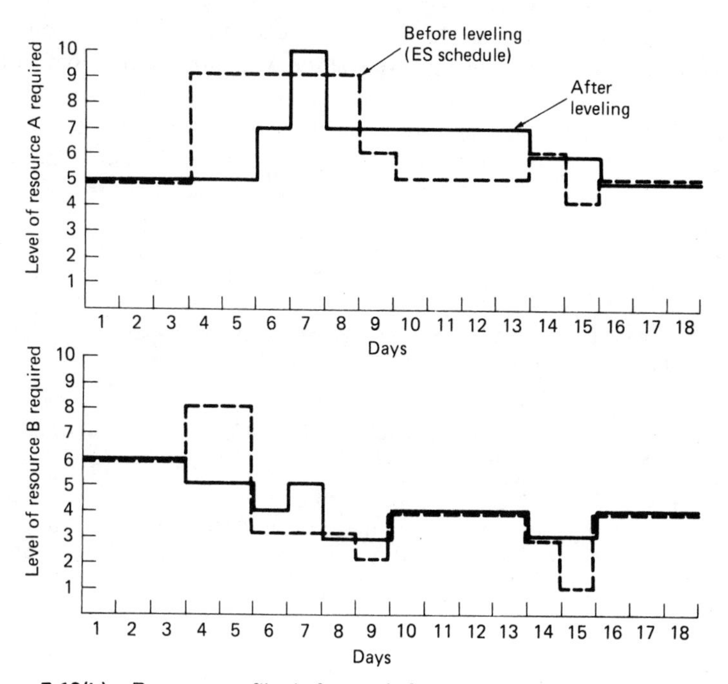


**Figure 7-10(b)** Resource profiles before and after application of Burgress procedure.

ever, this can be done only at the expense of increased variability in the level of resource B, with a corresponding increase in the total sum of squares.

Burgess' ideas are utilized in a recent paper by Woodworth and Willie,[23] who present a multiproject approach in which the sum of squares smoothing criterion is applied to each resource sequentially, across all projects. Although the authors do not make it explicit, the procedure would presumably be applied to resources in some descending order of criticality (i.e., most critical resource first, etc.).

## FIXED RESOURCE LIMITS SCHEDULING

The process of resource leveling as illustrated above will smooth the profiles of resource demand to the extent allowed by available activity float. But this process does not always produce satisfactory schedules if the amounts of available resources are tightly constrained. The final profiles of resource usage in Figure 7-10(b), for example, show levels that might still exceed the maximum amount of resources available. What would happen, for instance, if the amounts of Resource A and B were limited to, say, 6 and 7 units per day, respectively?

The answer to this question requires the use of the second category of procedures mentioned previously, called fixed resource limits scheduling. Also often called *constrained-resource scheduling*, or *limited resource allocation*, these are techniques designed to produce schedules that will not require more resources than are available in any given period, with project durations which are increased beyond the original critical path length as little as possible.

An example flow chart of this type of scheduling procedure is shown in Figure 7-11. The procedure will be applied to the same sample network of

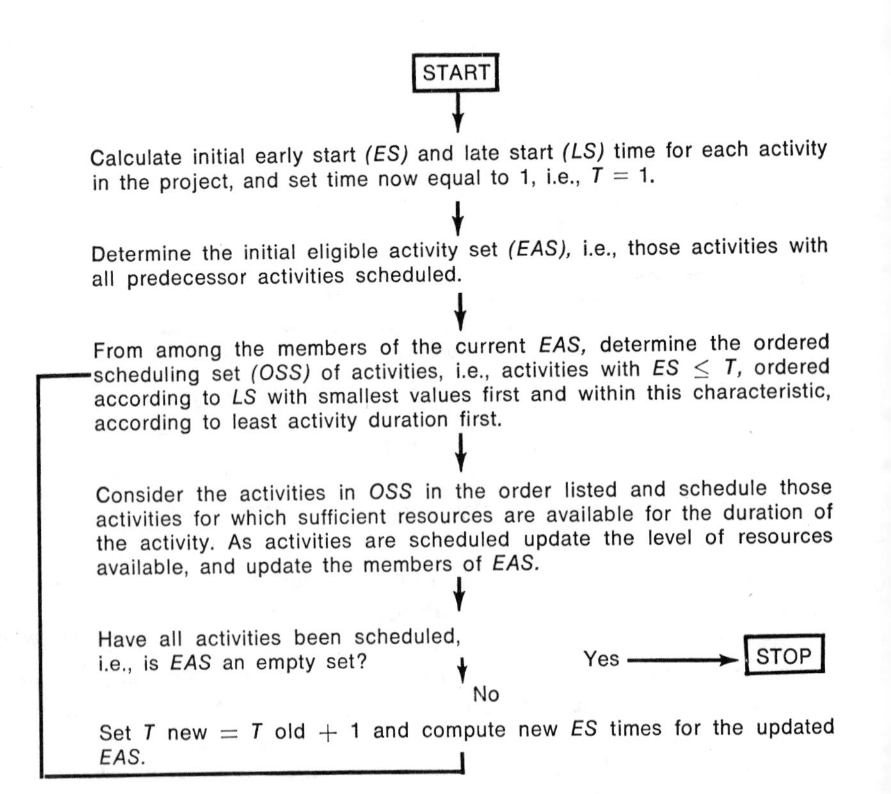


**Figure 7-11** Basic constrained-resource scheduling procedure.

Figure 7-4 that we have been using, but with the amounts of Resource types A and B available limited to 6 and 7 units per period, respectively.

The procedure involves a parallel approach (stepping through time, period-by-period). First a set of activities is defined whose predecessors are all scheduled; this is called the Eligible Activity Set (*EAS*). Then, only the activities in the *EAS* with *ES* less than or equal to some time value T are considered. These activities are ordered with least slack first and, within this criterion, with shortest duration first. This ordered list of activities is referred to as the Ordered Scheduling Set (*OSS*).

The bookkeeping for this scheduling procedure can be simplified by ordering the activities in the *OSS* according to their late start time (*LS*) instead of by slack. The minimum-*LS* ordering is identical to the ordering obtained by using minimum slack, as shown in Reference 4. However, an advantage is that the *LS* values do not change from time period to time period, whereas slack values continuously decrease for an activity that is ready to be scheduled in a given period but is not scheduled.

The complete schedule shown in Figure 7-12 can be obtained by continuing

| ACTIVITY | Time 1 | 2 | 3 | 4 | 5 | 6 | 7 | 8 | 9 | 10 | 11 | 12 | 13 | 14 | 15 | 16 | 17 | 18 | 19 | 20 | 21 | 22 |
|---|---|---|---|---|---|---|---|---|---|---|---|---|---|---|---|---|---|---|---|---|---|---|
| A | 3A<br>2B | 3A<br>2B | 3A<br>2B | | | | | | | | | | | | | | | | | | | |
| B | 2A<br>4B | 2A<br>4B | 2A<br>4B | 2A<br>4B | 2A<br>4B | | | | | | | | | | | | | | | | | |
| C | | | | 3A<br>1B | 3A<br>1B | 3A<br>1B | 3A<br>1B | 3A<br>1B | 3A<br>1B | | | | | | | | | | | | | |
| D | | | | | | | | | | 4A<br>3B | 4A<br>3B | | | | | | | | | | | |
| I | | | | | | | | | | | | 3A<br>2B | 3A<br>2B | 3A<br>2B | 3A<br>2B | | | | | | | |
| E | | | | | | | | | | | | 2A<br>0B | 2A<br>0B | 2A<br>0B | | | | | | | | |
| G | | | | | | 3A<br>1B | 3A<br>1B | 3A<br>1B | 3A<br>1B | | | | | | | | | | | | | |
| F | | | | | | | | | | | | 1A<br>1B | 1A<br>1B | 1A<br>1B | | | | | | | | |
| J | | | | | | | | | | | | | | | | 4A<br>1B | 4A<br>1B | | | | | |
| H | | | | | | | | | | | | | | | 2A<br>2B | 2A<br>2B | 2A<br>2B | 2A<br>2B | 2A<br>2B | | | |
| K | | | | | | | | | | | | | | | | | | | | 5A<br>4B | 5A<br>4B | 5A<br>4B |
| Level of A unassigned | ~~6~~<br>~~3~~<br>1 | ~~6~~<br>~~3~~<br>1 | ~~6~~<br>~~3~~<br>1 | ~~6~~<br>~~4~~<br>1 | ~~6~~<br>~~4~~<br>1 | ~~6~~<br>~~3~~<br>0 | ~~6~~<br>~~3~~<br>0 | ~~6~~<br>~~3~~<br>0 | ~~6~~<br>~~3~~<br>0 | ~~6~~<br>2 | ~~6~~<br>2 | ~~6~~<br>~~3~~<br>~~1~~<br>0 | ~~6~~<br>~~3~~<br>~~1~~<br>0 | ~~6~~<br>~~3~~<br>~~1~~<br>0 | ~~6~~<br>~~3~~<br>1 | ~~6~~<br>~~4~~<br>0 | ~~6~~<br>~~4~~<br>0 | ~~6~~<br>4 | ~~6~~<br>4 | ~~6~~<br>1 | ~~6~~<br>1 | ~~6~~<br>1 |
| Level of B unassigned | ~~7~~<br>~~5~~<br>1 | ~~7~~<br>~~5~~<br>1 | ~~7~~<br>~~5~~<br>1 | ~~7~~<br>~~3~~<br>2 | ~~7~~<br>~~3~~<br>2 | ~~7~~<br>~~6~~<br>5 | ~~7~~<br>~~6~~<br>5 | ~~7~~<br>~~6~~<br>5 | ~~7~~<br>~~6~~<br>5 | ~~7~~<br>4 | ~~7~~<br>4 | ~~7~~<br>~~5~~<br>4 | ~~7~~<br>~~5~~<br>4 | ~~7~~<br>~~5~~<br>4 | ~~7~~<br>~~5~~<br>3 | ~~7~~<br>~~5~~<br>4 | ~~7~~<br>~~5~~<br>4 | ~~7~~<br>5 | ~~7~~<br>5 | ~~7~~<br>3 | ~~7~~<br>3 | ~~7~~<br>3 |

**Figure 7-12** Application of basic scheduling procedure to network of Figure 7-4.

the above procedure as indicated in Figure 7-13. The final schedule shows that project completion is delayed 4 days past the early finish time of 18 days computed for this project without regard to limits on available resources.

The procedure illustrated by this example represents a potentially powerful *general approach* to a broad class of scheduling problems. The procedure can be applied to multiple project problems as well as single projects. For the multiproject case, project start and end dates for each project are necessary. This information is used to compute total activity slack (float), or the equivalent late start times, to permit determination of the order in which various activities are considered for scheduling. It should also be emphasized that the number of resources being considered is not limited by this procedure but only by the capacity of the computing system being used. The basic approach illustrated by this simple example is, in fact, essentially how the majority of heuristic-based constrained-resource scheduling systems operate, including some large commercially-offered computer programs. However, the basic steps indicated here are often embellished considerably to permit more realistic simulation of the numerous variations followed in the actual practice of scheduling project activities. Some examples of these embellishments will be given later.

### Relative Effectiveness of Heuristic Rules

If a heuristic other than minimum slack had been used for ordering the activities in the *OSS*, a different schedule might have been produced for the example network. The min-slack heuristic was used here because it has been found, in several studies, to generally produce the best results. For example, a 1975 study[4] comparing the 8 rules shown in Table 7-1 showed that the minimum slack rule generally outperformed the 7 other rules, as shown in Figure 7-14. These results were based on a sample of small (i.e., 27-activity) multiresource problems for which the optimal solution, in terms of minimum schedule duration, could be calculated. A different study which involved larger networks from actual practice ranging in size up to 180 activities showed that either a Late-Start (equivalent to min-slack) or Late-Finish heuristic was most effective for both parallel or serial allocation procedures.[10] However, that study also concluded that the choice of allocation procedure (i.e., parallel or serial) was of greater significance in some cases than the choice of scheduling heuristic.

Although individual studies such as the two cited above have indicated the *general* best-effectiveness of a particular heuristic, or type of heuristic, it must be emphasized that all such studies have also shown that no one heuristic—or combination of heuristics—*always* produces the best results on *every* problem. This is perhaps the greatest disadvantage of heuristics: Rules that perform well on one problem may perform poorly on another, and vice-versa. In practice, even with more sophisticated procedures, it is not possible to absolutely guarantee in ad-

T = 1
EAS: A B
ES: 1 1
LS: 1 5
OSS: A 8
Schedule A in periods 1-3 and remove it from EAS.
Schedule B in periods 1-5 and remove it from EAS.
Add C, D, and G to EAS.

T = 2
EAS: C D G
ES: 4 4 6
LS: 4 9 10
OSS: None—No activities can be scheduled.
Since minimum ES = 4, skip T = 3 and go to T = 4.

T = 4
EAS: C D G
ES: 4 4 6
LS: 4 9 10
OSS: C D
Schedule C in periods 4-9 and remove it from EAS.
ADD I to EAS.

T = 5
EAS: D G I
ES: 5 6 10
LS: 9 10 10
OSS: D
No activities can be scheduled.

T = 6
EAS: D G I
ES: 6 6 9
LS: 9 10 10
OSS: D G
Schedule G in priods 6-9 and remove it from EAS.
Since D can be scheduled no sooner than period 10 because of resource constraints, to go T = 10 = ES of I.

T = 10
EAS: D I
ES: 10 10
LS: 9 10
OSS: D I
Schedule D in periods 10-11 and remove it from EAS.
Add E and F to EAS.

T = 11
EAS: I E F
ES: 11 12 12
LS: 10 11 11
OSS: I
No activities can be scheduled.

T = 12
EAS: I E F
ES: 12 12 12
LS: 10 11 11
OSS: I E F
Schedule I in periods 12-15 and remove it from EAS.
Schedule E and F in periods 12-14 and remove them from EAS.
Add J and H to EAS.

T = 13
EAS: J H
ES: 16 15
LS: 14 14
OSS: None—No activities can be scheduled.
Since minimum ES = 15, skip T = 14 and go to T = 15.

T = 15
EAS: J H
ES: 16 15
LS: 14 14
OSS: H
Schedule H in periods 15-19 and remove it from EAS.

T = 16
EAS: J
ES: 16
LS: 14
OSS: J
Schedule J in periods 16-17 and remove it from EAS.
Add K to EAS.

T = 17
EAS: K
ES: 18
LS: 16
OSS: None—no activities can be scheduled.

T = 18
EAS: K
ES: 18
LS: 16
OSS: K no activities can be scheduled since K can be scheduled no sooner than period 20 because of resource constraints, go to T = 20.

T = 20
EAS: K
ES: 20
LS: 16
OSS: K
Schedule K in periods 20-22 and remove it from EAS.
All activities have been scheduled. STOP.

**Figure 7-13** Details of basic scheduling procedure applied to network of Figure 7-4.

**Table 7-1. Some Heuristics Used in Constrained-Resource Scheduling.**

| *Heuristic Scheduling Rules Evaluated* | | |
|---|---|---|
| *Rule* | *Notation* | *Operating Features* |
| Minimum Activity Slack | MINSLK | Schedules first those activities with lowest activity slack time (total float). |
| Minimum Late Finish Time | LFT | Schedules first those activities with the earliest values of late finish time. |
| Resource Scheduling Method | RSM | Priority index calculated on basis of pair-wise comparison of activity early finish and late start times. Gives priority to activities roughly in order of increasing late finish time. |
| Greatest Resource Demand | GRD | Schedules first those activities with greatest resource demand in order to complete potential bottleneck activities. |
| Greatest Resource Utilization | GRU | Gives priority to that *group* of activities which results in the minimum amount of idle resources in each scheduling interval. Involves an integer linear programming logarithm. |
| Shortest Imminent Operations | SIO | Schedules first those activities with shortest durations in an attempt to complete the greatest number of activities within a given time-span. |
| Most Jobs Possible | MJP | Gives priority to the largest possible *group* of jobs which can be scheduled in an interval. Involves an integer linear programming logarithm. |
| Random Activity Selection | RAN | Priority given to jobs selected at random, subject to resource availability limits. |

Note: The MINSLK rule has been shown equivalent to a minimum-late-start-time rule.

vance which particular heuristic—or combination of heuristics—will produce best results for a given problem.

In spite of this drawback, heuristic procedures are used very widely in practice. The schedules produced by these procedures may not be the theoretically best possible, but they are usually good enough to use for planning purposes in view of the uncertainties typically associated with activity durations and resource constraints and requirements. Furthermore, some very powerful computer-based solution procedures incorporating a variety of imaginative heuristics have been developed which will produce schedules for large, complex projects under a variety of special assumptions such as job splitting, "crashing," stretching out of jobs by varying the rate of resource application, etc. General informa-

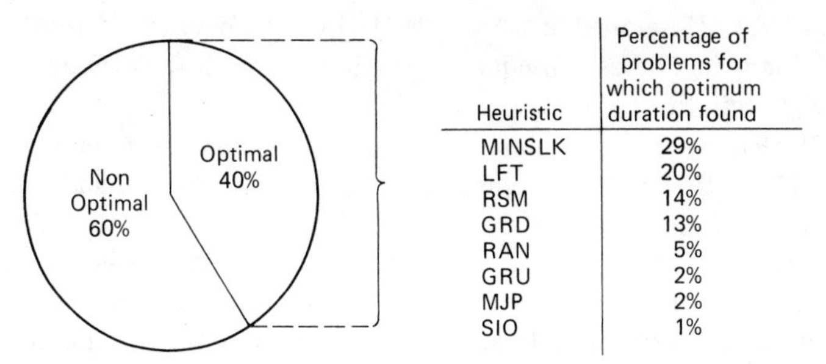


| Heuristic | Percentage of problems for which optimum duration found |
|---|---|
| MINSLK | 29% |
| LFT | 20% |
| RSM | 14% |
| GRD | 13% |
| RAN | 5% |
| GRU | 2% |
| MJP | 2% |
| SIO | 1% |

(a) Percent of problems for which optimal minimum duration solution was obtained. (Optimal vs. nonoptimal totals include problems in which one *or more* heuristic sequencing rules produced a minimum duration solution.)

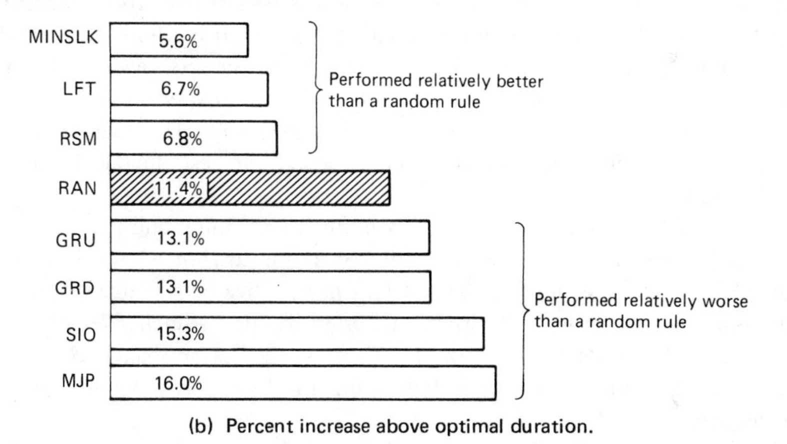


(b) Percent increase above optimal duration.

**Figure 7-14** Heuristic rules results.[5]

tion about a number of such commercially-available computer programs is given in Chapter 11. Many other such programs have been developed by individual companies for proprietary use. The details of these approaches are not generally available for publication.* However one program, representative of this class of comprehensive heuristic procedures, that has been described in the open literature is described briefly below.

## Wiest's SPAR-1 Model

One of the more comprehensive heuristic scheduling procedures developed by an independent researcher is called SPAR-1 (for Scheduling Program for Allocat-

*These programs vary widely with regard to approaches used in scheduling under resource constraints. Some, such as the McAuto "MSCS" package, allow the user to choose from among several available heuristic rules or even create his own. Others utilize one or more built-in heuristic rules without allowing any user choice.

ing Resources). Developed in the mid-1960s by J. D. Wiest,[20,22] it has served as a model for many other subsequent procedures. A flow diagram of the procedure is given in Figure 7-15.

The SPAR-1 model is similar to the basic procedure diagrammed earlier in Figure 7-11, in that a parallel scheduling approach is followed and activities are sorted for scheduling within each time period in order of their total slack. However, the SPAR-1 model includes a number of useful embellishments, as will be seen in the following description. One unusual feature not apparent from the flow diagram is a probability-based selection procedure for determining whether a job from the list of jobs eligible for scheduling in each period is actually scheduled. Under this approach the first job on the list is not automatically selected for scheduling, but is selected with some probability less than 100%. Jobs from the top of the list which are not selected are moved to the bottom of the list. Eventually all jobs in the list will be scheduled, but the sequence of scheduling will vary randomly, and hence the project schedule will vary on successive runs of the model, even though inputs are exactly the same. This feature allows a number of different schedules to be generated and the best one selected according to a specific criterion.

In the SPAR-1 model each job has a normal, maximum, and minimum level of resources which can be assigned to it. Since the total man-days required for job performance is assumed to be constant irrespective of resource loading, job duration can be normal, "crashed" (shortened), or "stretched" (increased), according to whether the level of resources assigned is normal, maximum, or minimum. Resource allocation is done automatically by the program, according to whether the job is critical or noncritical.

Referring to Figure 7-15, it can be seen that the SPAR-1 model iterates day by day starting with day $d = 1$. Jobs with $ES = 1$ are selected and ordered according to their slack (the "augment critical job routine" actually comes into play later and will be explained below). If the job to be scheduled is not critical, attempts are made to schedule it with a normal level of resources ("normal crew size"). If this is not possible, the minimum crew size is considered before the job is postponed until day $d + 1$. If the job to be scheduled is critical a different procedure is followed. First, if sufficient resources are available the job is scheduled at its maximum allocation; that is, it is crashed. If this is not possible a normal crew size is considered. If neither of these attempts is successful two special subroutines are tried. In the first of these, efforts are made to *borrow needed resources* from currently active jobs. Resources are borrowed from a job only when the resultant "stretching" of the job will not delay the entire project.

The second subroutine that may be tried for critical jobs *reschedules active jobs* to a later start date. That is, jobs that can be postponed without delaying the project are delayed for scheduling until a later period; the resources thus

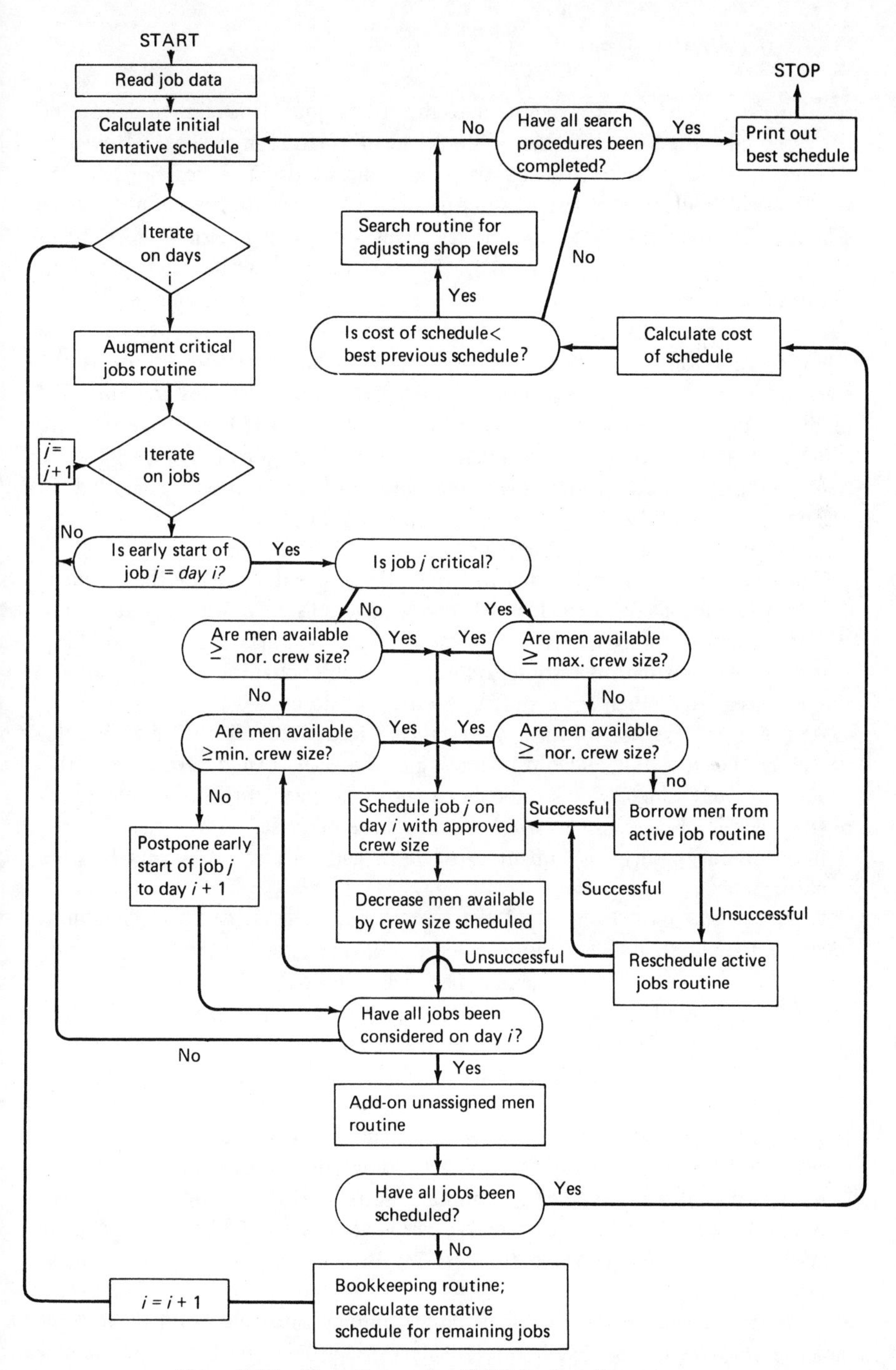


**Figure 7-15** Flow diagram for SPAR-1 scheduling program.

freed up are available for assignment to the critical job. This reschedule routine has essentially the same effect as a "look ahead" feature might have. However, instead of actually looking ahead to predict the needs of critical jobs (which would be difficult in the limited resources cases, since activities are not always scheduled at their *ES* date), the model changes previous decisions as it moves along from day to day by "taking back" resources that were assigned to jobs in earlier periods which do not need them as much as the critical jobs.

If neither of the special subroutines (borrow resources or reschedule active jobs) is successful, a minimum resource level schedule is attempted before the start of the critical job is postponed. Then, after all possible jobs are scheduled on day *d*, the *add-on unassigned resources* subroutine is called. In this procedure a list of unused resources is compiled. A list of active jobs requiring these resources that are not at their maximum allocation levels is also compiled, in ascending order of slack. The assigned levels of these jobs is then increased, until either jobs or resources is exhausted.

Whenever the program iterates to a new day (i.e., day $d + 1$), the *augment critical jobs* subroutine is tried. With this feature jobs previously scheduled and still active are examined. If any of these is critical and has a resource level less than its maximum—and if resources are still available—that job's resource allocation is increased as much as possible up to the maximum.

As the program moves from day to day it also updates all critical path data, i.e., early and late start and finish times, and job slack. The importance of this step is apparent, since a job's criticality may be changed through actions taken by the model. For example, when resources are increased to a critical job the resultant duration decrease might cause it to gain positive slack and become noncritical.

When all jobs have been scheduled over all days, the cost of the schedule is computed, including daily overhead expenses, due date bonuses or penalties, and resource costs based on the assumed available levels. Depending upon the cost of the schedule compared with previous schedules, the procedure is either terminated or a new set of available resources is tested.

SPAR-1 is able to accommodate projects of several thousand jobs, and several dozen resource types, under a variety of special assumptions, such as job splitting, shift or nonshift scheduling, variable resource limits, and single or multiple projects. The results obtained from SPAR-1 on small tests problems have shown that in some cases it gave nearly optimal results. However, no heuristic-based program, irrespective of its complexity, can guarantee to yield an optimal project schedule, or even consistently best results from problem to problem, as noted earlier.

Analytic procedures, such as linear programming, can yield optimal solutions consistently, but are restricted to relatively small projects with few special conditions. For this reason, and in spite of their limitations, heuristic procedures appear to offer, for the foreseeable future, the best approach for scheduling realistic-sized projects under conditions of limited resources.

## OPTIMIZATION PROCEDURES

Another major category of constrained-resource scheduling procedures—those for producing optimal solutions—has been characterized by relatively less practical progress than the heuristic methods. The optimization procedures which have been developed can be divided into two subcategories:

1. procedures based on linear programming (LP);
2. procedures based on enumerative and other mathematical techniques.

*Linear programming* (LP) was first proposed as a method for solving the constrained-resource scheduling problem in the early 1960s. Numerous formulations of the problem were proposed but found to be impractical except for solving small problems of only a few activities. Wiest described the use of those early LP formulations on the constrained-resource problem as "akin to using a bulldozer to move a pebble."[20] Wiest showed that a 55-activity network with four resource types would require more than 6,000 equations and 1,600 variables. Later researchers improved on these early formulations and solved somewhat larger problems. In recent years, with the increasing capability and decreasing cost of computers, some researchers have once again begun to investigate the use of linear programming procedures, either alone or in combination with other approaches. An example of the latter type of approach is that of Patterson and Huber,[14] who combine a minimum bounding procedure with integer LP to reduce the computation time required in arriving at a minimum project duration. Essentially, their approach involves starting the optimization procedure off with a "good" lower bound solution, to reduce the domain of possible solutions over which the LP algorithm must search. The "minimum bound" they use to initiate the search procedure is simply an estimate of the minimum project duration implied by the tightest resource constraint, i.e., if the cumulative resource requirement of that resource is 80 units, and if 8 units per day are the maximum available, then the minimum project duration = 80/8 = 10 days. They tested this approach on a series of small problems ranging in size up to about 20 activities and 3 different resource types, with encouraging results, compared to those of other LP approaches.

Another recent approach is that of Talbot.[19] He uses both integer linear programming and implicit enumeration in formulating and solving resource-constrained project scheduling problems in which job performance time is a function of resource allocation, and total cost is used as an objective function. His reported computational experience indicates that the methodology can provide optimal solutions to small problems and heuristic solutions to larger problems.

The LP approaches that have been proposed to date appear to share the common weakness of unpredictability of effectiveness. That is, they produce an optimal solution quickly on some problems but not on others. The Patterson-Huber algorithm mentioned above, for example, operating on an IBM 360-67 computer solved some 20-activity problems optimally in seconds while other, essentially similar, problems required 10–20 minutes of computer time. If the characteristics of those problems which are amenable to solution by LP could be identified, the future for LP approaches might appear brighter than it now does.

In summary, in spite of some encouraging recent development, LP approaches still have not progressed to the point where they are capable of routinely solving the type of large, complex problems easily handled by heuristic procedures. They have been used with heuristic procedures to determine the best sequence of activities within the *OSS* at individual time periods. However, they remain today primarily an interesting research topic for academicians.

*Enumeration Techniques.* The second category of optimization procedures consists of techniques that have appeared on the scene fairly recently. These are based on enumeration of all possible activity sequencing combinations (i.e., schedules) and include the so-called "branch and bound" procedures. These methods were first applied to constrained-resource project scheduling in the late 1960s, when researchers in the USA and Germany independently demonstrated that such procedures were capable of optimally solving networks of up to about 100 single-resource activities or 30 multiresource activities, under assumption of nonconstant resource availabilities and activity splitting.[3,11,12] More recent developments include procedures which appear capable of consistently producing optimal solutions to larger problems much more efficiently.

The term "branch and bound" (B&B) refers to a generic type of optimization procedure which involves partitioning a problem into subproblems (branching), and evaluating these subproblems (computing bounds). The procedure can be conveniently modeled by the nodes and branches of a tree, to enumerate possible alternatives in arriving at the best solution. This solutions tree is sometimes referred to as the "B&B tree." The power of the procedure stems from the fact that the enumeration process is *implicit*, not explicit, since some incomplete schedule alternatives generated early in the process can be identified immediately as not capable of leading to an improved complete schedule (compared to some initial starting value), and thus are not pursued in the solution development process. B&B procedures that have been developed by different researchers differ in the branching rules used (for deciding which portion of the tree to explore next) and in the bounding rules used (for identification of nodes least likely to produce an improved solution).

**Table 7-2. Computation Results For Optimal Solution of 233 Problems.**[16]

| *Network Size (No. Activities)* | *Number Solved* | *Mean CPU Time (min.)** | *Variance in CPU (min.)** |
|---|---|---|---|
| 23 | 60 | 0.20 | 0.07 |
| 27 | 60 | 0.76 | 1.02 |
| 35 | 57 | 2.61 | 19.17 |
| 43 | 56 | 5.84 | 60.75 |
| | 233 | | |

*IBM 370-155.

A number of different branch and bound procedures for the project scheduling problem have been developed by various researchers. One of the more effective to date is that of Stinson, et al., described in Reference 17. The details of this procedure, which is designed for minimizing project duration, are summarized in Appendix 7-1, and illustrated on a small sample problem. An optimal solution for the familiar example problem used throughout this chapter (network of Figure 7-4) is also presented there.

To give some indication of the capabilities of the B&B approach, Stinson's procedure produced the results shown in Table 7-2 for a sample of 240 artificially-generated test problems. Out of the total of 240 project-type problems attempted (job-shop scheduling problems were also solved), optimal solutions were found for 233, or 97%. The program was terminated short of optimality for the 7 unsolved problems. In each case computer memory demand, not computation time, was the limiting factor, with 500K bytes of memory storage on an IBM 370-155 allocated. A representative 4-resource problem in the 43-activity class is shown in Figure 7-16, with the optimal activity start times.

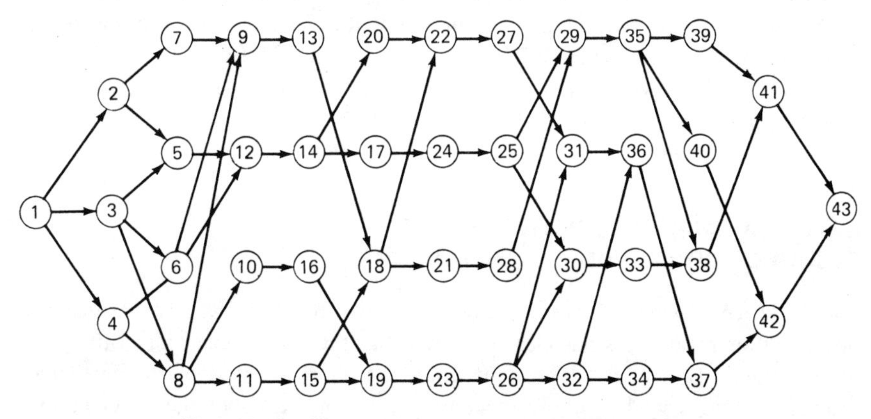


**Figure 7-16** Network for representative B&B problem.

**ACTIVITY DURATIONS, RESOURCES REQUIRED, AND OPTIMUM START TIMES**

| Act. No. | Dur. | Res. Req. | | | | Opt. Start | Act. No. | Dur. | Res. Req. | | | | Opt. Start |
|---|---|---|---|---|---|---|---|---|---|---|---|---|---|
| 1 | 0 | 0 | 0 | 0 | 0 | 0 | 22 | 5 | 2 | 6 | 6 | 1 | 48 |
| 2 | 4 | 3 | 2 | 4 | 2 | 0 | 23 | 1 | 3 | 2 | 4 | 1 | 39 |
| 3 | 6 | 6 | 2 | 3 | 1 | 0 | 24 | 1 | 2 | 6 | 4 | 3 | 43 |
| 4 | 6 | 5 | 4 | 3 | 2 | 6 | 25 | 1 | 1 | 1 | 2 | 4 | 44 |
| 5 | 2 | 1 | 3 | 3 | 1 | 6 | 26 | 5 | 6 | 3 | 6 | 1 | 43 |
| 6 | 2 | 4 | 1 | 6 | 2 | 12 | 27 | 6 | 6 | 2 | 2 | 6 | 55 |
| 7 | 3 | 4 | 1 | 4 | 5 | 4 | 28 | 3 | 5 | 1 | 5 | 1 | 40 |
| 8 | 5 | 1 | 2 | 2 | 2 | 12 | 29 | 3 | 3 | 3 | 2 | 4 | 45 |
| 9 | 7 | 1 | 5 | 2 | 3 | 17 | 30 | 7 | 4 | 3 | 4 | 3 | 48 |
| 10 | 6 | 6 | 1 | 5 | 3 | 25 | 31 | 5 | 2 | 5 | 3 | 3 | 61 |
| 11 | 8 | 6 | 1 | 6 | 3 | 17 | 32 | 2 | 1 | 3 | 5 | 5 | 53 |
| 12 | 2 | 5 | 1 | 6 | 6 | 14 | 33 | 3 | 2 | 3 | 3 | 2 | 62 |
| 13 | 1 | 3 | 4 | 4 | 3 | 24 | 34 | 6 | 4 | 1 | 4 | 1 | 62 |
| 14 | 7 | 1 | 1 | 1 | 1 | 16 | 35 | 7 | 3 | 5 | 5 | 4 | 55 |
| 15 | 2 | 1 | 1 | 4 | 1 | 25 | 36 | 1 | 1 | 4 | 5 | 4 | 67 |
| 16 | 1 | 1 | 1 | 2 | 3 | 31 | 37 | 3 | 1 | 5 | 1 | 2 | 68 |
| 17 | 9 | 1 | 3 | 5 | 2 | 27 | 38 | 2 | 5 | 3 | 1 | 3 | 67 |
| 18 | 5 | 4 | 2 | 3 | 1 | 31 | 39 | 2 | 2 | 4 | 3 | 3 | 65 |
| 19 | 3 | 5 | 1 | 4 | 3 | 36 | 40 | 1 | 6 | 3 | 2 | 3 | 69 |
| 20 | 3 | 2 | 2 | 5 | 3 | 40 | 41 | 2 | 2 | 6 | 5 | 4 | 71 |
| 21 | 4 | 3 | 5 | 6 | 5 | 36 | 42 | 3 | 3 | 3 | 2 | 4 | 71 |
| | | | | | | | 43 | 0 | 0 | 0 | 0 | 0 | 74 |

Resource Limits: 10 units each type.

**Figure 7-16** Continued

All of the implicit enumeration approaches developed to date appear to share the same disadvantage common to LP optimization procedures: unpredictability of computation time from problem to problem. For example, as Table 7-2 shows, the variance in CPU time required by the Stinson approach was quite large, ranging up to as much as 10 times the mean for a given class of problems. While some efforts were made in the testing of the procedure to isolate possible problem characteristics contributing to such large variance, no significant conclusions were drawn.

## NETWORK CHARACTERISTICS: USE IN RESOURCE-CONSTRAINED SCHEDULING

The experience gained in resource-constrained scheduling with heuristic and optimization procedures has clearly shown that the effectiveness of both types of procedures is strongly dependent upon the characteristics of the particular problems being solved. Yet surprisingly little is known about the exact nature of this relationship.

In contrast to the considerable effort that has been expended over the past

20 years on the development of solution *techniques*, comparatively little effort has gone into study of the *problems themselves*, i.e., the size, shape, complexity, etc., of project network scheduling problems. This is particularly true of problems encountered in actual practice.

The lack of an accepted methodology for simply describing networks in meaningful terms has hindered the development of a useful classification scheme for resource-constrained scheduling applications. For example, except for the number of activities and resource types, there is today no readily accepted way for describing the differences/similarities between networks, except in very general descriptive terms, such as "short and fat," "long and thin," etc.

The need for a system or methodology for describing or measuring the difference between networks, or changes to a given network problem, has been long recognized. One use of such a scheme would be in estimating the computational requirements or general effectiveness associated with a particular optimization procedure. Another use, in the case of heuristics, would be in predicting which of several heuristics might be most effective on a particular problem, or in estimating the impact on schedule duration of changes in network structure and/or resource constraints, when scheduling given networks with a selected heuristic.

Various quantitative factors or summary measures have been tested by differ-

**Table 7-3. Examples of Network Summary Measures.**[7]

I. *Measures that characterize network size, shape, and logic*

Examples: *Length:* max. no. of consecutive nodes from beginning to end
*Width:* max. no of nodes in parallel
*Complexity:* $\frac{\text{No. of arcs}}{\text{No. of nodes}}$

II. *Measures that indicate time characteristics of the network*

Examples: Sum of Activity Durations
Average Activity Duration
Variance in Activity Durations
Critical Path Duration
Total Network Slack
Density: $\frac{\text{Sum of activity durations}}{\text{Sum of durations + total free slack}}$

III. *Measures that characterize resource demands/availabilities*

Examples: Total Work Content (equivalent to cumulative resource requirements)

Average Resource Requirement Per Activity

Average Resource Requirement Per Period

Product Moment (a measure of location of predominant influence of the resource type, in terms of first or second half of original project duration)

Resource Utilization Factor (a measure of resource "tightness," in terms of requirement versus availabilities)

ent researchers in attempts to explain the relationship between problem characteristics and effectiveness of a particular solution technique. In general these attempts have not been successful. However, Patterson[13] developed a regression model which satisfactorily predicted the relative effectiveness of selected heuristic rules for project duration minimization on a sample of both single and multiproject problems.

A somewhat different scheme of network summary measures is described in Reference 7. One of the more complete such schemes yet proposed, it is intended to provide both a general scheme of project network descriptive measures as well as factors for use in resource-constrained scheduling. This approach divides the domain of possible summary measures into 3 categories. Table 7-3 shows these categories, along with selected examples of summary measures in each category.

One potential use of such summary measures is for estimating the impact of

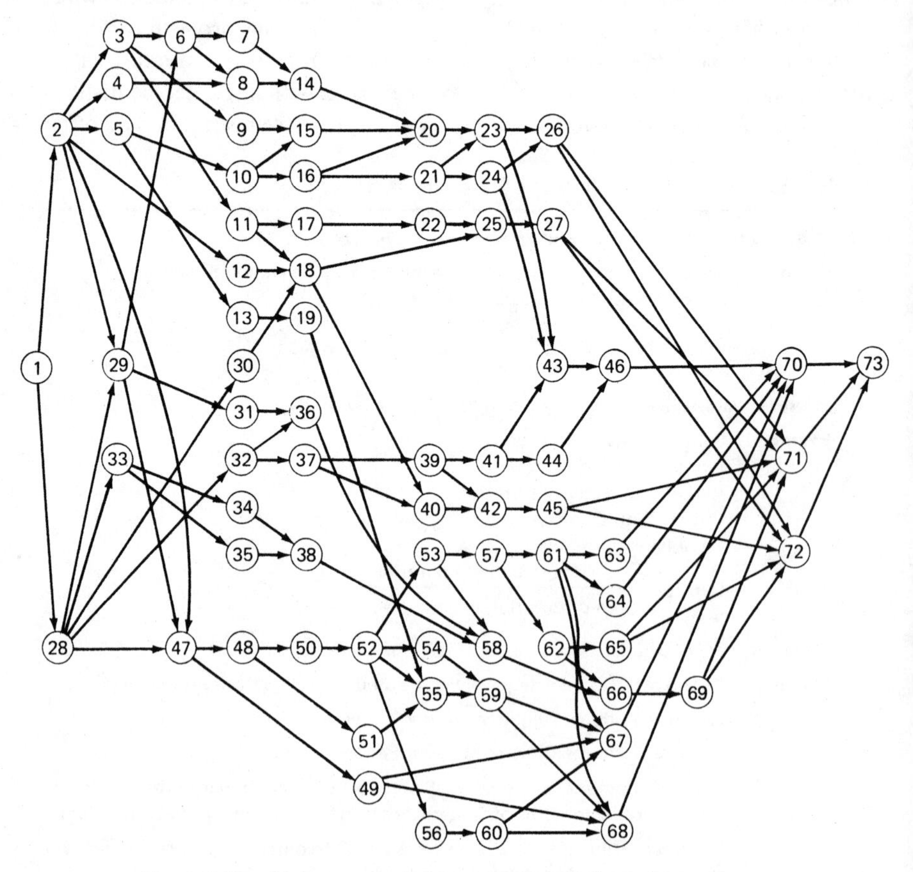


**Figure 7-17** Project network for management information system.

**ACTIVITY DURATIONS AND UNITS OF EACH RESOURCE TYPE REQUIRED**

| *Activity No.* | *Duration* | *Type 1 Resource* | *Type 2 Resource* | *Activity No.* | *Duration* | *Type 1 Resource* | *Type 2 Resource* |
|---|---|---|---|---|---|---|---|
| 1 | 4 | 0 | 0 | 38 | 8 | 0 | 1 |
| 2 | 2 | 1 | 0 | 39 | 4 | 1 | 0 |
| 3 | 3 | 1 | 0 | 40 | 5 | 0 | 1 |
| 4 | 6 | 0 | 1 | 41 | 8 | 0 | 1 |
| 5 | 5 | 1 | 0 | 42 | 11 | 0 | 1 |
| 6 | 4 | 1 | 0 | 43 | 6 | 0 | 3 |
| 7 | 4 | 1 | 0 | 44 | 6 | 1 | 0 |
| 8 | 4 | 0 | 1 | 45 | 7 | 0 | 1 |
| 9 | 7 | 0 | 1 | 46 | 3 | 0 | 3 |
| 10 | 8 | 1 | 0 | 47 | 3 | 1 | 0 |
| 11 | 6 | 1 | 0 | 48 | 1 | 1 | 0 |
| 12 | 7 | 0 | 1 | 49 | 14 | 0 | 1 |
| 13 | 3 | 1 | 0 | 50 | 3 | 1 | 0 |
| 14 | 2 | 0 | 1 | 51 | 7 | 0 | 1 |
| 15 | 4 | 0 | 1 | 52 | 10 | 2 | 0 |
| 16 | 12 | 1 | 0 | 53 | 9 | 2 | 0 |
| 17 | 5 | 1 | 0 | 54 | 6 | 1 | 0 |
| 18 | 9 | 0 | 2 | 55 | 10 | 0 | 2 |
| 19 | 8 | 0 | 1 | 56 | 3 | 2 | 0 |
| 20 | 8 | 0 | 2 | 57 | 6 | 2 | 0 |
| 21 | 4 | 1 | 0 | 58 | 13 | 0 | 2 |
| 22 | 13 | 1 | 0 | 59 | 6 | 0 | 2 |
| 23 | 10 | 0 | 1 | 60 | 10 | 0 | 1 |
| 24 | 3 | 1 | 0 | 61 | 4 | 1 | 0 |
| 25 | 13 | 0 | 1 | 62 | 3 | 1 | 0 |
| 26 | 7 | 0 | 1 | 63 | 3 | 1 | 0 |
| 27 | 8 | 0 | 1 | 64 | 8 | 1 | 0 |
| 28 | 1 | 1 | 0 | 65 | 8 | 1 | 0 |
| 29 | 3 | 1 | 0 | 66 | 12 | 0 | 3 |
| 30 | 9 | 0 | 1 | 67 | 7 | 0 | 1 |
| 31 | 6 | 0 | 1 | 68 | 6 | 0 | 1 |
| 32 | 9 | 1 | 0 | 69 | 7 | 0 | 3 |
| 33 | 2 | 1 | 0 | 70 | 13 | 0 | 0 |
| 34 | 8 | 1 | 0 | 71 | 5 | 0 | 3 |
| 35 | 9 | 0 | 1 | 72 | 5 | 0 | 3 |
| 36 | 6 | 0 | 1 | 73 | 4 | 0 | 0 |
| 37 | 12 | 1 | 0 | | | | |

**Figure 7-17** Continued

alternative resource availability levels on project duration, to avoid the necessity of actual solution by computer. The network shown in Figure 7-17, for example, is for an actual project involving the design and implementation of software for a computer-based management information system. The resources required in this case were 2 different types of computer programmers. Project duration, computed without regard to resource limits, was 75 days. However, the number of programmers of each type was severely limited. Possible combinations of each type of personnel ranged from a low of 3 type 1 and 7 type 2 to a maximum of 5 type 1 and 9 type 2 (activity durations were fixed and did not change with resource loading).

When the network was scheduled successively under each of these 9 levels of manpower availability (using a minimum Late Finish Time heuristic), the project durations indicated by the solid curve ("scheduled" values) in Figure 7-18 was obtained.

A regression equation was developed for estimating these scheduled values, using selected combinations of the types of summary measures illustrated in Table 7-3. This regression equation was used as the basis for a forecasting model in the form of an interactive computer program. The program, when given the standard network data (i.e., durations and logic), resource requirements, and possible combinations of resource limits of each type, automatically calculated the necessary summary measures and produced the "forecast" durations shown in Figure 7-18.

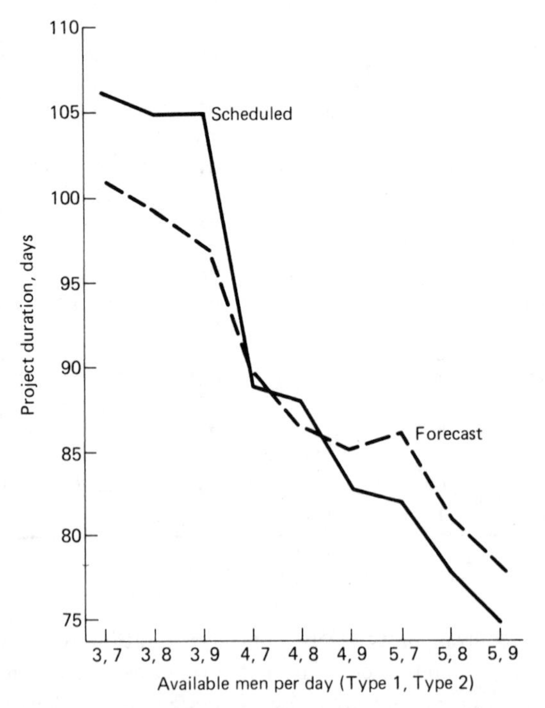


**Figure 7-18** Forecast and scheduled durations for network of Figure 7-17.

## SUMMARY

Techniques for scheduling activities to satisfy resource constraints have developed significantly in both number and capability over the past 20 years. The past 10 years, in particular, have seen increased activity in this area.

Considerable progress has been made in the past 10 years in the development of implicit enumeration procedures for producing optimal resource-constrained solutions. And because of the continued improvements in computer technology, these procedures appear within the realm of practical application to modest-sized problems in special cases. However, the extreme unpredictability of computation time associated with these procedures will most likely prevent their widespread use until either improved algorithms are developed or ways are found to exploit problem characteristics by applying them only in instances where they will be effective.

Heuristic scheduling procedures have also experienced significant progress over the past 10 years. Today there are literally dozens of large-scale, computer-based project scheduling packages available, which offer varied and imaginative heuristics that will handle the largest networks under an extreme variety of special conditions. Such heuristic procedures are the mainstay of commercial applications, and will likely continue to be for the immediate future.

## REFERENCES


1. Agin, Norman, "Optimum Seeking with Branch and Bound," *Management Science,* December 1966.
2. Burgess, A. R. and J. B. Killebrew, "Variation in Activity Level on a Cyclic Arrow Diagram," *Journal of Industrial Engineering,* March–April 1962.
3. Davis, Edward W. and George E. Heidorn, "An Algorithm for Optimal Project Scheduling Under Multiple Resource Constraints," *Management Science*, August 1971.
4. Davis, E. W and J. H. Patterson, "A Comparison of Heuristic and Optimum Solutions in Resource-Constrained Project Scheduling," *Management Science,* April 1975.
5. Davis, E. W. and J. H. Patterson, "Resource-Based Project Scheduling: Which Rules Perform Best?" *Project Management Quarterly*, Sept. 1976.
6. Davis, E. W., "Project Scheduling Under Resource Constraints: Historical Review and Categorization of Procedures," *AIIE Transactions*, Dec. 1973.
7. Davis, E. W., "Project Network Summary Measures and Constrained Resource Scheduling," *AIIE Transactions*, June 1975.
8. Elmaghraby, S. E., *Activity Networks: Project Planning and Control by Network Models*, John Wiley & Sons, New York, 1977.
9. Fendley, Larry G., "Toward the Development of a Complete Multi-Project

Scheduling System," *Journal of Industrial Engineering*, October 1968; unpublished PhD thesis, Arizona State University, 1966, same title.
10. Gordon J. H., "Heuristic Methods In Resource Allocation," *Proceedings of the 4th Internet Conference*, Paris, 1974.
11. Johnson, T. J. R., "An Algorithm for the Resource Constrained Project Scheduling Problem," unpublished PhD Dissertation, Massachusetts Institute of Technology (1967).
12. Mueller-Mehrbach, H., "Ein Verfahren zur Planung des Optimalen Betriebsmitteleinsatzes bei der Terminierung von Grossprojekten," *Zeitschrift fuer Wirtschaftliche Fertingueng*, Heft 2 and 3, Feb./March 1967.
13. Patterson, James, "Project Scheduling: The Effects of Problem Structure on Heuristics Performance," *Naval Research Logistics Quarterly,* 1976, Vol. 23, pp. 95.
14. Patterson, James H. and Walter D. Huber, "A Horizon-Varying, Zero-One Approach to Project Scheduling," *Management Science*, February 1974.
15. Schrage, L., "Solving Resource-Constrained Network Problems by Implicit Enumeration-Non-Preemptive Case," *Operations Research*, 10 (1970).
16. Stinson, Joel P., "A Branch and Bound Algorithm for a General Class of Multiple Resource-Constrained Scheduling Problems," unpublished PhD Dissertation, University of North Carolina (1976).
17. Stinson, Joel P., E. W. Davis, and B. Khumawala, "Multiple Resource—Constrained Scheduling Using Branch and Bound," *AIIE Transactions*, September 1978.
18. Talbot, F. B. and J. H. Patterson, "An Efficient Integer Programming Algorithm With Network Cuts for Solving Resource-Constrained Scheduling Problems," *Management Science*, July 1978.
19. Talbot, F. B., "Project Scheduling with Resource-Duration Interactions: The Nonpreemptive Case," Working paper No. 200, Graduate School of Business Administration, University of Michigan, January 1980.
20. Weist, Jerome D., "Heuristic Model for Scheduling Large Projects with Limited Resources," *Management Science*, February 1967.
21. Weist, Jerome D., "Heuristic Programs for Decision Making," *Harvard Business Review*, September-October 1965.
22. Weist, Jerome D. and Ferdinand K. Levy, *A Management Guide to PERT/CPM,* Prentice-Hall, Inc., Englewood Cliffs, NJ, 1969.
23. Woodworth, Bruce M. and C. T. Willie, "A Heuristic Algorithm for Resource Levelling in Multi-Project, Multi-Resource Scheduling," *Decision Sciences*, 1975, Vol. 6, pp. 525–540.

## EXERCISES

**1.** Suppose the resource requirements associated with each activity in the network of Figure 7-4 have changed to the following levels:

| Activity | Resource Requirement Type A | Type B |
|---|---|---|
| A | 3 | 2 |
| B | 2 | 4 |
| C | 3 | 1 |
| D | 4 | 3 |
| E | 4 | 0 |
| F | 2 | 2 |
| G | 3 | 1 |
| H | 4 | 4 |
| I | 6 | 4 |
| J | 4 | 1 |
| K | 5 | 4 |

a. Draw the ES and LS resource profiles.
b. Construct the cumulative resource requirements curves.
c. Calculate the lower-bound estimate of the average daily resource requirements of each type; use the internal tangent method where appropriate.
d. What is the criticality index for each resource if the following maximum levels of each are available?

| | Type A | Type B |
|---|---|---|
| Case 1 | 8 | 8 |
| Case 2 | 7 | 7 |

e. What is the expected minimum project duration with Case 1 resource limits of d above?

2. Using the Burgess leveling procedure and the revised resource requirements above for the network of Figure 7-4, level the resource requirements as much as possible.
3. Using the shortest-job-first heuristic (SIO rule in Table 7-1) instead of the min-LS rule to order activities in the *OSS*, reschedule the network of Figure 7-4 with the basic procedure of Figure 7-13. Use lowest job number in case of ties.
4. Using the new resource requirements given in (1) above for the network of Figure 7-4 and resources available of 8 type A, 8 type B, apply the basic scheduling procedure of Figure 7-11.
5. Repeat exercise 4 above, but instead of using the Min-LS rule to order activities in the *OSS*, use the "Greatest Resource Demand" (GRD) rule shown in Table 7-1. Use lowest job number in case of ties.
6. Schedule the example of Figure 7-4 using the Wiest procedure outlined in Figure 7-15. Assume that only resource type A is required, that the activity resource requirements of type A shown in Figure 7-4 are the "Normal crew size" and that the maximum and minimum loadings are one unit more or one unit less, respectively. Assume type A resources available equal 8 units. Ignore schedule costs.
7. Verify the project schedule shown in Figure 7-9 using the Burgess algorithm.

# APPENDIX 7-1
# APPLICATION OF BRANCH AND BOUND METHODS TO RESOURCE CONSTRAINED PROJECT SCHEDULING

The description of the branch and bound method here as applied to resource-constrained project scheduling is based on Stinson, Davis, and Khumawala.[17]* The version of the problem addressed is minimization of project duration, i.e., establishing a schedule of feasible activity start times such that the entire project is completed in a minimum span of time. It is assumed that (1) resource requirements of each activity are at a constant level for each resource type during the entire interval in which the activity is in progress, (2) the level of availability of each resource type is constant over the entire span of project duration, and (3) activities once started are not interrupted.

The B&B procedure for the resource-constrained scheduling problem consists of creating nodes in the decision tree, which represent unique "partial schedules." Each partial schedule represents scheduling decisions for some subset of the total number of activities. The partial schedules created are always feasible (i.e., they satisfy both precedence and resource constraints) and nonredundant (no two schedules are exactly alike). Furthermore the solutions tree is exhaustive in that every possible feasible project schedule will exist at the terminal nodes of the tree.

The tree generation process starts with creation of an initial node representing the set of activities which can be started at the beginning of the project (time 0). From this point a family of partial schedules is created by branches with new

*Reference numbers are for the references to Chapter 7.

nodes to the tree; each node in the family created from a particular node has in common with the others all scheduling decisions made previously in creating the common parent node. However each is unique from the others in that it includes one new decision involving the scheduling of one or more activities previously unscheduled. Each branching operation therefore creates only as many new partial schedules as there are feasible combinations ("feasible subsets") of activities that may enter the schedule at some point in time, $t_n$. Thus a partial schedule, $PS_n$, can be visualized as a real project in progress at time $t_n$: Some activities (the "complete set," $C_n$) will have been completed at $t_n$, others (the "active set," $A_n$) may be actively in progress, to be finished at a later date, and still others may be ready to be scheduled when all predecessor activities have been completed.

The complete B&B procedure involves simultaneous tree generation and "pruning," to prevent growth beyond unmanageable bounds. However the procedure is best understood by first showing the complete (i.e., unpruned) tree for a sample problem; then techniques for pruning the tree will be described.

The small 4-activity project of Figure 7-19 will be used as an example. The B&B solutions tree generated for this project is shown in Figure 7-20. The first node (partial schedule) represents the only feasible scheduling decision at time zero, which is to place activity 1, the dummy start activity, in an "active" status. When activity 1 is completed (also at time zero) there are three possible scheduling decisions which are feasible (i.e., 3 possible feasible subsets): Schedule activity 2 by itself (node 2), schedule activity 3 by itself (node 3), or schedule both activities 2 and 3 together (node 4). The latter is allowable only because the resource requirements of both activities together do not exceed the amount available.

To illustrate continuation of the process, consider branching from node 2, where activity 2 is scheduled to start at time zero. The next set of scheduling decisions will occur when activity 2 is completed (time 1). Thus all scheduling

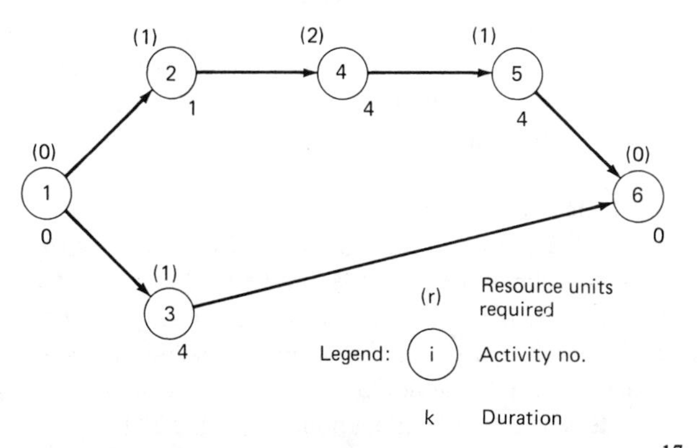


**Figure 7-19** Simple network for branch and bound illustration.[17]

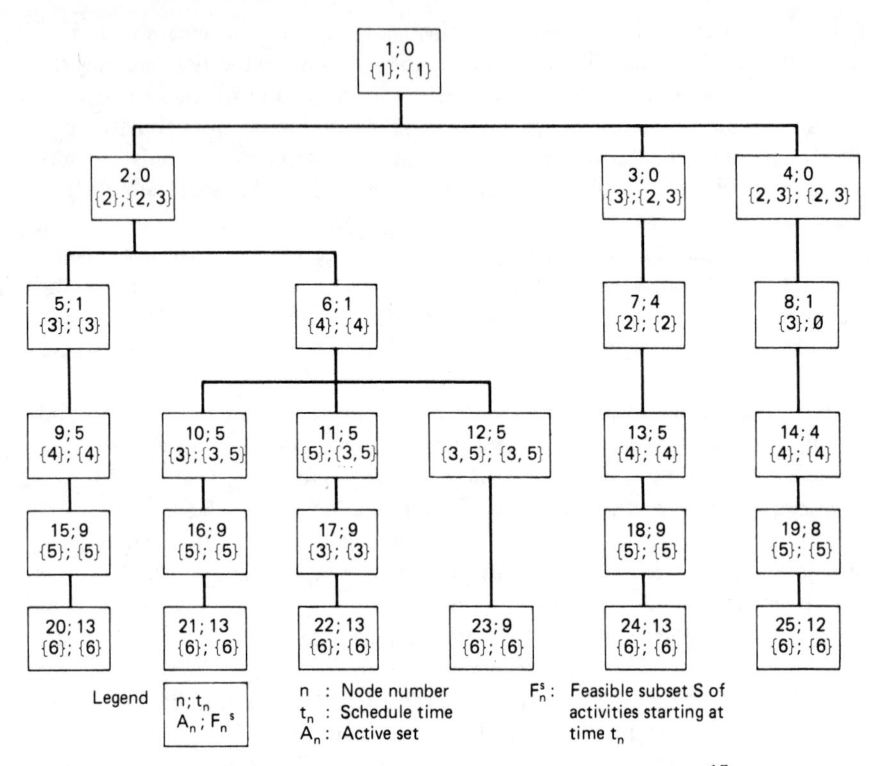


**Figure 7-20** Complete decision tree for illustrative network.[17]

decisions in the new family of nodes emanating from node 2 will occur at time 1. The feasible alternatives are: Schedule activity 3 alone (node 5) or schedule activity 4 alone (node 6). The alternative of scheduling both activities 3 and 4 together is not feasible because of resource constraints.

Continuing from node 6 (activity 4 selected to start at time 1), there are three possible alternatives: Schedule activity 3 alone (node 10), schedule activity 5 alone (node 11), or schedule activities 3 and 5 together (node 12). All three new nodes have the same "decision time" of day 5, since activity 4 is complete at time 5.

If node 12 is selected to continue from, it can be seen that both activities 3 and 5 will be completed simultaneously at the end of day 9. At this point the only possible new decision is to schedule the final (dummy) activity 6. Thus node 23, a terminal node, is a complete schedule since all project activities have been scheduled. Node 23 is, in fact, the optimal schedule; its completion time is 9 days. The Gantt chart for node 23 is shown in Figure 7-21 along with all nonoptimal complete schedules (terminal nodes 20, 21, 22, 24, and 25).

The tree of solutions created for this sample problem can be seen to be ex-

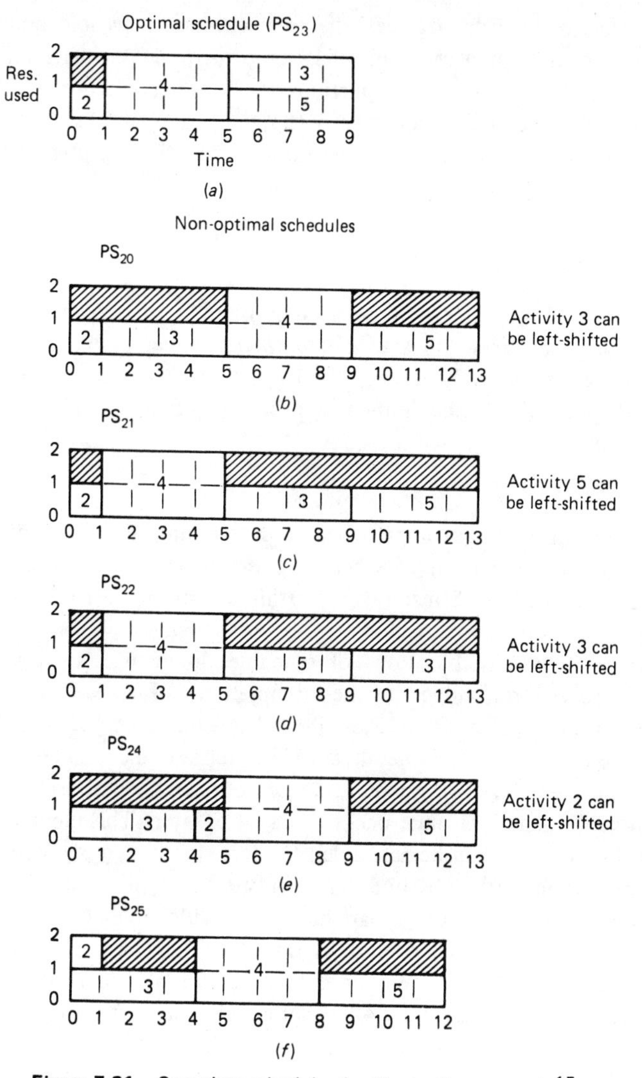

**Figure 7-21** Complete schedules for illustrative network.[17]

haustive, and feasible with respect to both precedence and resource constraints. It is also nonredundant since no two complete schedules are exactly alike. It offers a framework for evaluating all possible feasible schedules. However, following this process is quite impractical for even modest-sized problems, because the number of complete schedules rapidly becomes extremely large as the number of activities increases. The framework offered is thus useful only to the extent that portions of the solutions tree can be pruned away during the tree creation process, in order to keep the tree within manageable bounds. Such pruning is allowable, however, only if it can be proved conclusively that further branching from the pruned-away portions of the tree could not possibly lead to a complete schedule which would be better than some other complete schedule that already exists or that can be developed from the nonpruned portions of the tree.

## PRUNING THE TREE

Pruning the solutions tree can be accomplished with two basically different categories of procedures: (1) *schedule dominance* and (2) *lower bounds*. Schedule dominance is relatively more powerful, but operationally more difficult to implement. We will describe dominance procedures first.

### Schedule Dominance

There are several different approaches to determining schedule dominance; all of them involve some form of schedule comparison. The approach described here was developed by Stinson[16] and utilizes some proofs of Schrage[15] and Johnson.[11]

To grasp the essential notions of schedule dominance, look again at the branching process from node 1 of the example tree in Figure 7-20. In branching from node 1 to create nodes 2, 3, and 4, all possible combinations of activities 2 and 3 were scheduled. Note that node 3 represented scheduling activity 3 alone. Since activity 3 has a duration of four days, all scheduling decisions "downstream" of node 3 must occur at day 4 or later. This includes activity 2, which technologically can be scheduled to start simultaneously with activity 3, is resource-feasible with activity 3, but is only 1 day in duration. Thus, irrespective of when activity 2 later appears in any schedule emanating from node 3, it can always be left-shifted to start at time 0. The Gantt chart for node 24 (shown in Figure 7-21), the only complete schedule downstream from node 3, shows activity 2 starting at day 4. If activity 2 were left-shifted, the schedule could be shortened by one day and hence replicate the complete schedule of node 25. Consequently it can be said that the partial schedule of node 3 is *dominated* by another partial schedule (node 4) and node 3 could have been eliminated from consideration in expanding the tree (thereby eliminating nodes 7, 13, 18, and 24). Nodes 10 and 11 could also have been pruned in similar fashion.

Both nodes 2 and 3 in the sample tree involve scheduling only one activity while both simultaneously are feasible. Yet node 3 is dominated while node 2 is not (clearly, since the optimal path emanates from it).

While node 2 obviously cannot be dominated directly, any immediate successor which incorporates activity 3 into the schedule *may* be dominated. Here node 5 is dominated because activity 3 can be left-shifted one day in the complete schedule shown downstream as node 21, making it also equivalent to the complete schedule of node 25. Thus node 5 is also dominated by node 4, the parent of node 25.

At this point in the pruning process the surviving portions of the tree are shown in Figure 7-22. Only one path other than the optimal path remains. This path cannot be eliminated by the type of schedule dominance illustrated above, but may be pruned by use of a lower bound.

### Lower Bound Pruning

Note that at any point in the project scheduling (i.e., for any partial schedule) a minimum length schedule for the remaining (unscheduled) activities can be calculated *ignoring possible resource conflicts.* In other words, there is an inescapable duration of the remaining work implied soley by the duration and precedence requirements of the remaining activities. The completion time of this path constitutes a lower bound on completion time of any partial schedule emanating from this partial schedule. The authors of Reference 17 call this a *precedence-based lower bound* (LBP). If it is not less than the completion time for a known complete schedule (called the current upper bound), the partial schedule may be pruned.

Another type of lower bound can be calculated by ignoring precedence con-

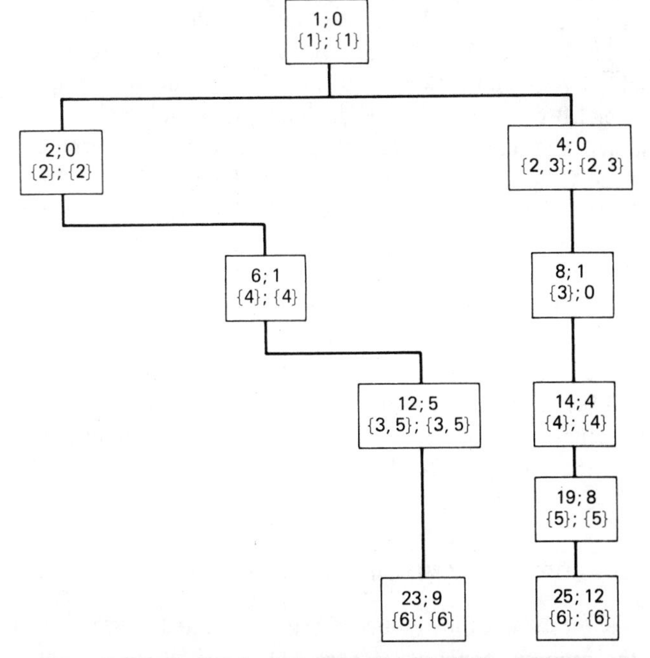


**Figure 7-22** Reduced solutions tree for illustrative problem.

straints and looking at resource requirements. Any project has a fixed work content of each resource type, equal to the sum over all activities of the daily resource requirement times activity duration. With a known limit on resource availability in each time period, a *resource-based lower bound* (LBR) is clearly implied. For example, if the project in question requires 100 man-days of work of some resource type and 7 men of that type are available each day, then before the project starts we know that its duration must equal or exceed 15 days ($\frac{100}{7}$ = 14.3 ≈ 15). Similarly, suppose some partial schedule with a decision time of 5 days is being considered, with a total of 75 man-days of unscheduled work remaining. If there is an existing possible solution of 15 days and the resource limit is 7 men per day, then in the 10 days between and including day 5 and day 14 (i.e., one day less than the existing schedule) there are only 70 man-days available. Thus this partial schedule could not possibly be part of a shorter complete solution and could be eliminated.

In practice both the precedence-based LB and the resource-based LB can be computationally implemented much more efficiently than the above descriptions suggest. For example, a more convenient way of using critical path information is to calculate a late start time (*LST*) for each activity, based on the best complete schedule to date. Any partial schedule which leaves an activity in the set of remaining unscheduled activities on a date greater than or equal to its *LST* cannot lead to an improved solution.

The two LB's mentioned above have been found through computational experience to be relatively weak because they consider *either* resource constraints and ignore precedence constraints, or vice versa. Use of these two LB's will not, for example, permit elimination of node 4 in our sample problem.

A stronger LB can be developed which involves simultaneous consideration of both resource and precedence constraints. This bound, termed a *critical sequence lower bound* (LBC) is described in References 16 and 17. In practice, all three lower bounds (i.e., LBC, LBR, and LBP) are calculated and the largest of the three taken as the lower bound for the partial schedule.

Figure 7-23 shows a reduced solutions tree for the familiar 2-resource problem of figure 7-4. The initial lower bound is 19 (LBR based on resource 1: $\frac{111}{6}$ = 18.5 ≈ 19); however, a lower bound of 22, based on the heuristic solution of Figure 7-12, can also be used. Cumulative resource idleness (used in branching, as described below) is indicated for each partial schedule. The partial schedule alternatives eliminated by dominance or lower bound pruning are indicated. As can be seen, the optimum schedule duration is 20 days. Figure 7-24 shows the resulting bar-chart schedule.

## TREE DEVELOPMENT AND BRANCHING RULES

The process of solving the resouce-constrained scheduling problem with B&B involves the primary sequential steps of (1) branching, or building up the tree,

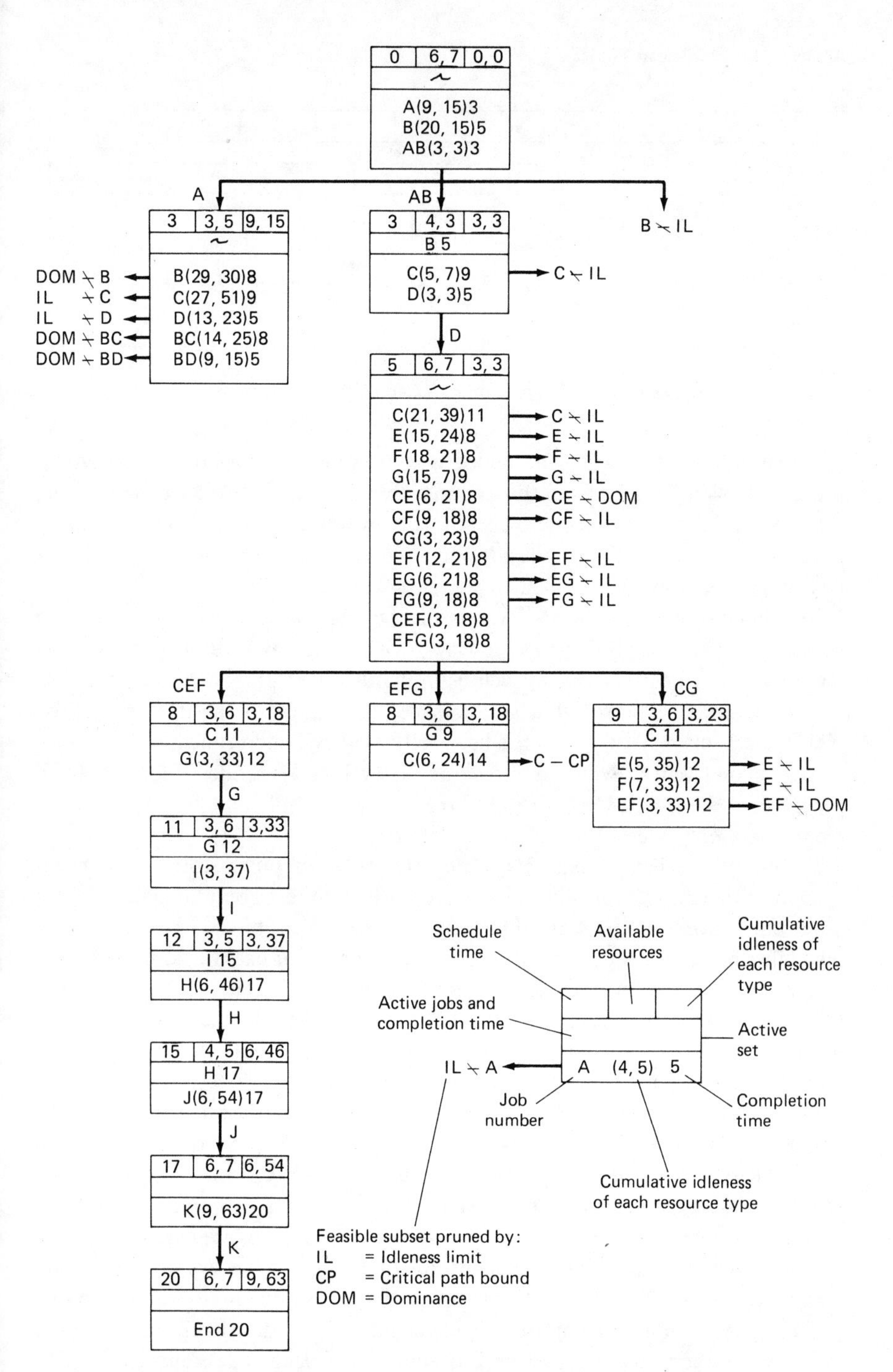


**Figure 7-23** Reduced solutions tree for network of Figure 7-4 (see legend, next page).

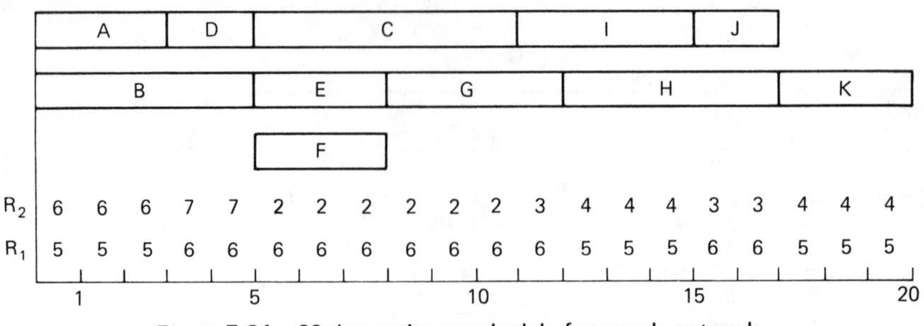


**Figure 7-24** 20-day optimum schedule for sample network.

(2) bounding, or evaluating the nodes, and (3) pruning nonoptimal portions of the tree. As can be seen from the example above, the more effective the pruning, the more efficient the procedure becomes. However pruning is also enhanced by the manner in which the tree is allowed to grow, i.e., the manner in which the next unpruned node from which to branch is selected.

The node selection rule used for branching can cause the solutions tree to develop in two strongly contrasting ways. One such way is termed "backtracking." In this scheme the tree develops in depth very rapidly in a restricted area until a complete new solution is obtained; lateral areas of the tree are not explored until this restricted area has been fully explored and pruned.

The second way in which the tree can be developed is termed "skiptracking." In this case the tree tends to grow in breadth more rapidly and downward movement is in more or less uniform fashion laterally.

Skiptracking schemes require significantly more computer storage than backtracking procedures, but generally can be solved in less computer time. More importantly for the constrained-resource scheduling problem, tree pruning by schedule dominance is possible with skiptracking but not backtracking.

Two node selection rules which result in skiptracking are: Select the node having the least lower bound, or select the node which has the least total accumulated resource idleness up to the current partial schedule time. Use of any one node rule alone, however, is not as effective on very large problems as several rules together; this is because many nodes could have the same lower bounds. A series of partial schedule "attributes," including the three lower bounds described earlier, can be grouped into a "decision vector," i.e., a series of tie-breaking rules, for selecting the next node from which to branch. If no ties exist for the first attribute in the vector, the other attributes are ignored.

B&B procedures represent perhaps the most promising avenue of mathematically rigorous alternatives for attacking the constrained-resource scheduling problem. As computer memories have expanded along with significant decreases in the cost of computer time, these procedures have become less unreasonable to consider, at least for some small-medium sized problems.

# 8

# TIME-COST TRADE-OFF PROCEDURES

The results of the planning and scheduling stages of the critical path method provide a network plan for the activities making up the project and a set of earliest and latest start and finish times for each activity. In particular, the earliest occurrence time for the network terminal event is the estimated "normal" project duration time, based on "normal" activity time estimates. This state of the overall project planning and control procedure is depicted in box (3) of Figure 1-4 (Chapter 1). The purpose of this chapter is to consider the question raised in the next step, i.e., whether the current plan satisfies time constraints placed on the project, or in general, to consider the relationship between project duration and total project costs.

Time constraints arise in a number of ways. First, the customer might contractually require a scheduled completion time for the project. Then, the original time constraint might change after a project has started, requiring new project planning. These changes arise because of changes in the customer's plans; or, when delays occur in the early stages of a project, the new expected completion time of the project may be too late. The most interesting time constraint application, and the one which was the basis for the development of the "CPM" time-cost trade-off procedure by Kelley and Walker,[10,11] arises when we ask for the project schedule that minimizes *total project costs*, direct plus indirect. This is equivalent to the schedule that just balances the (indirect) marginal value of time saved (in completing the project one time unit earlier) against the (direct) marginal cost of saving it. This situation occurs frequently, for example, in the major overhaul of large systems, such as chemical plants, paper machines, aircraft, etc. Here the value of time saved is very high, and furthermore it is known quite accurately. In this application, the crux of the problem amounts to developing a procedure to find the minimum (marginal) cost of saving time. This assumes, of course, that some jobs can be done more quickly if more resources are allocated to them. The resources may be men, machinery, and/or materials. We will assume that these resources can be measured and estimated, reduced to monetary units, and summarized as a direct cost per unit time.

Thus, the main purpose of this chapter can be stated as the development of a procedure to determine activity schedules *to reduce the project duration time with a minimum increase in the project direct costs, by buying time along the critical path(s) where it can be obtained at least cost.*

This procedure can be applied informally in a very simple manner. For example, consider the network presented by Davis,[4] and shown in Figure 8-1. The critical path is composed of activities 0-1, 1-2, 2-4, and 4-5, with normal duration times of 4, 6, 5, and 7, respectively, giving a project duration of 22 days. The other paths through the network have positive float values indicated along the arrows. Also, the third value inside the parentheses indicates the cost to reduce the duration of the activity by one day. For example, activity 0-1 has a normal duration of 4 days, and it can be reduced to 3 days at a cost of $70. It is easy to see in this example that the cheapest way to compress this project is to add "resources" to activity 1-2 and reduce its duration by 1 or 2 days. The net result in this case would be a project duration of 21 days at an extra cost of $50, or a 20 day duration at an extra cost of $100. The effect of a 2 day reduction in activity 1-2 is noted directly on the network by the changes made in the forward pass computations only. At this point it is interesting to note that there are now three critical paths; 0-2-4-5 and 0-1-4-5 have been added to the original critical path 0-1-2-4-5. Further reductions in the project duration are still possible, but now one must determine the economical way to cut all three paths simultaneously.

Systematic methods of carrying out the above procedure will be taken up in this chapter. First, the original CPM approach of Kelley and Walker,[10,11] based on simple linear time-cost trade-off curves for each activity, will be presented. Then, a heuristic hand computational procedure will be presented, based on these same assumptions. Finally, an extensive treatment of the use of mathematical programming to give exact solutions to this problem is presented in Appendix 8-1.

It is assumed in all of the procedures to be developed in this chapter that unlimited resources are available. If this is not the case, or if personnel must be paid up to some fixed resource requirement level, whether they are needed or not on a particular day, then the procedures described in Chapter 7 on resource allocation may be more appropriate than those described here. Although this

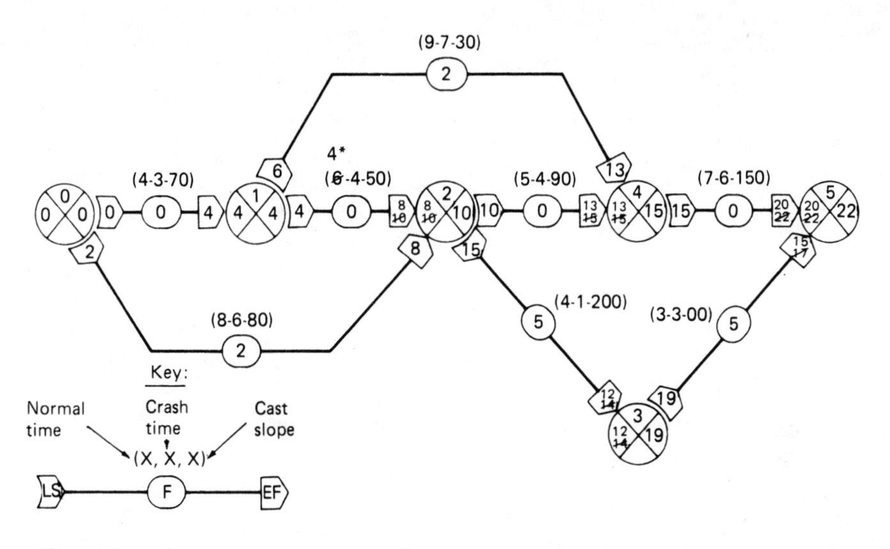


*Changes in the forward pass computations reflect a compression of activity 1-2 from 6 to 4 days.

| Activity | Normal | | "Crash" | | Cost slope |
|---|---|---|---|---|---|
| | Time | Cost | Time | Cost | |
| (0, 1) | 4 days | $210 | 3 days | $280 | $70 |
| (0, 2) | 8 days | 400 | 6 days | 560 | 80 |
| (1, 2) | 6 days | 500 | 4 days | 600 | 50 |
| (1, 4) | 9 days | 540 | 7 days | 600 | 30 |
| (2, 3) | 4 days | 500 | 1 day | 1,100 | 200 |
| (2, 4) | 5 days | 150 | 4 days | 240 | 90 |
| (3, 5) | 3 days | 150 | 3 days | 150 | ** |
| (4, 5) | 7 days | 600 | 6 days | 750 | 150 |
| | | $3,050 | | $4,280 | |

**This activity cannot be expedited.

**Figure 8-1** Example of elementary time-cost trade-off procedure showing compression of project from 22 to 20 days.

assumption of unlimited resources is not often completely satisfied, there are situations where it is satisified to the extent required here. For example, the project in question may be a high priority project that will draw personnel from a large number of low priority or deferrable work activities, so that there are effectively unlimited resources.

## THE CRITICAL PATH METHOD (CPM) OF TIME-COST TRADE-OFFS

The development of the basic CPM time-cost trade-off procedure is based on a number of special terms which are defined below and are further shown in Figure 8-2.

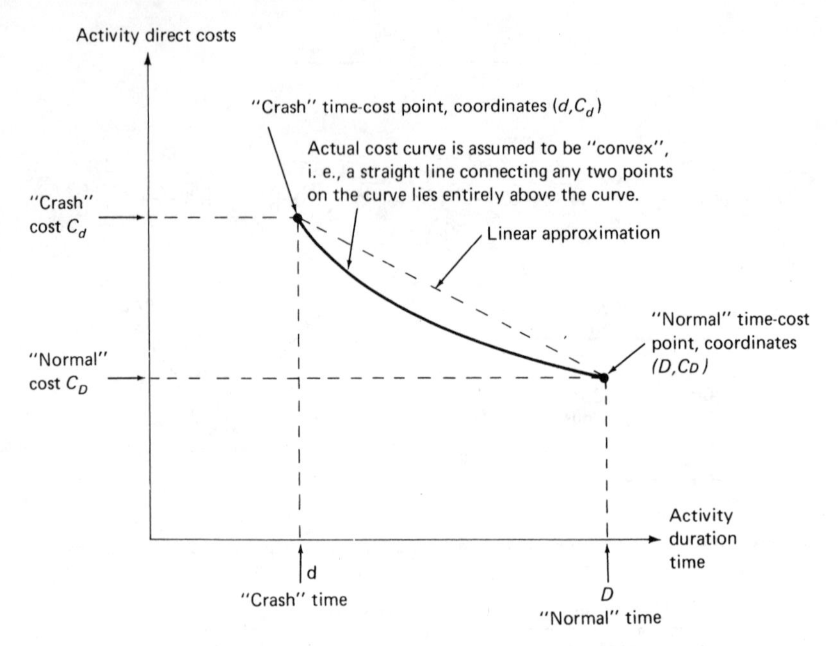


**Figure 8-2** Activity time-cost trade-off input for the CPM procedure.

*Definition:*

*Activity direct costs* include the costs of the material, equipment, and direct labor required to perform the activity in question. If the activity is being performed in its entirety by a subcontractor, then the activity direct cost is equal to the price of the subcontract, plus any fee that may be added.

*Definition:*

*Project indirect costs* may include, in addition to supervision and other customary overhead costs, the interest charges on the cumulative project investment, penalty costs for completing the project after a specified date, and bonuses for early project completion.

*Definition:*

*Normal activity time-cost point.* The normal activity cost is equal to the minimum of direct costs required to perform the activity, and the corresponding activity duration is called the normal time. (It is this normal time that is used in the basic critical path planning and scheduling, and the normal cost is the one usually supplied if the activity is being subcontracted.) The normal time is actually the shortest time required to perform the activity under the minimum direct cost constraint, which rules out the use of overtime labor or special time saving (but more costly) materials or equipment.

*Definition:*

*Crash activity time-cost point.* The crash time is the fully expedited or minimum activity duration time that is technically possible, and the crash cost is assumed to be the minimum direct cost required to achieve the crash performance time.

The normal and crash time-cost points are denoted by the coordinates $(D, C_D)$ and $(d, C_d)$, respectively, in Figure 8-2. For the present, it will be assumed that the resources are infinitely divisible, so that all times between $d$ and $D$ are feasible, and the time-cost relationship is given by the solid line. It will also be assumed that this curve is convex, (defined below), and can be adequately approximated by the dashed straight line. These assumptions are relaxed in the treatment of this problem given in Appendix 8-1.

## ACTIVITY TIME-COST TRADE-OFF INPUTS FOR THE CPM PROCEDURE

The basic activity inputs to the CPM procedure have been illustrated in Figure 8-2, where it is assumed that the time-cost trade-off points lie on a *continuous linear or piece-wise linear* decreasing curve. A piece-wise linear curve is illustrated in Figure 8-3. It is further assumed that the activities are independent, in the sense that buying time on one activity does not affect in any way the availability, cost, or need to buy time on some other activity. This assumption would, for example, be violated if a special resource could be obtained to speed up simultaneously two separate activities in the network, since this would mean that buying time on one activity would automatically include the other.

If the *actual* time-cost relationship departs significantly from the assumed straight line, but is "convex," then it may be necessary to fit the actual cost curve with a series of straight lines, as shown in Figure 8-3. A convex curve is one for which a straight line connecting any two points on the curve lies entirely above the curve; similarly, if the straight line lies entirely below the curve, it is said to be "concave." These definitions are illustrated in Figure 8-4; the need for the convexity assumption will be described below.

In Figure 8-3, the actual cost curve, which is convex, is approximated by a piece-wise linear curve, each piece being treated as a separate activity or pseudo-activity. In the project network, the actual activity, $A$, is replaced by the pseudo-activities, $A_1$, $A_2$, and $A_3$ drawn in series as shown in Figure 8-3. In this illustration, the approximation is with three pieces; however, the procedure can be

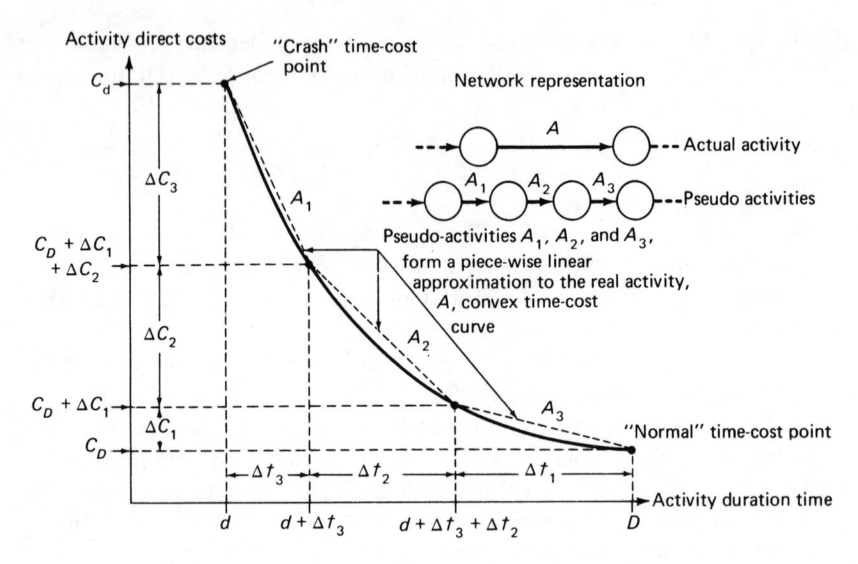


|Cost slope| = ("crash" cost - "normal" cost) / ("normal" time - "crash" time)

| Pseudo-activities | "Normal" | | "Crash" | | Time-cost slope |
|---|---|---|---|---|---|
| | Time | Cost,$ | Time | Cost,$ | |
| $A_1$ | $d + \Delta t_3$ | 0 | $d$ | $\Delta C_3$ | $(\Delta C_3/\Delta t_3)$ |
| $A_2$ | $\Delta t_2$ | 0 | 0 | $\Delta C_2$ | $(\Delta C_2/\Delta t_2)$ |
| $A_3$ | $\Delta t_1$ | $C_D$ | 0 | $(C_D + \Delta C_1)$ | $(\Delta C_1/\Delta t_1)$ |
| Total: $A$ | $D =$ | $C_D$ | $d$ | $C_d =$ | $(C_d - C_D)/(D - d)$ |

$D = d + \Delta t_3 + \Delta t_2 + \Delta t_1$ $\qquad C_d = \$ (C_D + \Delta C_1 + \Delta C_2 + \Delta C_3)$

**Figure 8-3** Piece-wise linear approximation to convex time-cost curves using pseudo-activities.

extended in an obvious way to any number of pieces. The coordinates of the normal and crash time-cost points for each pseudo-activity are given in the table at the bottom of the figure, where it can be noted that the sum of the three pseudo-activities, $A_1$, $A_2$, and $A_3$, gives the whole activity, $A$, and the sum of the coordinates of the normal and crash points for the pseudo-activities gives

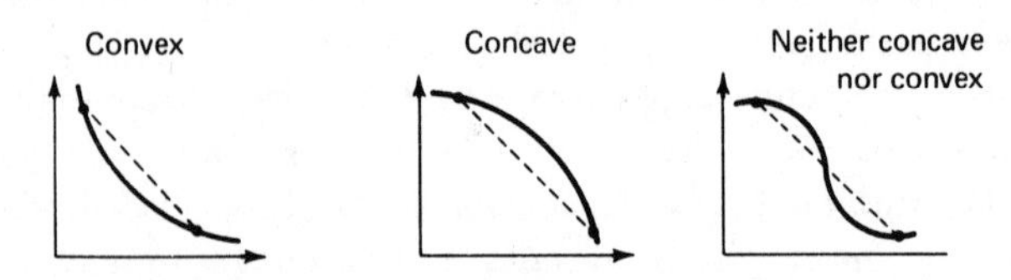


**Figure 8-4** Illustration of convexity and concavity.

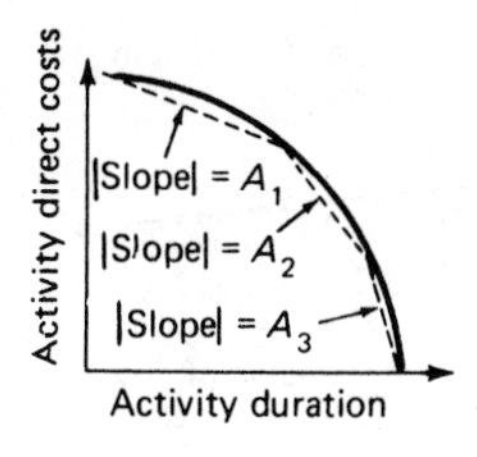


**Figure 8-5** Example of a concave activity time-cost trade-off curve.

the coordinates of the same points for the whole activity, i.e., $(D, C_D)$ and $(d, C_d)$. The reason for the convexity requirement can be explained heuristically in terms of Figure 8-3. If the activity is currently scheduled at its normal time, $D$, then physically, pseudo-activity $A_3$ must be augmented first, then $A_2$, and finally $A_1$. Since the CPM computational procedure effectively searches the critical activities to find the one that can be augmented the cheapest, it will naturally choose the pseudo-activities in the proper order, i.e., $A_3$, $A_2$, and finally $A_1$, since the cost slopes increase as one goes from $A_3$ to $A_2$ to $A_1$, for any *convex* curve. However, if the time-cost curve was not convex, then the cost slopes may be lowest for $A_1$, and highest for $A_3$, as shown in Figure 8-5. In this case, the CPM computational procedure would augment the activities in a sequence that would not be physically meaningful, i.e., in the order $A_1$, $A_2$, and finally $A_3$.

### The CPM Computational Procedure

The CPM computational procedure chooses the duration times for each activity so as to minimize the total project direct costs and at the same time satisfy the constraints on the total project completion time and on the individual activities, the latter being dictated by both the logic of the project network and the performance time intervals $(d, D)$ established for each activity.

For example, again consider the simple project presented in Figure 8-1. Applying the "CPM" procedure, we would start by setting all activity durations at their "normal" value. This gives a project duration of 22 days as determined by the critical path consisting of activities 0-1, 1-2, 2-4, and 4-5. The associated total direct cost of project performance is $3050, as indicated in Figure 8-6. Note that this cost could be increased to $3,870 through unintelligent decision making by "crashing" all activities not on the critical path, with no decrease in project duration. Between these upper and lower cost values for a project duration of 22 days there are several other possible values, depending upon the number of noncritical activities crashed.

If all activity durations are set at "crash" values, the project duration can be decreased to 17 days, with a total cost of $4,280, as shown by the extreme upper left point of Figure 8-6. This corresponds to the practice of speeding up a

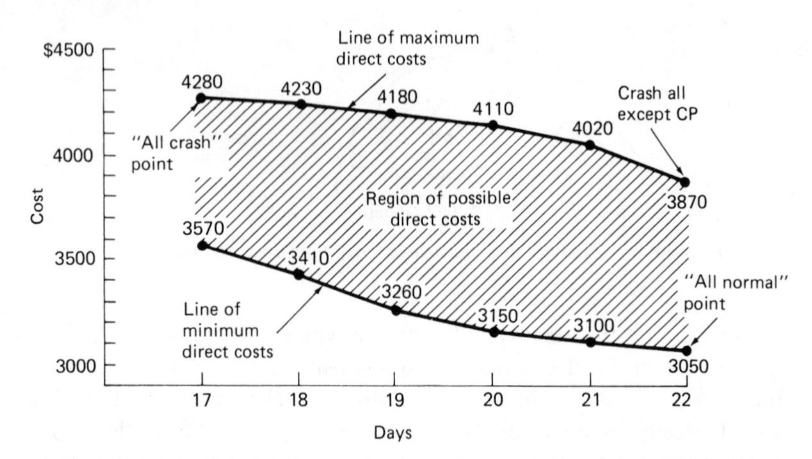


**Figure 8-6** Project duration vs. direct cost for sample network in Figure 8-1.

project by requiring crash times *across the board* (everybody works overtime), instead of along the critical paths only. In place of this, a duration of 17 days could also be achieved at lower cost by not "crashing" activities unnecessarily. Thus, activity 0-2 can be set at 7 instead of 6, activity 1-4 at 8 instead of 7, and activity 2-3 at 4 instead of 1. With all other activities set at crash values, the associated cost of performance for 17-day project duration is reduced to $3,570. This value is the *lowest possible* value for 17-day project duration, as will be shown below. It is the optimal "CPM" solution for a 17-day schedule.

Between 22 and 17 days there are several possible values of project duration, as shown in Figure 8-6. For each such duration there is a range of possible cost values depending upon the durations of individual activities and whether activities are crashed unnecessarily or not. Figure 8-6 shows the curve of both maximum and minimum costs and the region of possible costs for each duration between these curves.

In this simple example the minimum direct cost curve is easily determined by trial and error, or by the hand application of a heuristic procedure such as that given below. But in more realistic cases consisting of several dozen or more activities, such trial and error determination becomes extremely tedious, if not impossible. In these situations, systematic computation schemes, usually mathematical programming, must be used to determine the value of the minimum cost curve for every possible level of project duration. The details of these procedures are discussed in the appendix to this chapter, while their application is illustrated in Figure 8-7.

Suppose the project duration is under management control, such as in the case in a plant overhaul. Assuming adequate resources are available, the rational choice of a project schedule would be made so as to minimize the total project costs. Then, one would select the project duration schedule indicated in Figure

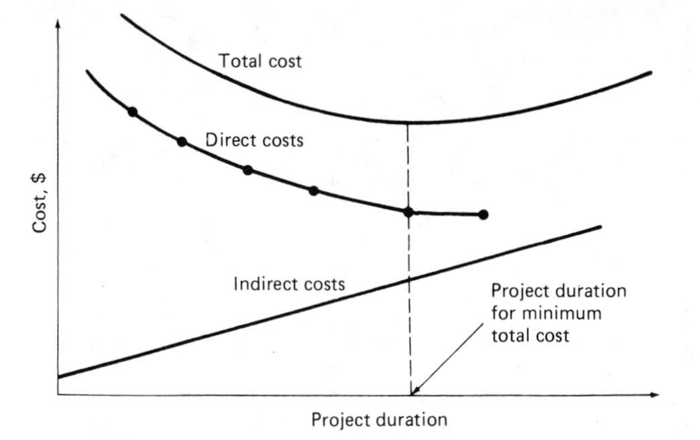


**Figure 8-7** Determining project schedule for minimum total cost.

8-7, corresponding to the minimum value of the total cost curve. The critical ingredient of this process is the minimum direct cost curve, which was not really available prior to the inception of this "CPM" optimization procedure.

## HAND COMPUTATIONAL PROCEDURE

The introduction of this chapter included a discussion of the hand compression of the project shown in Figure 8-1. The activity to be shortened was merely selected by visual inspection of the network. In this case the most economical alternative was quite obvious. However, as the compression process continues, the number of critical paths increases, and it very quickly becomes difficult to evaluate all of the possible alternatives. For this reason, it is useful to adopt some procedure which can reasonably be followed by hand. While these heuristic hand procedures do not always lead to optimal solutions, they usually come quite close.

The 8 step hand procedure presented below is a slight modification of the method developed by Siemens.[15] The key element of this procedure is the cost slope and time available, illustrated in Figure 8-3. The cost slope will be denoted by $C_{ij}$ for an arbitrary activity $(i - j)$.

$$\text{Cost Slope} = C_{ij} = (C_d - C_D)_{ij}/(D - d)_{ij}$$

$$\text{Time Available} = TA_{ij} = (D - d)_{ij}$$

An "effective" cost slope, $EC_{ij}$, is then defined by Siemens as the cost slope divided by the number of inadequately shortened paths, $N_{ij}$, which contain activity $(i - j)$.

$$EC_{ij} = C_{ij}/N_{ij}$$

| Activity | Paths requiring reduction | | | Cost slope | Time Reduction Available | Iteration | | | | | | |
|---|---|---|---|---|---|---|---|---|---|---|---|---|
| | 0–1–2–4–5 | 0–1–4–5 | 0–2–4–5 | | | 1 | 2 | 3 | 4 | 5 | 6 | |
| 0–1 | 1 | 1 | | 70 | ~~1~~ 0 | 35 | X | X | X | X | X | Effective cost slopes $EC_{ij}$ |
| 0–2 | | | 1 | 80 | ~~2~~ 1 | 80 | 80 | 80 | 80 | 80 | 80 | |
| 1–2 | 2 | | | 50 | ~~2~~ 0 | 50 | 50 | 50 | 50 | X | X | |
| 1–4 | | 1 | | 30 | ~~2~~ 1 | 30 | 30 | 30 | 30 | 30 | X | |
| 2–4 | 1 | | 1 | 90 | ~~1~~ 0 | 45 | 45 | X | X | X | X | |
| 4–5 | 1 | 1 | 1 | 150 | ~~1~~ 0 | 50 | 50 | 50 | X | X | X | |

| Initial path length | 22 | 20 | 20 | Iteration | Action | Iteration cost | Cumulative cost |
|---|---|---|---|---|---|---|---|
| | 5 | 3 | 3 | 0 | – | – | 3050 |
| Remaining time reduction required (17 day project duration) | 4 | 2 | ↓ | 1 | Cut 0–1 by 1 day | 70 | 3120 |
| | 3 | ↓ | 2 | 2 | Cut 2–4 by 1 day | 90 | 3210 |
| | 2 | 1 | 1 | 3 | Cut 4–5 by 1 day | 150 | 3360 |
| | 0 | ↓ | | 4 | Cut 1–2 by 2 days | 100 | 3460 |
| | ↓ | 0 | ↓ | 5 | Cut 1–4 by 1 day | 30 | 3490 |
| | ↓ | ↓ | 0 | 6 | Cut 0–2 by 1 day | 80 | 3570 |

**Figure 8-8** Application of modified Seimens algorithm to the network problem shown in Figure 8-1, for the compression of the project duration from 22 to 17 days.

The procedure described below chooses from among all available activities to be shortened, the one with the lowest *effective* cost slope. This heuristic will usually come quite close to the optimal solution. Each step of this procedure is illustrated in Figure 8-8.

### Modified Siemens Algorithm

1. Prepare the project network and time estimates, and list in columns all paths through the network whose expected lengths are *greater* than the desired (scheduled) project duration, $T_s$. The length of a path is merely the sum of the durations of all activities on the path in question. Also note at the bottom of each path column (row marked iteration 0), the time reduction that is required, i.e., expected path length minus $T_s$.*

---

*A systematic procedure for finding the $K$ shortest paths through a network is given by Yen[16]. A useful alternative procedure would be to modify a slack sort of the project activities so that each slack path extends from the initial to the terminal network event.

2. List (in rows) all activities present in at least one of the listed paths noting for each activity its cost slope, $C_{ij}$, and time reduction available, $TA_{ij}$.*
3. Compute the effective cost slopes, $EC_{ij}$, and record them in the column headed iteration 1.
4. For the path(s) with the most remaining time reduction required, select the activity with the lowest effective cost slope. Break ties by considering the following ordered list:
   4.1. Give preference to the activity which lies on the greatest number of inadequately shortened paths.
   4.2. Give preference to the activity which permits the greatest amount of shortening.
   4.3. Choose an activity at random.
5. Shorten the selected activity $(i - j)$ as much as possible, which will be equal to the minimum of the following:
   5.1. the unallocated time remaining for the selected activity $(i - j)$, or
   5.2. the smallest demand of those inadequately shortened paths containing the activity $(i - j)$.
6. Sell back, or deshorten, as much time as possible on paths that have been overcut, as long as this action does not cause any new paths to become inadequately shortened.
7. Stop if all paths have been adequately shortened. If not, recalculate those effective cost-slopes where any of the following have occurred:
   7.1. a path which was inadequately shortened prior to this iteration, has been adequately shortened, or
   7.2. all unallocated time for the activity just shortened has been consumed and there are one or more additional cost-slope/supply pairs for this activity (see the footnote to Step 2).
8. Return to Step 4.

### Example Problem

The sample problem shown in Figure 8-1 will be used to illustrate the above algorithm. The results of Steps 1, 2, and 3 are shown in Figure 8-8 for the case where a 17-day project duration is required. There are 5 paths through the network, but only the 3 listed in Figure 8-8 exceed the desired project duration of 17 days. Also, activities 2-3 and 3-5 have been omitted because

*If a piecewise linear approximation is being used, as in Figure 8-3, then the data pairs ($C_{ij}$, $TA_{ij}$) must be given for each linear segment, with increasing cost slopes in columns going from left to right. Only one cost-slope/supply pair for each activity can be considered during any iteration of the following recurring steps. Also, these pairs are considered sequentially left-to-right as they are used up in the shortening process, and from right-to-left if deshortening occurs.

they are not present in any of the 3 paths which require shortening. The cross-hatching in Figure 8-8 is used to denote the activities that are *not* present in the path indicated at the head of the column.

The repetitive portion of this algorithm then begins by applying step 4, which leads to cutting activity 0-1 by 1 day in the first iteration. This activity was chosen because it has the lowest *effective* cost slope of the 4 activities on path 0-1-2-4-5, which is the path currently requiring the greatest amount of shortening. This change is shown by placing a 1 in the two cells corresponding to the row for activity 0-1 and the two path columns which contain this activity; the "Time Reduction Available" for this activity is updated from 1 to 0; and, finally, the "Remaining Time Reduction Required" is updated in the row for iteration 1, to reflect that paths 0-1-2-4-5 and 0-1-4-5 have been reduced from 5 and 3 to 4 and 2, respectively. A total of 6 iterations are required to reach a final solution, which requires an augmentation of the direct activity costs of \$520, for a total of \$3570. It should also be noted in Figure 8-8, that 17 days is the minimum or "crash" project duration, because all activities on the first path (0-1-2-4-5) have zero "Time Reduction Available."

A summary of repeated applications of this algorithm to cover the entire range of possible project compressions is shown in Table 8-1. The results shown here essentially give the project schedule, activity by activity, for a given project duration. It can be shown that each of these solutions is "optimal," in the sense that no other solution will give lower total direct costs. This will not always occur with this heuristic, based on lowest *effective* cost slope, but solutions obtained this way will usually be very good. These results were plotted in Figure 8-6 above, where they form the line of minimum direct costs. As stated before, this is the signal contribution of the "CPM" procedure. It permits the specification of the optimal (total cost minimization) schedule, as shown in Figure 8-7.

## ECONOMIC IMPLICATIONS OF PROJECT TIME-COST TRADE-OFF

The concept of project time-cost trade-off could have considerable economic importance, considering the extremely large and costly systems being developed today, over extended periods of time. For example, facilities for electric power generation, mass transportation, sophisticated weapons systems, oil production, etc. In each of these cases, the system costs are in the billions of dollars, and the development times are in excess of 5 years. Alternative project schedules could have considerably different total costs. An analysis of this problem by Cukierman and Shiffer[3] is given below.

Consider the project duration denoted by $T$, as a variable to be selected by management on a rational basis. Factors to be considered include the following,

**Table 8-1 Summary of Hand Computation Solutions to the Example in Figure 8-1.**

| | *Project Duration (Days)* | | | | | | | | | |
|---|---|---|---|---|---|---|---|---|---|---|
| | *17* | | *18* | | *19* | | *20* | | *22* | |
| *Activity* | *Amount Shortened* | *Cost* | *Amount Shortened* | *Cost* | *Amount Shortened* | *Cost* | *Amount Shortened* | *Cost* | *Amount Shortened* | *Cost* |
| 0–1 | 1 | 70 | 1 | 70 | 1 | 70 | 0 | 0 | 0 | 0 |
| 0–2 | 1 | 80 | 0 | 0 | 0 | 0 | 0 | 0 | 0 | 0 |
| 1–2 | 2 | 100 | 1 | 50 | 1 | 50 | 2 | 100 | 0 | 0 |
| 1–4 | 1 | 30 | 0 | 0 | 0 | 0 | 0 | 0 | 0 | 0 |
| 2–3 | 0 | 0 | 0 | 0 | 0 | 0 | 0 | 0 | 0 | 0 |
| 2–4 | 1 | 90 | 1 | 90 | 1 | 90 | 0 | 0 | 0 | 0 |
| 3–5 | 0 | 0 | 0 | 0 | 0 | 0 | 0 | 0 | 0 | 0 |
| 4–5 | 1 | 150 | 1 | 150 | 0 | 0 | 0 | 0 | 0 | 0 |
| Total Excess Direct Costs | – | 520 | – | 360 | – | 210 | – | 100 | – | 0 |
| Base Direct Costs | – | 3050 | – | 3050 | – | 3050 | – | 3050 | – | 3050 |
| Total Direct Costs | – | 3570 | – | 3410 | – | 3260 | – | 3150 | – | 3050 |

which denote present value or discounted values based on an appropriate rate of return, $r$.

$X(T)$ = *minimal* discounted costs of a project plan with delivery date $T$.

$Z(T)$ = discounted value to the buyer, of the project completed at time $T$.

$V(T, \ldots)$ = discounted value of net payments from buyer to seller, which depend on $T$, as well as on the detailed payment schedule; hence the multivariate notation $(T, \ldots)$ including $T$ and other variables.

Now, the net benefits to the buyer, contractor, and their total to society can be defined, respectively, as follows:

$$B_B = Z(T) - V(T, \ldots)$$

$$B_C = V(T, \ldots) - X(T)$$

$$B_S = B_B + B_C = Z(T) - X(T)$$

If you assume that the functions $Z(T)$ and $X(T)$ reflect, respectively, the projects true present value of benefits and costs to society, then society wants to maximize $B_S$. This implies that the socially optimal completion time, $T_S$, will be determined by the (first-order) condition $dZ/dT = dX/dT$, independently of the $V$ function.* This point $(T_S)$ is noted in Figure 8-9, where the slopes of $Z(T)$ and $X(T)$ are the same. That is, at the point where the marginal benefits equal the marginal costs.

From the social point of view, the $V$ function is only a device for transferring income between the buyer and the seller. But its specification is important to all parties if we are to insure that the contractor has the incentive to also deliver the project at time $T_S$, instead of some other time, such as $T_C$ shown in Figure 8-9. A sufficient condition to achieve the desired result is to have $\partial V/\partial T = dZ/dT$ for all $T$. Cukierman and Shiffer[3] point out that this will not be accomplished by most project payment plans, $V(T, \ldots)$. For example, periodic progress payments proportional to contractor expenditures, or agreed upon payments to be made upon achievement of project milestones, will not achieve the required incentive. A plan that could achieve the desired result, for example, would be payments executed at predetermined dates, independently of the progress of the project, plus an annual fine equal to the discount rate $r$, times

*$dZ/dT$ is the derivative of $Z$ with respect to $T$, that is, the rate of change of $Z$ with respect to $T$.

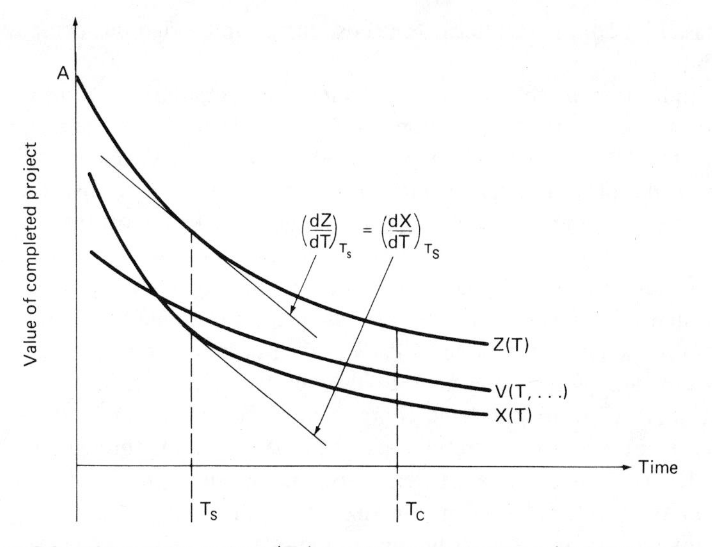


**Figure 8-9** The socially optimal ($T_S$) and the contractor's best ($T_C$) delivery times.

the value to the buyer of the completed project at the time of completion, paid over the life of the project.

This is an interesting, perhaps academic, consideration of an extremely important problem from an economic point of view. Of the three functions in Figure 8-9, $X(T)$ is the most difficult to obtain. The "CPM" time-cost trade-off concept could be very helpful in this endeavor to determine a socially optimal project schedule. It should be added that in the determination of $X(T)$, the availability of resources must also be considered. This could be done, for example, by trial and error simulation. First, assume a trial set of activity time-cost trade-off function inputs, and determine the corresponding "CPM" project schedules for various project durations. The resource level vs. time profiles for these schedules can then be smoothed by the methods in Chapter 7, and finally, this smoothed level can be judged as to whether it is feasible or not from a resource staffing point of view.

It appears that the project time-cost trade-off concept has more to offer than the applications to date indicate. A real understanding of this concept may lead to improved solutions to large scale systems problems currently facing society.

## SUMMARY

The general philosophy of the time-cost trade-off problem has been presented in this chapter, along with a hand computation procedure, and the "CPM" procedure based on continuous linear time-cost curves and a rigorous computational algorithm.

An application of the "CPM" procedure in a computerized project management system is described by Benson and Sewall.[1] Their system, called PLANIT, includes the "CPM" procedure as a subroutine, which they call FASNET. It is applied in the project planning stage to determine project time-cost alternatives, and also in the project control stage, as needed to make up lost time. It should be added that this program also incorporates input constraints on resource availability as described in Chapter 7. It represents one of the few project management systems that incorporate both of these project planning tools.

Embellishments of the basic "CPM" problem are presented in the appendix to this chapter. First, extensions to the objective function include the work of Elmaghraby,[5] who considered completion schedules, not only on the project end event but also on an arbitrary set of milestone events throughout the network. He has developed an efficient network flow algorithm to determine the project/activity schedule which minimizes the sum of the total activity direct costs plus the penalties for tardy completion of all of the milestone events. Another extension of this type was by Moore.[13] He used goal programming to consider multiple objectives, such as completion times for milestone events, operation within a specified budget, etc.

In addition to these embellishments of the objective function, a number of authors have devised ways of handling more complex activity time-cost tradeoff relationships. These include concave curves in place of the usual convex assumption. Also, time-cost relationships consisting of a set of feasible time-cost points, or a combination of feasible points and continuous curves are described in the appendix to this chapter.

## REFERENCES

1. Bensen, L. A. and R. F. Sewall, "Dynamic Crashing Keeps Projects Moving," *Computer Decisions*, Feb. 1972, pp. 14–18.
2. Charnes, A. and W. W. Cooper, "A Network Interpretation and a Directed Subdual Algorithm for Critical Path Scheduling," *Journal of Industrial Engineering*, Vol. 13, No. 4 (1962), pp. 213–218.
3. Cukierman, A. and Z. F. Shiffer, "Contracting for Optimal Delivery Time in Long-Term Projects," *The Bell Journal of Economics*, Vol. 7, No. 1, Spring 1976, pp. 132–149.
4. Davis, E. W., "Networks: Resource Allocation," *Industrial Engineering*, Vol. 6, No. 4, April 1974, pp. 117–120.
5. Elmaghraby, S. E. and P. S. Pulat, "Optimal Project Compression with Due-Dated Events," *Naval Research Logistics Quarterly*, Vol. 26, No. 2, June 1979, pp. 331–348.
6. Elmaghraby, S. E., *Activity Networks: Project Planning and Control by Network Models*, John Wiley & Sons, New York, 1977.

7. Fondahl, J. W., "Can Contractors Own Personnel Apply CPM Without Computers," *The Constructor*, November 1961, pp. 56–60, and December 1961, pp. 30–35.
8. Fulkerson, D. R., "A Network Flow Computation for Project Cost Curves," *Management Science*, Vol. 7, No. 2, January 1961, pp. 167–179.
9. Hindelang, T. J. and J. F. Muth, "A Dynamic Programming Algorithm for Decision CPM Networks," *Operations Research*, Vol. 27, No. 2, March–April 1979, pp. 225–241.
10. Kelley, J. E. and M. R. Walker, *Critical Path Planning and Scheduling*, Proceedings of the Eastern Joint Computer Conference, December 1959, pp. 160–173.
11. Kelley, J. E., Jr., "Critical Path Planning and Scheduling: Mathematical Basis," *Operations Research*, Vol. 9, No. 3 (1961), pp. 296–320.
12. Meyer, W. L. and L. R. Shaffer, *Extensions of the Critical Path Method Through the Application of Integer Programming*, Report issued by the Dept. of Civil Engineering, University of Illinois, Urbana, Ill., July 1963.
13. Moore, L. J., B. W. Taylor III, E. R. Clayton, and S. M. Lee, "Analysis of a Multi-Criteria Project Crashing Model," *American Institute of Industrial Engineering Trans.*, Vol. 10, No. 2, June 1978, pp. 163–169.
14. Panagiotakopoulos, D., "A CPM Time-Cost Computational Algorithm for Arbitrary Activity Cost Functions," *INFOR*, Vol. 15, No. 2, June 1977, pp. 183–195.
15. Siemens, N., "A Simple CPM Time-Cost Trade-off Algorithm," *Management Science*, Vol. 17, No. 6, Feb. 1971, pp. B-354-363.
16. Yen, J. Y., "Finding the K Shortest Loopless Paths in a Network," *Management Science*, Vol. 17, No. 11, July 1971, pp. 712–716.


## EXERCISES

1. Given below are the network data and the time-cost trade-off data for a small maintenance project.

**Table 8-2**

| *Job (Activity)* | *Predecessor Jobs* | *Normal (days)* | *Normal (dollars)* | *Crash (days)* | *Cost Slope (dollars/day)* |
|---|---|---|---|---|---|
| *A* | none | 3 | 50 | 2 | 50 |
| *B* | none | 6 | 140 | 4 | 60 |
| *C* | none | 2 | 50 | 1 | 30 |
| *D* | *A* | 5 | 100 | 3 | 40 |
| *E* | *C* | 2 | 55 | 2 | — |
| *F* | *A* | 7 | 115 | 5 | 30 |
| *G* | *B, D* | 4 | 100 | 2 | 70 |
| Total | | | 610 | | |

ndirect costs, including the cost of lost production, and asso-
sts, supervision, etc., to be as follows.

| (days) | 12 | 11 | 10 | 9 | 8 | 7 |
|---|---|---|---|---|---|---|
| ollars) | 900 | 820 | 740 | 700 | 660 | 620 |

**Table 8-3**

| Activity No. | Predecessor Activity Nos. | Description of Activity | Time (hours) a | m | b | Cost (dollars) Normal | Crash |
|---|---|---|---|---|---|---|---|
| 101 | — | inspect & measure pipe | 2 | 4 | 5 | 16 | 22 |
| 102 | — | devlp. cal. mtls. list | 3 | 6 | 8 | 18 | 25 |
| 103 | 101 | make drawings of pipe | 2 | 3 | 5 | 12 | 18 |
| 104 | — | deactivate line | 7 | 8 | 10 | 8 | 14 |
| 105 | 102 | procure calender parts | 120 | 244 | 320 | 12 | 35 |
| 106 | 102 | assemble calender work crew | 6 | 8 | 9 | 20 | 30 |
| 107 | 103 | devlp. matl. list (pipe) | 3 | 4 | 7 | 10 | 13 |
| 108 | 104, 105, 106 | deactivate calender | 4 | 4 | 5 | 3 | 3 |
| 109 | 107 | procure valves | 136 | 220 | 280 | 10 | 20 |
| 110 | 107 | procure pipe | 136 | 200 | 240 | 10 | 22 |
| 111 | 107 | assemble work crew (pipe) | 4 | 6 | 7 | 16 | 20 |
| 112 | 108 | tie off warps | 1 | 2 | 3 | 3 | 8 |
| 113 | 110, 111 | prefab pipe sections | 20 | 40 | 50 | 120 | 240 |
| 114 | 111 | erect scaffold | 6 | 12 | 15 | 30 | 65 |
| 115 | 112 | disassemble calender | 4 | 10 | 14 | 90 | 210 |
| 116 | 112 | empty & scour vats | 2 | 3 | 5 | 6 | 9 |
| 117 | 104, 114 | remove old pipe | 18 | 30 | 38 | 180 | 300 |
| 118 | 115 | repair calender | 35 | 70 | 98 | 650 | 1500 |
| 119 | 113, 117 | position new pipe | 6 | 8 | 12 | 50 | 110 |
| 120 | 118 | lubricate calender | 3 | 5 | 6 | 10 | 22 |
| 121 | 109, 119 | position new valves | 5 | 7 | 10 | 66 | 100 |
| 122 | 119 | weld new pipe | 6 | 8 | 11 | 50 | 60 |
| 123 | 120 | reassemble calender | 18 | 22 | 24 | 200 | 270 |
| 124 | 123 | adjust & balance calender | 6 | 8 | 14 | 80 | 95 |
| 125 | 121, 122 | insulate pipes | 15 | 20 | 30 | 60 | 75 |
| 126 | 121, 122 | connect pipes to boiler | 3 | 4 | 5 | 24 | 30 |
| 127 | 121, 122, 123 | connect pipes to calender | 7 | 8 | 11 | 48 | 50 |
| 128 | 116, 124 | refill vats | 1 | 1 | 1 | 2 | 2 |
| 129 | 125, 126 | remove scaffold | 3 | 4 | 4 | 16 | 18 |
| 130 | 126 | pressure test | 5 | 6 | 9 | 15 | 16 |
| 131 | 128 | tie in warps | 3 | 4 | 5 | 8 | 10 |
| 132 | 127, 131 | activate calender | 2 | 2 | 3 | 14 | 14 |
| 133 | 129, 130, 132 | clean up | 3 | 4 | 5 | 15 | 18 |

Assuming any integer times between the normal and crash activity times are feasible, use the hand computational time-cost trade-off procedure to show that the total costs are $1510, 1470, 1430, 1470, 1530 and 1620 for 12, 11, 10, 9, 8, and 7 day project durations, respectively.

2. Given in Table 8-3 are the data for a steam calendar and pipeline maintenance project. Three sets of times are given under columns headed $a$, $m$, and $b$. In this problem only the first two columns will be used: the times under the column headed $a$ are the "crash" activity performance times while those under the column headed $m$ are the "normal" performance times.
   a. Draw the network and make the basic scheduling computations using activities-on-arrows or activities-on-nodes.
   b. Identify the critical path.
   c. Indicate, in step-by-step detail, the activity augmentations required to reduce the project duration to 248 hours while keeping the total project direct costs at a minimum. Assume that only the normal and crash times are feasible.

3.*In the appendix to this chapter, the linear programming formulations are given for activity time-cost trade-off curves that are either continuous, or a collection of feasible time-cost points. These formulations can be extended to cover the case where combinations of these two types of trade-off functions occur. An analysis of this problem indicates that the transitions can occur in three distinct ways as shown in Figure 8-10, i.e., any combinations of continuous curves and discrete points can be handled by appropriate combinations of these three transitions. Write out the linear programming formulation for each of these transitions.

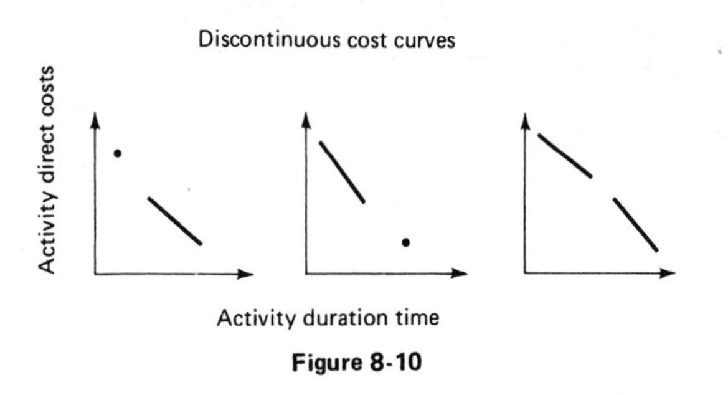


**Figure 8-10**

4.*Consider the simple project network shown in Figure 8-11, along with the accompanying activity time-cost trade-off curves. Using the linear (integer)

*This problem pertains to material taken up in the appendix to this chapter.

programming techniques described in the appendix to this chapter, write the linear programming formulation of this problem, and determine the schedule of activity duration times ($y_{ij}$) which will minimize the total direct project costs for a project duration constraint of $\lambda = 10$ time units.

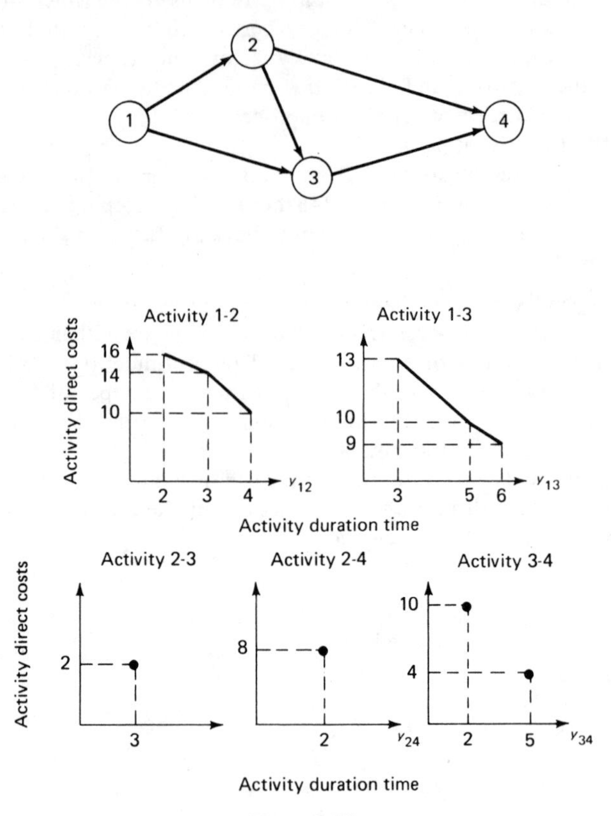


**Figure 8-11**

# APPENDIX 8-1 APPLICATIONS OF MATHEMATICAL PROGRAMMING TO THE DEVELOPMENT OF PROJECT COST CURVES

The purpose of this appendix is to present a review of the current applications of mathematical programming techniques to the problem of generating project cost curves and the various extensions of this basic problem. The treatment given here will assume that the reader is familiar with the basic linear programming formulation of an optimization problem.

First, the basic network scheduling (forward pass) computations will be viewed as a problem in linear programming; this approach was published by Charnes and Cooper.[2] The time-cost trade-off problem will then be introduced along with several logical extensions. Finally, this will be followed by the treatment of more complex activity time-cost trade-off functions, along with solution algorithms.

## LINEAR PROGRAMMING FORMULATION OF BASIC SCHEDULING COMPUTATIONS

The various topics treated in this appendix will be illustrated using the simple network* shown in Figure 8-12. The numbers appearing along each activity in this figure give $(d_{ij}, D_{ij}, C_{ij})$. For example, $(d_{12}, D_{12}, C_{12}) = (1, 3, 3)$ indi-

*By permission from L. R. Ford, Jr., and D. R. Fulkerson, "Flows in Network," The Rand Corporation, 1962. Published, 1962, by Princeton University Press.

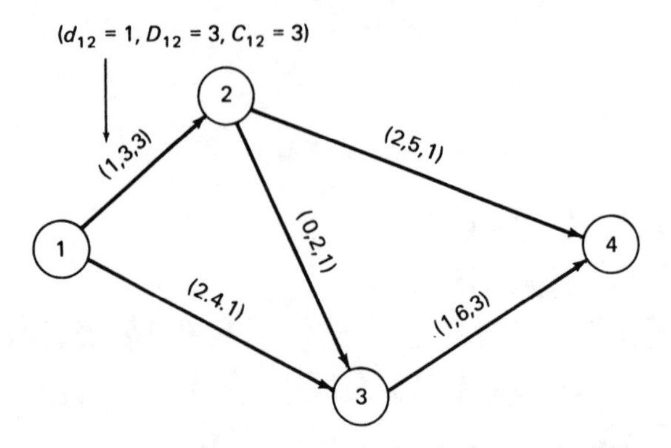


**Figure 8-12** Illustrative network with data giving crash time, normal time, and cost slope for each activity.

cates that for activity 1-2, the crash and normal performance times are 1 and 3 time units, respectively, and the slope of the linear time-cost trade-off curve is 3 monetary units per time unit. In the applications, the time unit is chosen so that the $d_{ij}$'s and $D_{ij}$'s are all integer valued, and the monetary unit is chosen so that the $C_{ij}$'s are also integer valued.

For purposes of illustrating how the basic forward pass computations can be formulated as a linear programming problem, assume that all activities in Figure 8-12 are scheduled to be performed at their normal times, i.e., the middle number of the three numbers given for each activity. Now the project network may be viewed as a *flow network*, in which a hypothetical unit of flow leaves the source, node (event) 1, and enters the sink, node 4. Also, nodes 2 and 3 play the role of "transhipment" points and, thus, at these nodes there must be a *conservation of flow*, i.e., the total flow into node 2 must equal the total flow away from node 2, and similarly for node 3. The performance time of each activity, $y_{ij}$, is then interpreted as the time (or cost) of transporting *a unit of flow* from node $i$ to node $j$. Viewing a project network in this way reduces the problem of finding the critical path(s) to the determination of the network path(s) or route(s) from the source, node 1, to the sink, node 4, which requires a maximum time (or cost) to traverse. The reader is no doubt well aware of the fact that there are easier ways of locating the network critical path. The main purpose of this *network flow* interpretation of the problem is to show that it can be formulated as a linear programming problem, a technique which will prove to be useful later in solving problems which cannot be easily solved in other ways.

Applying the above *network flow* interpretation to Figure 8-12 results in the linear programming formulation given in equation (1), which will be referred to as the *primal* problem. In this formulation, $y_{ij} = 0$ or 1 denotes the absence or presence of a unit flow from node $i$ to node $j$, a flow which is said to be along activity $i$-$j$.

PRIMAL PROBLEM

$$\text{Maximize } f[Y] = 3y_{12} + 4y_{13} + 2y_{23} + 5y_{24} + 6y_{34} \quad (1a)$$

$$\text{Subject to} \quad y_{12} + y_{13} = 1 \quad (1b)$$

$$-y_{12} + y_{23} + y_{24} = 0 \quad (1c)$$

$$-y_{13} - y_{23} + y_{34} = 0 \quad (1d)$$

$$-y_{24} - y_{34} = -1 \quad (1e)$$

$$y_{12} \geqq 0, \quad y_{13} \geqq 0, \quad y_{23} \geqq 0, \quad y_{24} \geqq 0, \quad y_{34} \geqq 0$$

Constraint equations (1b) and (1e) indicate that a unit of flow leaves the source, node 1, and enters the sink, node 4, while constraint equations (1c) and (1d) require a conservation of flow at the intermediate nodes 2 and 3. Thus, any set of $y_{ij}$'s which satisfy these constraints constitute a path from the source to the sink, as indicated by the $y_{ij}$'s that are equal to one. Then, since the $y_{ij}$'s are equal to one for the activities carrying the hypothetical unit flow, and zero otherwise, the objective function, $f[Y]$, gives the sum of the activity duration times for the chosen path from source to sink. When the objective function is maximized, the corresponding path, denoted by $Y^*$, is the longest or critical path through the network; the corresponding value of the objective function will be denoted by $f[Y^*]$. The solution of this problem is greatly facilitated by the use of the duality theorem of linear programming as described below.

## Formulation of the Dual Problem

From the well known duality theorem of linear programming, it can be shown that to every linear programming problem, call it the *primal*, there corresponds a related linear programming problem which is called the *dual* of the original problem. The connection between these two problems is stated in the following theorem.

### *Duality Theorem*

Given a linear programming problem, call it the primal, there is always a related linear programming problem, called the dual, defined as follows, in matrix notation.

PRIMAL

$$\text{Maximize } f[X] = \underset{1\times n}{[C]}\ \underset{n\times 1}{[X]}$$

$$\text{Subject to } \underset{m\times n}{[A]}\ \underset{n\times 1}{[X]} \leqq \underset{m\times 1}{[P]}$$

$$\underset{n\times 1}{[X]} \geqq \underset{n\times 1}{[O]}$$

DUAL

$$\text{Minimize } g[W] = \underset{1\times m}{[P']}\ \underset{m\times 1}{[W]}$$

$$\text{Subject to } \underset{n\times m}{[A']}\ \underset{m\times 1}{[W]} \geqq \underset{n\times 1}{[C]}$$

$$\underset{m\times 1}{[W]} \geqq \underset{m\times 1}{[O]}$$

If there exists a solution, $\underset{n\times 1}{[\mathbf{X}^*]}$, which gives a finite maximum value to $f[\mathbf{X}]$, there is always a "coupled" solution $\underset{m\times 1}{[\mathbf{W}^*]}$, for which $g[\mathbf{W}]$ has a finite minimum value, equal to $f[\mathbf{X}]$. (A solution to one problem makes the solution to the other problem readily available.)

To the above statement of the theorem must be added that while the primal problem has $n$ variables and $m$ constraints, the dual problem has the reverse, $m$ variables and $n$ constraints. Also, if any of the $n$ primal variables are unrestricted in sign, then the corresponding dual constraints are equalities, and if the primal constraints are equalities, then the dual variables are unconstrained in sign. With regard to the solutions of the primal and dual problems, if in the solution of the dual one of its inequality constraints is satisfied as an equality, then the primal variable corresponding to this dual constraint may be positive, whereas it must be zero if the dual constraint is satisifed as an inequality.

Applying the Duality Theorem to the above example, one obtains the dual formulation given in equation (2). Recall that in the primal there were 5 variables and 4 constraints; hence in the dual problem there are 5 constraints (one corresponding to each of the primal variables) and 4 variables (one corresponding to each of the primal constraints.) Also, since the primal constraints were all equalities, the dual variables are all unconstrained in sign.

DUAL PROBLEM

$$\text{Minimize } g[\mathbf{W}] = w_1 - w_4 \tag{2a}$$

$$\text{Subject to } w_1 - w_2 \geqq 3 \tag{2b}$$

$$w_1 - w_3 \geqq 4 \tag{2c}$$

$$w_2 - w_3 \geqq 2 \tag{2d}$$

$$w_2 - w_4 \geqq 5 \tag{2e}$$

$$w_3 - w_4 \geqq 6 \tag{2f}$$

$$-\infty < w_1 < \infty, \quad -\infty < w_2 < \infty, \quad -\infty < w_3 < \infty, \quad -\infty < w_4 < \infty$$

The advantage of the dual formulation to this problem is now fairly obvious. Since each dual constraint involves only two variables, they can be solved by inspection, if a value is assigned to $w_1$. To see this, consider the constraints written in the following equivalent form.

$$w_2 \leqq w_1 - 3 \quad w_2^* = -3 \tag{2bb}$$

$$\left.\begin{array}{l} w_3 \leqq w_1 - 4 \quad (2cc) \\ w_3 \leqq w_2 - 2 \quad (2dd) \end{array}\right\} w_3^* = -5$$

$$\left.\begin{array}{l} w_4 \leqq w_2 - 5 \quad (2ee) \\ w_4 \leqq w_3 - 6 \quad (2ff) \end{array}\right\} w_4^* = -11$$

It will be shown below that $w_4$ varies directly with the value assigned to $w_1$, and since $g[\mathbf{W}]$ is merely the difference between $w_1$, and $w_4$, $w_1$ can be assigned an arbitrary value without affecting $g[\mathbf{W}]$. Thus, one may let $w_1 = 0$, a choice which makes all of the other $w_i$'s, and in particular $w_4$, take on negative values. Then, to minimize $g[\mathbf{W}] = w_1 - w_4 \equiv -w_4$, one must minimize the absolute value of $w_4$, and at the same time satisfy each of the above constraints. It is easy to see by inspection that this is achieved by the solution indicated above, i.e., $w_1^* = 0$, $w_2^* = -3$, $w_3^* = -5$ and $w_4^* = -11$. This will be denoted as $[\mathbf{W}^*] = (0, -3, -5, -11)$. Thus, the optimal value of the dual objective function is $g[\mathbf{W}^*] = w_1^* - w_4^* = 0 - (-11) = 11$. It is interesting to note the similarity between the solution of inequalities (2bb) through (2ff), and the conventional forward pass computations. For example the value of $w_3^* = -5$ is the largest absolute value of the two solutions obtained by treating (2cc) and (2dd) as equalities, or equivalently, the smallest absolute value of $w_3$ which satisfies both of these constraints. This is analogous to calculating the earliest event time at a merge point by taking the largest of the earliest finish times of the merging activities, which are 4 and 5 in this case.

Noting that the second and fourth dual constraints, i.e., those given by equations (2c) and (2e) are satisfied as inequalities, one knows from the duality theorem that the second and fourth primal variables, i.e., $y_{13}$ and $y_{24}$ are zero, while the others may be positive. By checking the primal constraints equations (1b) through (1e), one sees that the optimal solution to the primal problem is given by $[\mathbf{Y}^*] = (y_{12}^*, y_{13}^*, y_{23}^*, y_{24}^*, y_{34}^*) = (1, 0, 1, 0, 1)$, and that $f[\mathbf{Y}^*] = 3 \times 1 + 4 \times 0 + 2 \times 1 + 5 \times 0 + 6 \times 1 = 11$. (Note also that the requirement, $f[\mathbf{Y}^*] = g[\mathbf{W}^*] = 11$ is satisfied.) Thus, it is concluded that the network critical path is comprised of activities 1-2, 2-3, and 3-4, and has a total duration of 11 time units, a value which can easily be verified by the routine forward pass computation.

## LINEAR PROGRAMMING FORMULATION OF THE CPM TIME-COST TRADE-OFF PROBLEM AND EXTENSIONS

The basic formulation of the time-cost trade-off problem will be given first, assuming each activity in the network shown in Figure 8-12 has a trade-off curve of the form shown in Figure 8-13. The advantage of the notation used in this figure is that it replaces the normal and crash costs by a single cost slope, $C_{ij}$. Again letting $y_{ij}$ denote the scheduled duration of activity $i$-$j$, one can write the total direct project costs as a function of these variables as follows:

$$\text{Total direct project costs} = \sum_i \sum_j (K_{ij} - C_{ij} y_{ij}) = K - \sum_i \sum_j C_{ij} y_{ij} \tag{3}$$

where $\Sigma_i \, \Sigma_j$ is used to denote the summation over all activities in the project network. Since the $K_{ij}$'s are fixed constants, whose sum is denoted by $K$ in

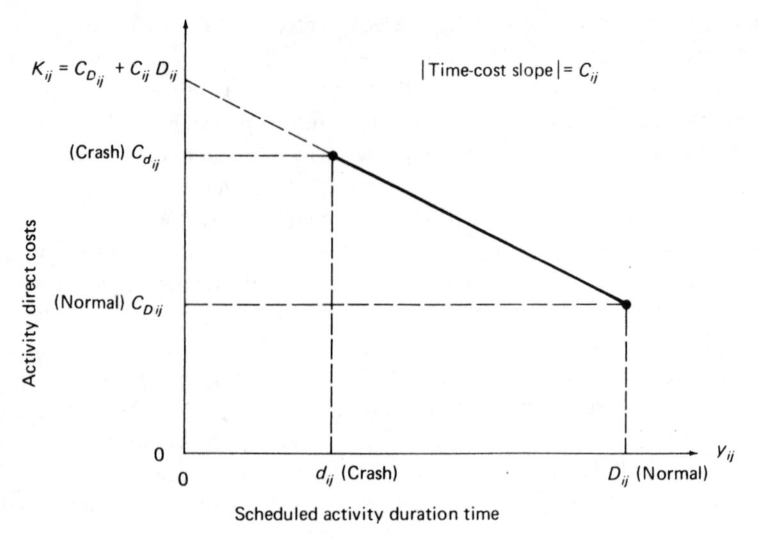


**Figure 8-13** Time-cost trade-off curve nomenclature.

equation (3), the total direct project costs are minimized if one will*

$$\text{Maximize } f[\mathbf{Y}] = \sum_i \sum_j C_{ij} y_{ij} \tag{4a}$$

$$\text{Subject to } T_i + y_{ij} - T_j \leqq 0, \quad \text{all } ij \tag{4b}$$

$$y_{ij} \leqq D_{ij}, \quad \text{all } ij \tag{4c}$$

$$-y_{ij} \leqq -d_{ij}, \quad \text{all } ij \tag{4d}$$

$$T_4 - T_1 \leqq \lambda \tag{4e}$$

where the $T_k$'s are (unknown) variables denoting the earliest expected time for node $k(k = 1, 2, 3, \text{or } 4)$, and $\lambda$ is the (constant) constraint placed on the total project duration, which is merely $T_4 - T_1$. Following the CPM convention of letting $T_1 = 0$, the last constraint, equation (4e) would merely become $T_4 \leqq \lambda$.

The first constraint equation (4b) applies to each activity, *i-j*, in the network, of which there are five in this example. These constraints merely state that for activity *i-j*, the difference between the earliest node times, $T_i$ and $T_j$, must be at least as great as $y_{ij}$, the scheduled duration of activity *i-j*. Similarly, equation (4c) applies to each of the five activities in this example; it constrains the scheduled activity duration time, $y_{ij}$, to be equal to or less than the normal activity time, $D_{ij}$. Finally, equation (4d) constains each $y_{ij}$ to be equal to or greater than the crash activity time, $d_{ij}$. The variables, $T_k$, have been omitted from the

*In this appendix, the symbol $T_i$ will be used in place of $E_i$, used elsewhere in this text.

objective function because their cost coefficients are zero; their role in this formulation is merely to insure that the scheduled values of $y_{ij}$ are feasible from the standpoint of network logic, and to insure that the project duration does not exceed $\lambda$.

While this problem could be solved using the simplex method to find the schedule $[\mathbf{Y}^*] = (y_{12}^*, y_{13}^*, y_{23}^*, y_{24}^*, y_{34}^*)$, which satisfies all of the constaints and at the same time maximizes $f[\mathbf{Y}]$, more efficient network flow algorithms have been developed. These procedures will not be given here. A detailed treatment can be found, for example, in the 2nd edition of this text, or in the original reference by Ford and Fulkerson.[8] An extensive treatment of this subject can also be found in Chapter 2 of the text by Elmaghraby.[6] The development of this procedure was motivated by considering the dual of the problem given above by equation (4). The special structure of the nonzero elements in the dual constraint matrix led to the discovery that this problem could be viewed as a network flow problem. A network flow algorithm was then devised to solve the dual problem, which is much simpler than the conventional simplex method. Thus, this time-cost trade-off problem, referred to as the "CPM" problem, can be solved very efficiently on a minicomputer, even for large networks. The output of these programs is essentially of the form shown above in Figure 8-8 and Table 8-1 for the modified Siemens hand computation procedure.

A natural extension of the "CPM" problem, as stated by eq. (4), was developed by Elmaghraby.[5] It allows scheduled completion time, $S_k$, to be placed on a set of milestone events in the network ($k = 1, 2, \ldots, K$), where $K$ is assumed to denote the project terminal event. This problem formulation also includes penalty rates, $p_k$, that are invoked for tardiness in meeting the schedules, $S_k$, placed on the project milestone events. This might be applied, for example, in the development of a mass transit system, a housing project, a shopping center, etc., where the facility can be completed and utilized in stages, as defined by the set of $K$ milestone events. The problem then is to determine the subset of activities whose durations are to be shortened, and the amount of that shortening, in order to incur the smallest total cost, i.e., cost of shortening plus the cost of tardiness. The linear programming formulation of this problem can be written as follows:

$$\text{Minimize } f(\mathbf{Y}, \mathbf{s}) = -\sum_i \sum_j c_{ij} y_{ij} + \sum_k p_k s_k \tag{5a}$$

$$\text{Subject to} \quad T_i + y_{ij} - T_j \leqq 0; \quad \text{all } ij \tag{5b}$$

$$y_{ij} \leqq D_{ij}; \quad \text{all } ij \tag{5c}$$

$$-y_{ij} \leqq -d_{ij}; \quad \text{all } ij \tag{5d}$$

$$T_k - T_1 - s_k \leqq S_k; \quad k = 1, \ldots, K \tag{5e}$$

$$s_k \geqq 0; \quad k = 1, \ldots, K \tag{5f}$$

In this formulation, the constraint set (5e) replaces the single constraint (4e) in the "CPM" problem. In these constraints, the variables, $s_k$, that are constrainted to be nonnegative, $s_k \geqq 0$, denote the tardiness in realizing event $k$, and the corresponding penalty, $p_k s_k$, appears in the objective function. Just as before, this problem can be solved by an efficient network flow algorithm (see Reference 5).

Another extension of the "CPM" problem was made by Moore[13] and others. They considered the determination of the optimal project schedule for the case where multiple objectives were considered. For example, the completion time of the entire project, as well as intermediate milestone events; duration of a particular set of activities that collectively should be allowed a specified minimum time; and an attempt to operate within a fixed budget for total direct project costs. This generalized model for the time-cost crashing problem is solved using goal programming, rather than linear programming. Modest sized networks could be handled by this procedure.

### Formulation for More Complex Time-Cost Functions

Chapter 8 discussed the use of a piece-wise linear approximation of a nonlinear but convex activity time-cost trade-off function. Suppose, for example, the trade-off curve for activity $i$-$j$ is as shown in Figure 8-14 below, which is continuous, convex, and nonincreasing. The actual tradeoff curve is shown at the left of Figure 8-14, with the two curves on the right depicting the separate segments of the piece-wise linear approximation to the actual curve. This representation of $y_{ij}$ in the objective function and the constraint equations is given below, where only that portion of the formulation pertaining to activity $i$-$j$ is given.

$$\text{Maximize } f[\mathbf{Y}] = \cdots + C_1 y_{1ij} + C_2 y_{2ij} + \cdots \tag{6a}$$

Subject to

$$\vdots$$

$$T_i + d_{ij} + y_{1ij} + y_{2ij} - T_j \leqq 0 \tag{6b}$$

$$0 \leqq y_{1ij} \leqq m - d \tag{6c}$$

$$0 \leqq y_{2ij} \leqq D - m \tag{6d}$$

$$\vdots$$

Using the simplex method to solve this problem will bring $y_{1ij}$ and $y_{2ij}$ into the solution in the proper order, i.e., $y_{1ij}$ will remain its maximum value until $y_{2ij}$ is reduced to zero, and then only will $y_{1ij}$ be reduced below its maximum value. This follows because $C_1 > C_2$, and the sum $C_1 y_{1ij} + C_2 y_{2ij}$ is being maximized. It is obvious that this would prevail for any number of straight line segments, as long as the actual trade-off is convex, which insures that $C_i > C_{i+1}$, for all $i$.

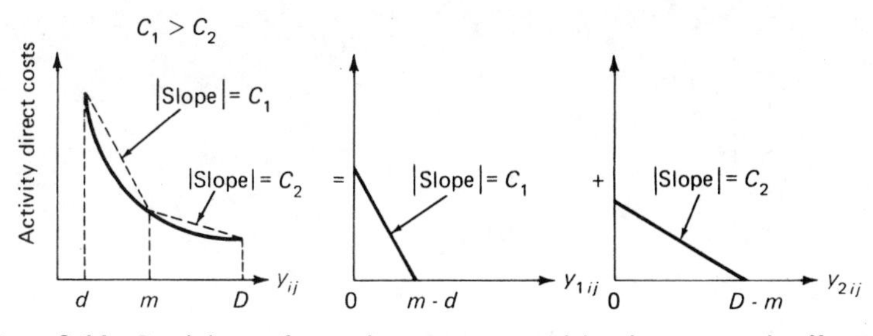


**Figure 8-14** Breakdown of a continuous convex activity time-cost trade-off curve.

### Formulation for Continuous Nonconvex Activity Time-Cost Trade-off Curves

If the trade-off curve is continuous, concave, and nonincreasing, the segments of a piece-wise linear approximation to the actual curve can be brought into the problem in their proper order by employing a nonnegative integer-valued variable. An approximation of this type is shown in Figure 8-15 below, where the actual cruve is shown on the left and the separate segments of this piece-wise linear approximation are shown on the right. In this case, the representation of $y_{ij}$ in the objective function and the constraint equations is given below, where again only that portion of the formulation pertaining to activity $i$-$j$ is given

$$\text{Maximize } f[\mathbf{Y}] = \cdots + C_1 y_{1ij} + C_2 y_{2ij} + \cdots \quad (7a)$$

Subject to $\vdots$

$$T_i + d_{ij} + y_{1ij} + y_{2ij} - T_j \leqq 0 \quad (7b)$$

$$0 \leqq y_{1ij} \leqq m - d \quad (7c)$$

$$0 \leqq y_{2ij} \leqq D - m \quad (7d)$$

$$\delta(m - d) \leqq y_{1ij} \quad (7e)$$

$$\delta(D - m) \geqq y_{1ij} \quad (7f)$$

$$\delta = \text{a non-negative integer} \quad (7g)$$

If one puts constraints (7e) and (7f) together, one finds that this system of constraints actually requires that $\delta$ be equal to either zero or one, as shown in equation (8).

$$y_{2ij}/(D - m) \leqq \delta \leqq y_{1ij}/(m - d) \quad (8)$$

If $y_{ij}$ is equal to its maximum value, then $\delta = 1$ since $y_{2ij} = D - m$, and because of the integer constraint on $\delta$, it must continue to equal 1 as long as $y_{2ij} > 0$.

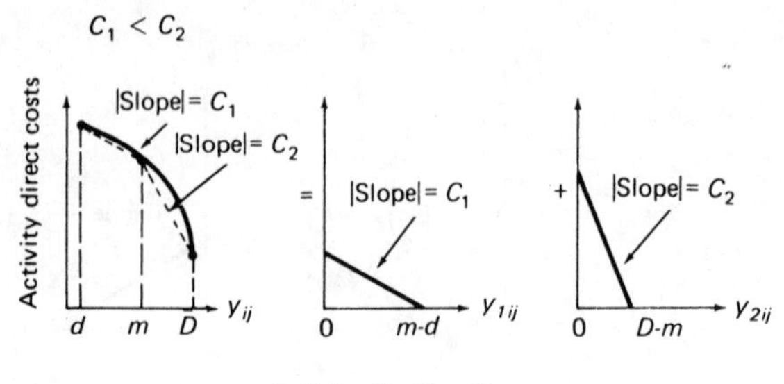


**Figure 8-15** Breakdown of a continuous concave activity time-cost trade-off curve.

Consequently, $y_{1ij}$ must, as it should, remain equal to its maximum value of $m - d$, as long as $y_{2ij} > 0$. When $y_{2ij} = 0$, $\delta$ can also equal zero, and only then can $y_{1ij}$ be less than $m - d$. Thus, one can see that the use of the integer-valued variable, $\delta$, forces the activity segments to vary in a manner dictated by the physical problem that they represent.

It is a simple matter to extend this formulation to more than two linear segments by introducing an additional integer-valued variable for each additional segment as shown in equation (9).

$$y_{k+1}/(d_{k+1} - d_k) \leqq \delta_k \leqq y_k/(d_k - d_{k-1}); \qquad k = 1, 2, \ldots, n \tag{9}$$

where in general the constraint, $0 \leqq y_i \leqq d_i - d_{i-1}$, holds. Also, it is important to note that this formulation of the problem will work on convex trade-off function as well as concave functions; however, it is not needed in the former case, as shown by equation (6). Because this formulation can be extended to any number of straight line segments, concave or convex, it follows that this integer-variable formulation can be used on any trade-off curve that is continuous and nonincreasing, such as the curve shown in Figure 8-16 which is neither convex nor concave.

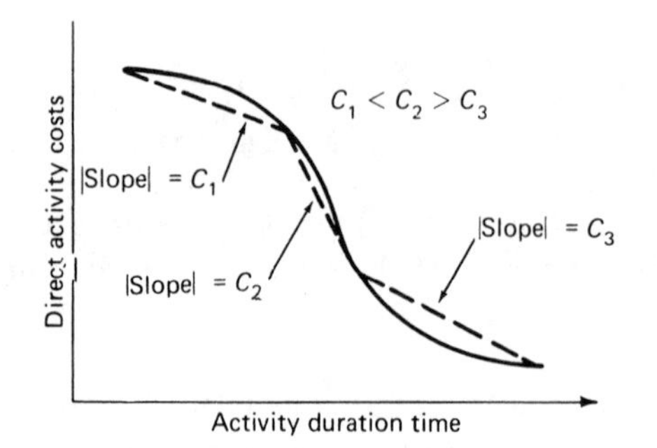

**Figure 8-16** Illustrative nonconvex, nonconcave trade-off function.

## Formulation for Feasible Point Time-Cost Trade-Off Functions

It often happens that an activity can only be performed in a small number of different times, which gives rise to a small set of feasible time-cost points. For example, consider an activity *i-j* that can only be performed at a normal time, $D_{ij}$, or a crash time, $d_{ij}$. In this formulation of the problem, one requires two* non-negative integer-valued variables as follows:

$$\begin{aligned} y_{D_{ij}} &= 1 \text{ if the activity duration is } D_{ij} \\ &= 0 \text{ if the activity duration is } d_{ij} \\ y_{d_{ij}} &= 1 \text{ if the activity duration is } d_{ij} \\ &= 0 \text{ if the activity duration is } D_{ij} \end{aligned} \tag{10}$$

Using these integer-valued variables, the linear programming formulation is as follows:

$$\text{Maximize } f[\mathbf{Y}] = \cdots - C_{D_{ij}} y_{D_{ij}} - C_{d_{ij}} y_{d_{ij}} + \cdots \tag{11a}$$

Subject to

$$\vdots$$

$$T_i + D_{ij} y_{D_{ij}} + d_{ij} y_{d_{ij}} - T_j \leqq 0 \tag{11b}$$

$$y_{D_{ij}} + y_{d_{ij}} = 1 \tag{11c}$$

$$y_{D_{ij}}, y_{d_{ij}} = \text{non-negative integers} \tag{11d}$$

$$\vdots$$

If the activity has $k$ different feasible time-cost points, the above formulation is extended by introducing one non-negative integer-valued variable for each feasible time-cost point and requiring that the sum of all of the variables be equal to one.

The use of integer variables in this manner is, indeed, a very powerful tool which could be used in a number of other ways. Consider, for example, the criticism put forth by Fondahl[7] that the CPM procedure does not account for the fact that activities are sometimes correlated in that a speed up of activity *A* must be accompanied by a speed up in say activity *B* as well, because it is accomplished by the use of a special resource which acts in common to both of these activities. In this case, one could accomplish the requirement that neither or both of the activities *A* and *B* are augmented by adding a pair of constraints

---

*One variable would suffice in this case, since $y_{D_{ij}} = 1 - y_{d_{ij}}$; however, the formulation given in equation (11) is used because it suggests the generalization to more than two feasible time-cost points.

to equation (11), i.e., $y_{d_A} - y_{d_B} \leqq 0$ and $y_{d_B} - y_{d_A} \leqq 0$, where $y_{d_A}$ and $y_{d_B}$ are already required to be non-negative integers by constraints like (11d). This pair of constraints accomplishes the desired result because they require, in effect, that both variables must be equal to zero or both must be equal to one.

### Solving Problems with Arbitrary Trade-Off Functions

It is not too difficult to combine the above formulations to handle more complicated trade-off functions which are combinations of discrete points and continuous curves. The details of this procedure can be found in the report by Meyer and Shaffer;[12] the three basic transition forms that must be handled are enumerated in problem 3 at the end of Chapter 8.

The application of the linear programming formulations given by equations (4), (6), (7), and (11), to the network given in Figure 8-11, but with more complicated activity trade-off functions, is presented in problem 4. The solution of this very small (five activities) problem requires 13 constraints, and the restriction that two of the variables (one for the concave cost curve for activity 1-2, and one for the discrete cost points for activity 3-4) be integer valued. To solve even this small problem using the simplex method, together with an integer programming algorithm, will require the use of a computer if the task is to be accomplished in a reasonable time and at a reasonable cost.

Current computer hardware and integer/programming software is such that these methods will handle networks of at most 100 activities. Other recent approaches to this problem have been more successful. For example, Panagiotakopoulos[14] has developed an efficient algorithm to determine optimal solutions to the "CPM" problem for the discrete case. That is, where the time-cost alternatives for each activity are specified by a set of feasible time-cost pairs, rather than continuous functions. The program for this algorithm is available from the author. It will handle networks well in excess of 100 activities.

A dynamic programming approach to the discrete case "CPM" problem was developed by Crowston, and extended by Hindelang and Muth.[9] Their approach is called decision CPM (DCPM) networks. It consists of conventional PERT/CPM *AND* nodes as well as *OR* nodes. The latter nodes are followed by a set of discrete time-cost activity performance alternatives. The computation time of this approach is said to grow linearly with the number of activities in the network. Thus it is potentially capable of handling very large networks.

In this appendix, the application of mathematical programming to the time-cost trade-off problem was introduced along with logical extensions to the objective function and the activity time-cost trade-off relationships. Finally, procedures to solve these various problems formulations were cited.

# 9

# THE PERT STATISTICAL APPROACH

In Chapter 1, PERT was described as being appropriate for scheduling and controlling research and development type projects, or others comprised primarily of activities whose actual duration times are subject to considerable chance variation. It is because of this variability that for projects of this type, the time element of project performance is usually of paramount importance. While the deterministic CPM approach, as described in Chapters 3 and 4, is quite frequently applied to programs of this type, the single estimate of the average activity performance time which it employs completely ignores the chance element associated with the conduct of the project activities. For example, an activity which is expected to take 10 days to perform, but might vary from 9 to 11 days would be treated no differently than an activity which is also expected to take 10 days to perform, but might vary from 2 to 25 days. The advantage of the PERT statistical approach, originally developed by D. G. Malcolm[9,11] and others, is that it offers a method of dealing with this chance variation, making it possible to allow for it in the scheduling calculations, and finally using it as a basis for computing the probability (index) that the project, or key milestones in the project, will be completed on or before their scheduled date(s).

This type of analysis could be very helpful in arriving at an acceptable project plan for implementation, as shown in box (5) of Figure 1-4, which illustrated the overall dynamic network-based planning and control procedure. For example, probabilities of meeting a required schedule, of greater than say 0.75, might signal a green light to go with the project plan. Values between 0.25 and 0.75

would signal a yellow caution light; careful attention to progress on this project is appropriate, and some project replanning may be anticipated. Finally, a probability less than 0.25 would signal red to immediately stop and replan to achieve a better chance of meeting the required schedule.

This type of early warning system has considerable merit when large uncertainty in activity duration time exists. For example, the major overhaul of aircraft involves considerable uncertainty that is not revealed until covers are removed from control cables, engines, etc., where corrosion and wear may be found to vary from minimal to extensive. In other projects, external elements, such as the weather, may interject uncertainty. For example, in the Alaskan pipeline project, risk analysis of the construction schedule required the inclusion of the effects of weather conditions within the project network plan.

## OVERVIEW OF PERT

To give an overview of the PERT statistical approach, consider the network originally presented in Figure 4-5, and shown here in modified form in Figure 9-1. In this network, the critical path consists of the three activities 0-3, 3-7, and 7-8. Now suppose that the performance of each of these activities is subject to a considerable number of chance sources of variation such as the weather, equipment failures, personnel or materials problems, or uncertainties in the methods or procedures to be used in carrying out the activity. It may be argued that if difficulties of one sort or another are encountered on a particular activity, that additional resources will immediately be applied to this activity, or subse-

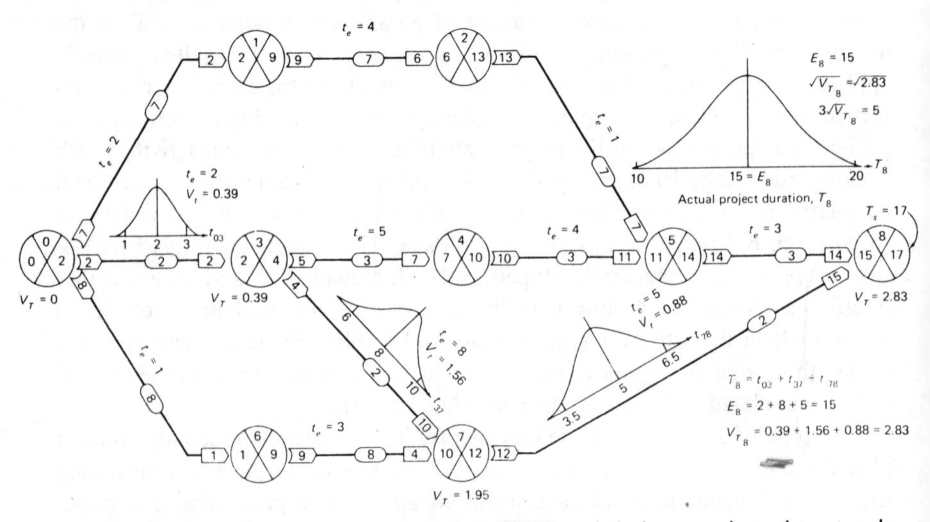


**Figure 9-1** Illustration of the "conventional" PERT statistical approach to the network originally presented in Figure 4-5.

quent activities on the critical path, so that the project will still be completed on time. This chapter is concerned with the problem of estimating *the probability of having to undertake such measures*. Hence, the uncertainties in performance time being referred to here are those associated with completing the originally defined activity with the originally specified resources.

Returning to the example, one notes that the actual performance times for the activities on the critical path, instead of being exactly 2, 8, and 5 days, are variables subject to random or chance variation, with *mean values* of 2, 8, and 5, respectively. Also, the actual time to perform the activities on the critical path is the sum of three random or chance variables, and (except for the slight possibility that activity 6-7 may be completed after activity 3-7 and activity 5-8 after activity 7-8) this sum is also the actual time to complete the project. Hence, to estimate the statistical distribution of project performance time, and in turn compute the probability of meeting a scheduled date for the completion of the project, it will be necessary to deal with the statistics of the sum of random variables. In the next section the theory of probability and statistics necessary to handle this problem will be considered. The following section will describe "conventional" PERT scheduling and probability calculation along with several practical applications. Several important refinements to the conventional PERT procedure are given at the end of this chapter.

The treatment of PERT given in this chapter is referred to as "conventional" because the calculation of earliest and latest activity start and finish times is made in the same way as described in Chapter 4, using expected activity performance times only. The variability associated with activity performance times is involved only in the computation of PERT probabilities. This simplifying assumption causes the PERT event times to be biased slightly, always on the low side, and the PERT probabilities considerably on the high side. For example, in Figure 9-1 the earliest expected complete time for activity 7-8 is 15. However, since the actual complete time for activity 5-8 may be greater than that for activity 7-8, even though its expected time of 14 is less (earlier) than that for activity 7-8, the correct *expected* time for the completion of both activities 5-8 and 7-8 must be greater than 15.

The fact that the conventional PERT procedure *always* underestimates the expected occurrence times of network merge events and their successors, disturbs users. The fact is, however, that this bias is *usually* relatively small; less than 5 percent. It turns out that the PERT probability estimates are more seriously biased, on the high side. This is due not only to the merge event time bias, but also to the relatively large error in the estimate of the variance of the actual project duration time. These biases will be illustrated in the latter sections of this chapter, along with two practical methods of circumventing them. One relatively simple method, called PNET, considers not only the critical path, but also a "sufficient" set of near critical paths so that accurate probability estimates can be made. The other method is Monte Carlo simulation. It is an order of magnitude more costly than PNET, but has the very important added advantage of estimating the criticality of all network activities and events. The latter is expressed as a probability that any activity in question will end up on the critical path when the project is actually conducted.

The PERT statistical approach to project management is not widely used today, even though its potential has grown over the past decade due to the proliferation of very costly and time consuming large scale systems projects. Crandall[4] has reported that the deterministic activity performance time approach has proven costly and invalid in projects such as the fabrication of offshore oil platforms and storage tanks, and projects in Alaska where weather is a factor in construction and logistic supply. Many other projects of this type could be cited, particularly large government hardware development projects whose history of large uncertainty in time and cost performance is well known. In view of this, it is hard to say why PERT is rarely applied. Certain contributing factors are known. The lack of validity of the basic assumptions underlying PERT have been cited in scores of technical papers. But by and large, most of these problems could be affordably circumvented by the two methods cited above. This reduces to the conjecture that top level project managers do not understand the basic principles of probability and statistics, or given that they do, they have just not learned how to use PERT to solve their management problems. For this reason, the next section is devoted to basic probability theory. While many readers of this text will already have a background in this material, they may find this material interesting as an example of how these concepts might be explained to managers who are to use the PERT probability approach to project planning and control. The material is presented with this objective in mind.

## BASIC PROBABILITY THEORY

### Probability as a Measure of Uncertainty

Some of the mathematical definitions of probability become highly abstract and the language somewhat complicated, although they will be discussed in less formidable language below. In a sense, however, one already knows what probability is all about. If one is told that an event is "almost certain," "highly probable," "about fifty-fifty," "highly unlikely," or "highly improbable," one has a good intuitive feel for the meaning of what is being said and, furthermore, this intuitive feeling is correct. All that probability theory attempts to do is to quantify these somewhat subjective statements in a precise and objective way.

In order to do this it has been found convenient to express probabilities on a scale that runs from 0 to 1. On this scale, zero represents impossibility and one represents certainty; the numbers in between represent varying degrees of likelihood. Instead of saying, for example, that it is "almost certain" that a device will continue operating for at least one hour and "highly improbable" that it will continue operating for more than one thousand hours, one can say that the respective probabilities are, say, 0.999 and 0.001. The definitions and mathematical procedures that enable one to go from qualitative to quantitative statements can become quite technical and highly specialized; the intent, however, is to enable one to make precise and valid statements about the degree of certainty or uncertainty associated with specific occurrences.

### The Managerial Function: Decision Making Under Uncertainty and Risk

The words "uncertainty" and "risk"† appear frequently in mathematical literature on probability concepts. The same two words or their synonyms are also a part of management's vocabulary, for the prime function of management is decision making under conditions of uncertainty with the objective of balancing the risks associated with a particular problem. Risk itself has two elements: the probability that something will happen, and the loss that will result if it does happen.

Consider the trivial example of deciding whether or not to wear a raincoat to work. If one decides to take a raincoat and it does not rain, there will be a loss, the effort or nuisance involved in carrying the raincoat. If one does not take the raincoat and it does rain; another kind of loss is involved, i.e., getting wet. The decision will, therefore, depend upon an evaluation of these possible losses and an assessment of the probability of rain. Current industrial applications of probability theory revolve around this concept of balancing risks. Probability theory is used, for example, in the determination of optimal inventory sizes, where the opposing risks are the costs of carrying too much stock, and the loss of sales that results when an out-of-stock condition occurs.

Probability has already been defined as a way of measuring uncertainty. Inasmuch as the problems facing management are the problems of uncertainty and risk, it is clear that probability has an important role to play in helping the manager to formulate and solve these problems.

### Empirical Frequency Distributions

To present a background in probability and statistics it is logical to begin by considering the basic raw material of statistics, i.e., observations of some measurable quantity subject to random or chance sources of variation. For discussion purposes consider a PERT activity which has been performed in the past a large number of times *under essentially the same conditions.* This assumes that no

†The definition of these two terms used here differs from that used in literature on decision theory, where the differentiation between uncertainty and risk is based on the presence or absence of a probability distribution associated with the variable in question.

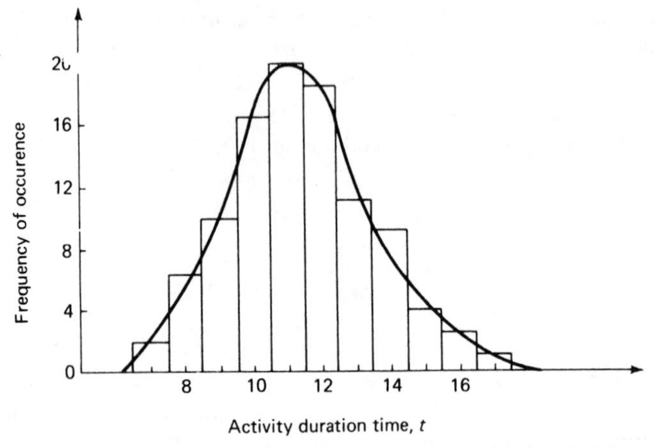


**Figure 9-2** Empirical frequency distribution of activity duration times.

learning, changes in working conditions, job description, etc., take place. Although PERT generally involves no statistical sampling of this sort, for the purpose of this discussion, one may suppose the duration times for this activity ranged from 7 to 17 days. Now suppose that one counts the number of times the activity required 7 days to perform, 8 days to perform, etc., and displays the resulting data in the form of an empirical frequency distribution or histogram as shown in Figure 9-2. If one had an infinite number of observations and made the width of the intervals in Figure 9-2 approach zero, the distribution would merge into some smooth curve; this type of curve will be referred to as the theoretical probability density of the random variable. The total area under such a curve is made to be exactly one, so that the area under the curve between any two values of $t$ is directly the probability that the random variable $t$ will fall in this interval.

## Characterization of an Empirical Distribution

To describe an empirical frequency distribution quantitatively, two measures are frequently employed—one which locates the point about which the distribution is centered, a measure of its central tendency or location, and the other which indicates the spread or dispersion in the distribution, a measure of its variability. These measures are illustrated in Figure 9-3. At the top and middle of the figure, the two distributions differ either in their mean values or dispersion, while at the bottom they differ in both respects. This same information is given by quantitative measures of these two characteristics of a frequency distribution.

In PERT computations, this text will use the familiar arithmetic average or

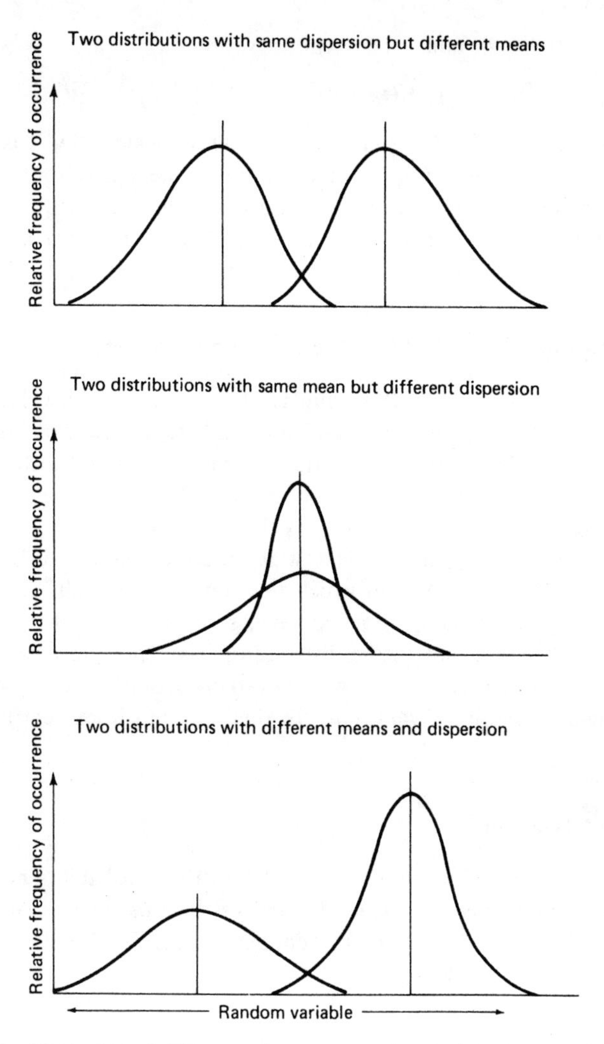


**Figure 9-3** Illustration of differences in measures of central tendency and dispersion.

mean as a measure of central tendency, and what is called the standard deviation as the measure of variability. These statistics will first be defined with respect to a sample of $n$ observations drawn from some distribution such as the one shown in Figure 9-2. If the $n$ observations are denoted by $t_1, t_2, \ldots, t_n$, these measures are computed as follows:

$$\text{Measure of central tendency} = \text{arithmetic mean}$$
$$= (t_1 + t_2 + \cdots + t_n)/n = \bar{t} \qquad (1)$$

Measure of variability = standard deviation = $s_t$

$$= \{[(t_1 - \bar{t})^2 + (t_2 - \bar{t})^2 + \cdots + (t_n - \bar{t})^2]/n\}^{1/2} \tag{2}$$

The above formula for the standard deviation indicates why it is sometimes referred to as the root-mean-square deviation; it is the square root of the mean of the squares of the deviations of the individual observations from their average. Computations will frequently use the square of the standard deviation, which, for convenience, is called the variance; $s_t^2$ = variance of $t$.

### Physical Interpretation of the Mean and Standard Deviation

The question usually asked at this point is what do $\bar{t}$ and $s_t$ (or $s_t^2$) mean? First of all, $t$ and $s_t$ both carry the same time units and are estimates of the true mean and standard deviation of the distribution shown by the smooth curve in Figure 9-2. These quantities will be denoted by $t_e$ and $(V_t)^{1/2}$, respectively; $\bar{t}$ approaches $t_e$ and $s_t$ approaches $(V_t)^{1/2}$ as the size of the sample, $n$, approaches infinity. If some assumption is made now about the theoretical distribution (the smooth curve in Figure 9-2) from which the sample was obtained, one can proceed with the interpretation. For example, suppose the random variable $t$ is "normally" distributed, that is, the distribution has a characteristic symmetrical bell shape which frequently occurs when a variable is acted upon by a multitude of random chance causes of variation. In this case, our interpretation is shown in Figure 9-4.

### Central Limit Theorem

The last bit of statistical machinery needed for PERT probability computations is the Central Limit Theorem, which is perhaps the most important theorem in all of mathematical statistics. In the context of PERT, this theorem may be stated in the following way.

### Central Limit Theorem

Suppose $m$ *independent* tasks are to be performed in order; (one might think of these as the $m$ tasks which lie on the critical path of a network). Let $t_1, t_2, \ldots, t_m$ be the times actually required to complete these tasks.

Note that these are random variables with true means $t_{e1}, t_{e2}, \ldots, t_{em}$, and true variances $V_{t1}, V_{t2}, \ldots, V_{tm}$, and these actual times are unknown until these specific tasks are actually performed. Now define $T$ to be the sum:

$$T = t_1 + t_2 + \cdots + t_m$$

and note that $T$ is also a random variable and thus has a distribution. The Central Limit Theorem states that if $m$ is large, say four or more, the distribution of

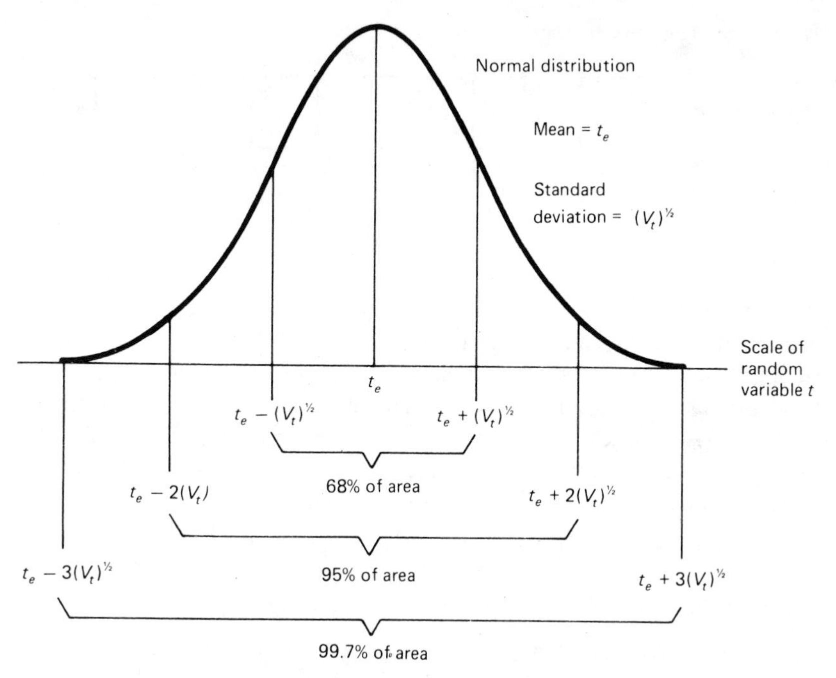


**Figure 9-4** Selected areas under the normal distribution curve.

$T$ is approximately normal with mean $E$ and variance $V_T$ given by

$$E = t_{e1} + t_{e2} + \cdots + t_{em}$$

$$V_T = V_{t_1} + V_{t_2} + \cdots + V_{t_m}$$

That is, the mean of the sum, is the sum of the means; the variance of the sum is the sum of the variances; and the distribution of the sum of activity times will be normal regardless of the shape of the distribution of actual activity performance times (such as given in Figure 9-2).

The normal distribution is extensively tabulated and therefore probability statements can be made regarding the random variable $T$ by using these tables. A table of normal curve areas is given in Appendix 9-1, and an example illustrating its use is given later.

## The Dice Tossing Experiment

To establish confidence in, and further understanding of the Central Limit Theorem, it is worthwhile to study its application to a familiar experiment—dice tossing. An experiment of tossing a single die can be described as shown in Figure 9-5.

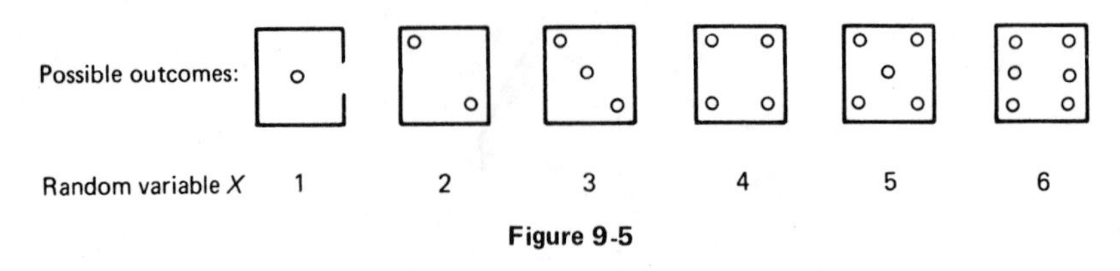


**Figure 9-5**

If the die being tossed is unbiased, the probability of each outcome of this experiment is equally likely, and hence has a probability of $\frac{1}{6}$, since there are just six possible outcomes. This is shown in Figure 9-6, which is called a theoretical probability distribution for the random variable $X$. The mean and variance of this theoretical distribution can be computed using equations (1) and (2) in which each possible value of the random variable $X$ is weighted by its theoretical probability. These results are indicated in Figure 9-6.

$$\text{Mean of } X = (1 \times \tfrac{1}{6}) + (2 \times \tfrac{1}{6}) + (3 \times \tfrac{1}{6}) + (4 \times \tfrac{1}{6}) + (5 \times \tfrac{1}{6}) + (6 \times \tfrac{1}{6}) = \tfrac{21}{6} = 3\tfrac{1}{2}$$

$$\begin{aligned}\text{Variance of } X &= (1 - 3\tfrac{1}{2})^2 \times \tfrac{1}{6} + (2 - 3\tfrac{1}{2})^2 \times \tfrac{1}{6} \\ &\quad + (3 - 3\tfrac{1}{2})^2 \times \tfrac{1}{6} + (4 - 3\tfrac{1}{2})^2 \times \tfrac{1}{6} \\ &\quad + (5 - 3\tfrac{1}{2})^2 \times \tfrac{1}{6} + (6 - 3\tfrac{1}{2})^2 \times \tfrac{1}{6} \\ &= \tfrac{70}{24} = 2\tfrac{11}{12}\end{aligned}$$

Now consider tossing two dice with the random variable $Y$ defined as the sum of the spots on both dice, i.e., $Y = X_1 + X_2$. The probability distribution for this example, shown in Figure 9-6, follows directly from the fact that there are 36 mutually exclusive and equally likely ways of tossing two unbiased dice, the outcomes of which are shown by the matrix in Figure 9-7, which gives the total number of spots for each of the 36 combinations.

The reader can verify that by using equations (1) and (2) in the same manner as shown above for the single die case, the mean and variance of this distribution are exactly twice the values for a single die, i.e., mean $= 2 \times 3\frac{1}{2} = 7$ and the variance $= 2 \times 2\frac{11}{12} = 5\frac{5}{6}$, as they should be according to the Central Limit Theorem. One should also note from Figure 9-6 that while the basic random variable $X$ has a rectangular distribution, the random variable $Y$ has a triangular distribution which represents a large step toward the theoretical normal distribution as dictated by the Central Limit Theorem.

To carry this experiment still one step further, consider tossing three dice with the random variable $Z$ defined as the sum of the spots on all three dice,

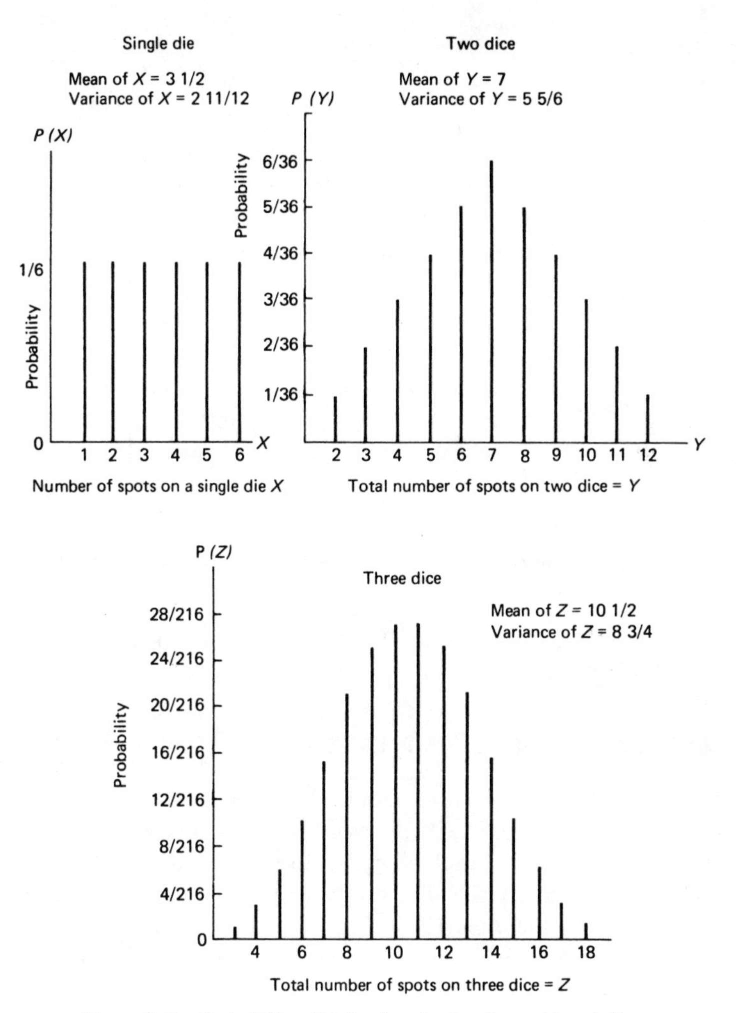


**Figure 9-6** Probability distribution for tossing unbiased dice.

i.e., $Z = X_1 + X_2 + X_3$. The reader can again verify that the mean and variance of this distribution follow the Central Limit Theorem, i.e.,

$$\text{Mean of } Z = 3 \times 3\tfrac{1}{2} = 10\tfrac{1}{2}$$

$$\text{Variance of } Z = 3 \times 2\tfrac{11}{12} = 8\tfrac{3}{4}$$

Examination of Figure 9-6 indicates that the shape of the distribution is now very close to the theoretical normal distribution. The Central Limit Theorem will be applied in the section below on the probability of meeting a scheduled

Second die

| First die | 1 | 2 | 3 | 4 | 5 | 6 |
|---|---|---|---|---|---|---|
| 1 | 2 | 3 | 4 | 5 | 6 | 7 |
| 2 | 3 | 4 | 5 | 6 | 7 | 8 |
| 3 | 4 | 5 | 6 | 7 | 8 | 9 |
| 4 | 5 | 6 | 7 | 8 | 9 | 10 |
| 5 | 6 | 7 | 8 | 9 | 10 | 11 |
| 6 | 7 | 8 | 9 | 10 | 11 | 12 |

**Figure 9-7** The thirty-six equiprobable ways of tossing two unbiased dice.

date, while an empirical verification of this theorem is taken up in exercise 1 at the end of this chapter.

## PERT SYSTEM OF THREE TIME ESTIMATES

PERT basic scheduling computations utilize the expected values, $t_e$, of the hypothetical distributions of actual activity performance times, as depicted in Figure 9-2 and 9-8 below. Since PERT addresses itself primarily to programs whose activities are subject to considerable random variation and to programs where time schedules are of the essence, it utilizes the standard deviations of the distributions shown in these figures in computing a measure of the chances of meeting scheduled dates of project milestones.

In making PERT computations, it must be realized that the activity performance time distribution is purely hypothetical, since one ordinarily is unable to do any statistical sampling whatever. [If historical (sample) activity duration data are available, they can be used to estimate $a$, $m$, and $b$, as indicated in Appendix 9-2]. After an activity has been performed, the observed actual performance time, denoted by $t$, for the activity can be considered as a single sample from this hypothetical distribution. However, all computations are made prior to the performance of the activity; hence, as stated above, the basis of PERT computations involves no statistical sampling, but rather it depends on the judgment of the person in charge of the activity in question. The latter judgement is, of course, based on a sampling of work experiences; however, this is not sampling in the strict statistical sense. In making these subjective estimates,

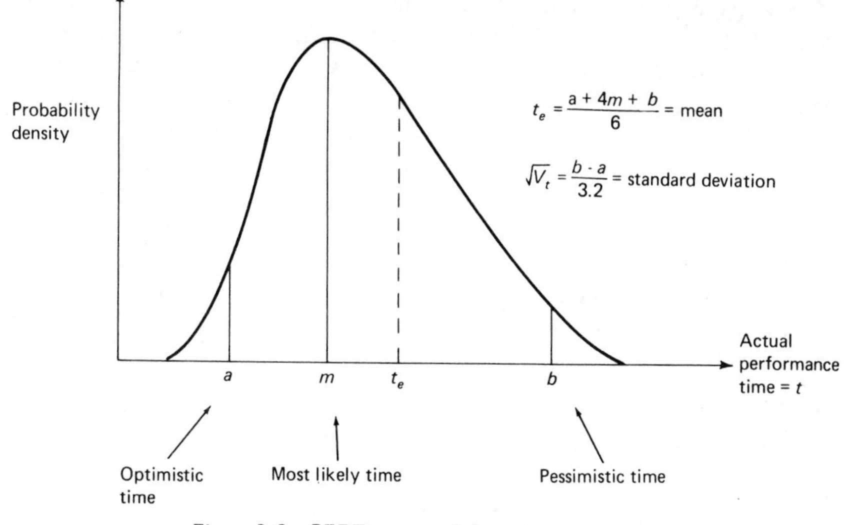


**Figure 9-8** PERT system of three time estimates.

one is asked to call on his general experience, and his knowledge of the requirements of the activity in question, to consider the personnel and facilities available to him, and then to estimate the three times shown in Figure 9-8. These times defined below, will then be used to estimate the mean and standard deviation of the hypothetical activity performance time distribution. The choice of a most likely time, $m$, and then a range of times from an optimistic estimate, $a$, to a pessimistic estimate, $b$, seems to be a natural choice of times.

*Definition:*

$a$ = optimistic performance time; the time which would be bettered only one time in twenty if the activity could be performed repeatedly under the same essential conditions.

*Definition:*

$m$ = most likely time; the modal value of the distribution, or the value which is likely to occur more often than any other value.

*Definition:*

$b$ = pessimistic performance time; the time which would be exceeded only one time in twenty if the activity could be performed repeatedly under the same essential conditions.

The above definitions of $a$ and $b$ are called the 5 and 95 percentiles, respectively, of the distribution of the performance time, $t$. These definitions are based on a study by Moder and Rodgers.[10] They differ from the original development of PERT,[9,11] where they were assumed to be the ultimate limits, or the 0 and 100 percentiles of the distribution of $t$. An intuitive argument for our definition is that since the estimates of $a$ and $b$ are based on past experience and judgment, the 0 and 100 percentiles would be very difficult to estimate, since they would never have been experienced. Further arguments of a statistical nature will be presented below.

Although the above definitions of $a$, $m$, and $b$ appear to be clear and workable, the following points will be helpful in obtaining *reliable* values for these time estimates.

1. One of the important assumptions in the Central Limit Theorem is the independence of the random variables in question. Since this theorem is the basis of PERT probability computations, the estimates of $a$, $m$, and $b$ should be obtained so that the assumption of independence is satisfied, that is, they should be made independently of what may occur in other activities in the project, which may in turn affect the availability of manpower and equipment planned for the activity in question. The estimator should submit values for $a$, $m$, and $b$ which are appropriate if the work is carried out with the initially assumed manpower and facilities, and under the assumed working conditions.
2. The estimates of $a$, $m$, and $b$ should not be influenced by the time available to complete the project, i.e., it is not logical to revise estimates by an across-the-board cut in times after learning that the project critical path is too long. This completely invalidates the PERT probabilities and destroys any positive contribution that they may be able to make in the planning function. Time estimates should be revised only when the scope of the activity is changed, or when the manpower and facilities assigned to it are changed.
3. To maintain an atmosphere conducive to obtaining unbiased estimates of $a$, $m$, and $b$, it should be made clear that these are estimates and not schedule commitments in the usual sense.
4. In general, the estimates of $a$, $m$, and $b$ should not include allowances for events which occur so infrequently that one does not ordinarily think of them as random variables. The estimates of $a$, $m$, and $b$ should not include allowances for acts of nature—fires, floods, hurricanes, etc.
5. In general, the estimates of $a$, $m$, and $b$ should include allowances for events normally classed as random variables. An example here would be the effects of weather. For activities whose performance is subject to weather conditions, it is appropriate to anticipate the time of the year when the activity will be performed and make suitable allowance for the anticipated prevailing weather in estimating $a$, $m$, and $b$.

### Estimation of the Mean and Variance of the Activity Performance Times

It is commonly known in statistics that for unimodal distributions the standard deviation can be estimated roughly as $\frac{1}{6}$ of the range of the distribution. This follows from the fact that at least 89 percent of any distribution lies within three standard deviations of the mean, and for the normal distribution this percentage is 99.7+ percent. Hence, one can use time estimates, $a$ and $b$, to estimate the standard deviation $(V_t)^{1/2}$ or the variance, $V_t$, as shown in equation (3):

$$(V_t)^{1/2} = (b - a)/3.2, \qquad \text{or} \qquad V_t = [(b - a)/3.2]^2 \tag{3}$$

As mentioned above, $a$ and $b$ were originally defined as the 0 and 100 percentiles of the distribution of $t$, and therefore, the divisor in equation (3) was 6 in place of the above value of 3.2, in the original development of PERT.[9,11] This is the basis of another argument in favor of our 5 and 95 percentile definitions of $a$ and $b$. In the paper by Moder and Rodgers,[10] it is shown that the difference $(b - a)$ varies from 3.1 to 3.3 (average of 3.2) standard deviations for a wide variety of distribution types ranging from the exponential distribution to the normal distribution, including rectangular, triangular, and beta type distributions. For this same set of distributions, the difference between the 0 and 100 percentiles varies, however, from 3.5 all the way to 6.0. Thus, the use of the 5 and 95 percentiles for $a$ and $b$ leads to an estimator of the standard deviation that is robust to variations in the shape of the distribution of $t$. This is of some importance, because in general we do not know the shape of the distribution of $t$, and further, we wish to avoid making any specific assumptions about it.

A simple formula for estimating the mean, $t_e$, of the activity time distribution has also been developed. It is the simple weighted average of the estimates $a$, $m$, and $b$ given in equation (4).

$$\text{Mean} = t_e = (a + 4m + b)/6 \tag{4}$$

To derive this formula for the mean, one must assume some functional form for the unknown distribution of $t$, such as shown in Figure 9-8. A likely candidate, chosen by Clark,[2] is the well known beta distribution, which has the desirable properties of being contained inside a finite interval, and can be symmetric or skew, depending on the location of the mode, $m$, relative to $a$ and $b$. Lacking an empirical basis for choosing a specific distribution, the beta distribution was historically accepted as a mathematical model for activity duration times, for purposes of deriving equation (4) only. Using this distribution as a model and assuming that equation (3) holds,* then $t_e$ is a cubic polynomial in $m$. Equation (4) is a linear approximation to the exact formula, whose accuracy is well within limits dictated by the accuracy of the estimates of $a$, $m$, and $b$.

---

*The 0 and 100 percentile definitions of $a$ and $b$ were used in this derivation. The same formula should hold, however, for the 5 and 95 percentile definitions used in this text.

It should also be pointed out that the mean is equal to the most likely or modal time ($t_e = m$), *only if the optimistic and pessimistic times are symmetrically placed about the most likely time*, i.e., only if $b - m = m - a$. Thus, in the CPM procedure the single time estimate, denoted by $D$, is an estimate of the *mean* activity duration time and is not necessarily the *most likely* time as defined here. This is essential, since according to the Central Limit Theorem, the expected total duration of a series of activities is the sum of their *mean* times and not a sum of their *most likely* times. In fact, since the distribution of the sum of random variables tends to the normal (symmetrical) distribution for which the mean and the mode are the same, the most likely (modal) duration time of a series of activities is not given by the sum of the individual activity most likely times, but rather by a sum of their mean times.

If a single time estimate system is being used, and an activity is encountered which has a skew distribution with a considerable amount of variation, then equation (4) might be of assistance in arriving at the single time estimate, $D$. In this case, a person might feel that he can estimate the mean activity duration time more accurately by estimating $a$, $m$, and $b$, and using equation (4) to convert these numbers to the required single time estimate, $D$. Some evidence to this effect was found in the study cited in Reference 10.

To illustrate the computation in the PERT statistical approach, consider the simple network given in Figure 9-9. Here, for the network originally presented in Figure 4-1, single time estimates have been replaced by estimates of $a$, $m$, and $b$. For example, $a = 1, m = 2$, and $b = 3$ for activity 1-2. In the middle diagram of Figure 9-9 are indicated the values of $t_e$, and $V_t$ computed from equations (3) and (4). These computations are illustrated below for activity 1-2.

$$t_e = (1 + 4 \times 2 + 3)/6 = 2$$

$$V_t = [(3 - 1)/3.2]^2 = 0.391$$

The results of the forward pass are indicated by the time scale at the bottom of Figure 9-9. The earliest expected event occurrence times, $E$, are computed in exactly the same manner (in conventional PERT) as outlined for the single time estimate systems, as shown in Figure 4-1.

The event variance, $V_T$, is computed in a manner quite similar to the computation of $E$. The rules are as follows:

RULE 1. $V_T$ for the initial network event is assumed to be zero.

RULE 2. The $V_T$ for the event following the activity in question in obtained by adding the activity's variance, $V_t$, to the variance of the predecessor event, except at merge events.

RULE 3. At merge events, $V_T$, is computed along the same path used to obtain $E$, i.e., the longest path. In case of ties, choose the path which gives the larger variance.

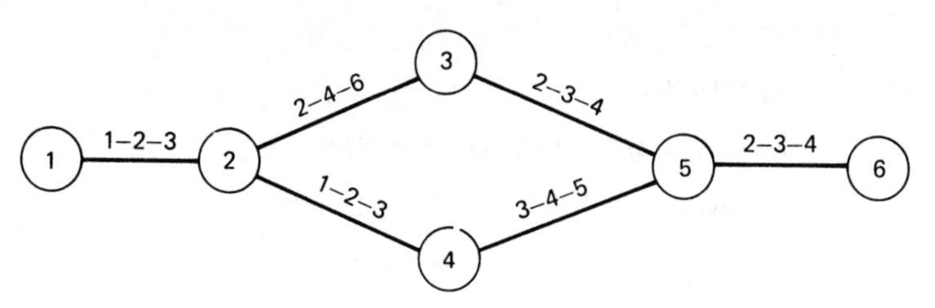


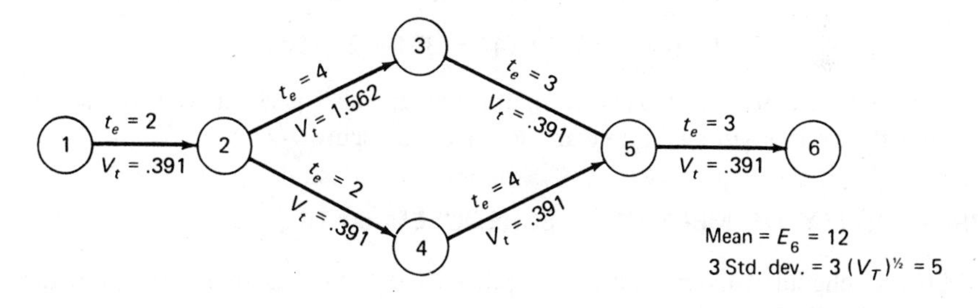


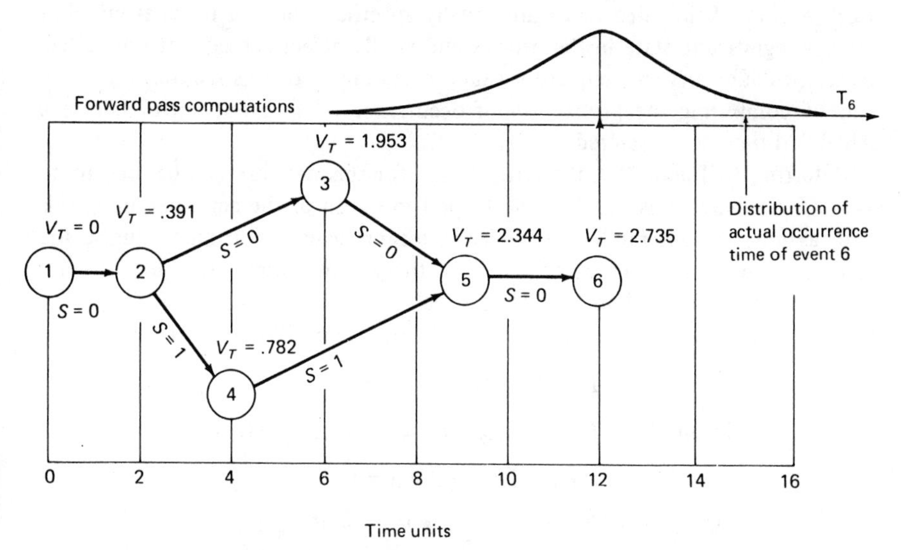


**Figure 9-9** PERT statistical computations.

Applying these rules to the network given in Figure 9-9, one obtains the following:

$$V_T \text{ (event 1)} = 0$$

$$V_T \text{ (event 2)} = 0 + 0.391 = 0.391$$

$$V_T \text{ (event 3)} = 0.391 + 1.562 = 1.953$$

$$V_T \text{ (event 4)} = 0.391 + 0.391 = 0.782$$

Since event 5 is a merge event, its variance according to Rule 3 above is computed along the path 1-2-3-5.

$$V_T \text{ (event 5)} = 1.953 + .391 = 2.344$$

$$V_T \text{ (event 6)} = 2.344 + .391 = 2.735$$

The backward pass and slack computations are also performed in the same manner shown in Figure 4-1; they are not given in Figure 9-9.

## PROBABILITY OF MEETING A SCHEDULED DATE

Although scheduled dates could be applied to the start or finish of a project activity, they have traditionally been applied to the time of occurrence of network events. Scheduled dates are usually specified only for those events that mark a significant state in the project and vitally affect subsequent project activities; such events are frequently called milestones. In this section, the problem of computing the probability of occurrence of an event, on or before a scheduled time, is considered.

Referring to Figure 9-9, the critical path for this network can be seen to be 1-2-3-5-6. Now consider the time to perform each of the activities along this path as independent random variables, the same assumption made during the process of collecting the $a$, $m$, and $b$ activity time estimates. Furthermore, the sum of these random variables, which shall be denoted by $T$, is itself a random variable which is governed by the Central Limit Theorem. Therefore

$$T = t_{1-2} + t_{2-3} + t_{3-5} + t_{5-6}$$

$$\text{Mean of } T = E_6 = (t_e)_{1-2} + (t_e)_{2-3} + (t_e)_{3-5} + (t_e)_{5-6}$$

$$E_6 = 2 + 4 + 3 + 3 = 12$$

$$\text{Variance of } T = V_T = V_{t1-2} + V_{t2-3} + V_{t3-5} + V_{t5-6}$$

$$V_T = 0.391 + 1.562 + 0.391 + 0.391 = 2.735$$

and finally, the Central Limit Theorem enables one to assume that the shape of the distribution of $T$ is approximately normal. This information is summarized

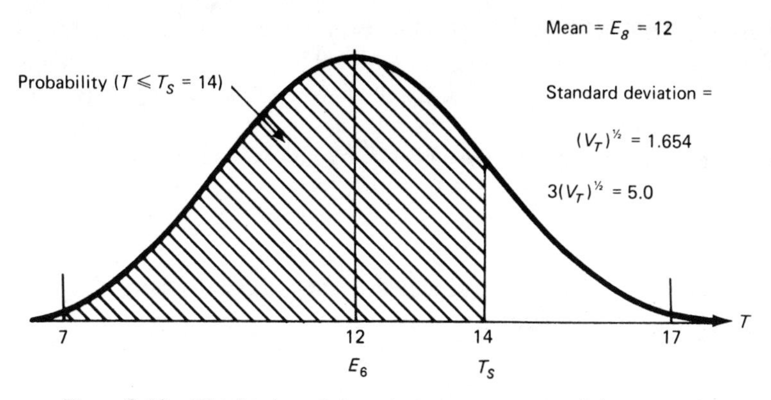


**Figure 9-10** Distribution of the actual occurrence time (T) of event 6.

in Figure 9-10, where the distribution is shown to "touch down" on the abscissa at three standard deviations on either side of the mean, i.e., at 12 ± 5.

Now, the problem of computing the probability of meeting an arbitrary scheduled date, such as 14 shown in Figure 9-10, is quite simple. Since the total area under the normal curve is exactly one, the crosshatched area under the normal curve is directly the probability that the actual event occurrence time, $T$, will be equal to or less than 14, which is the probability that the scheduled date will be met. This probability can be read from the table of normal curve areas, given in Appendix 9-1 at the end of this chapter. In order for this table to apply to any normal curve, it is based on the deviation of the scheduled date in question, $T_S$, from the mean of the distribution, $E_6$, in units of standard deviations, $(V_T)^{1/2}$. Calling this value $Z$, one obtains

$$P\{T \leqq T_S\} = P\{Z \leqq z = [T_S - E(T)]/\sqrt{V_T}\} \tag{5}$$

$$z = (14 - 12)/1.654 = 1.21$$

$$P\{T \leqq 14\} = P\{Z \leqq 1.21\} = 0.89$$

A value of $Z = 1.21$ indicates that the scheduled time, $T_S$, is 1.21 standard deviations greater than the expected time, $E_6 = 12$. Reference to Appendix 9-1B indicates that this value of $Z$ corresponds to a probability of approximately 0.89. Thus, assuming that "time now" is zero, one may *expect* this project to end at time 12, and the probability that it will end on or before the scheduled time of 14, *without expediting the project*, is approximately 0.89. It should be pointed out that if $T_S$ had been two days less than $E_6$ instead of being greater, i.e., $T_S = 12 - 2 = 10$, then $Z = -1.21$, and the corresponding probability would be 0.11. Hence, it is essential that the correct sign is placed on the $Z$.

The above phrase, "without expediting," *is very important*. In certain projects, schedules always may be met by some means or another, for example, by changing the schedule, by changing the project requirements, by adding additional personnel or facilities, etc. The probability being computed here is the probability that the original schedule will be met *without having to expedite the work* in some way or another. For this reason, the following rules should be adopted in dealing with networks having two or more scheduled dates.

*Definition:*

The probability of meeting a scheduled date is the probability of occurrence of an event on or before a specified date (time).

*Rule:*

To compute the probability of meeting a scheduled date, the variance of the initial project event should be set equal to zero, and all scheduled dates other than the one being considered should be ignored in making the variance and probability computations.

*Definition:*

The *conditional* probability of meeting a scheduled date is the probability of the occurrence of an event on or before a specified time, assuming that all prior scheduled events occur on their scheduled dates.

*Rule:*

To compute the *conditional* probability of meeting a schedule date, set the variances of the initial project event and all scheduled events equal to zero, and then make the usual variance and probability computations.

The above definitions and rules suggest where each of these two probabilities might be applied. If one is concerned primarily with the planning of a subnetwork consisting of the activities between two scheduled events, then the conditional probability is pertinent. However, if one is concerned with the entire project, then the unconditional probability of meeting a scheduled date seems pertinent, since it gives the probability of having to expedite a project somewhere in order to meet each of the scheduled event times.

## COMPUTATION OF CUMULATIVE PROBABILITY VS. PROJECT DURATION

Suppose the scheduled time for the completion of the project shown in Figure 9-9 had been $T_s = 10$; then the probability of meeting this schedule would have

**Table 9-1. Computation of Cumulative Probability of Meeting Alternative Project Scheduled Times**

| $T_S$ | $Z = (T_S - 12)/1.654$ | $P\{T \leqq T_S\}$ |
|---|---|---|
| 8 | −2.43 | 0.008 |
| 10 | −1.22 | 0.111 |
| 12 | 0 | 0.500 |
| 14 | 1.22 | 0.889 |
| 16 | 2.43 | 0.992 |
| 18 | 3.65 | 1.000 |

been the relatively low figure of 0.11, as discussed in the previous paragraph. One might then ask the question, what is the probability of meeting various revised project completion times? A convenient way to answer this question is with a graph (or table) giving cumulative probability vs. project duration time. Applying equation (5) above, this probability can be written as

$$\left\{\begin{matrix}\text{Cumulative Probability of}\\ \text{Completing Project on or}\\ \text{Before Time } T_s\end{matrix}\right\} = P\{T \leqq T_s\} = P\{Z \leqq [T_s - E(T)]/\sqrt{V_T}\}$$

Treating $T_s$ as a variable, a cumulative probability curve, such as shown in Table 9-1 and Figure 9-11, can be obtained from the following expression, together with Appendix 9-1.

$$P\{T \leqq T_s\} = P\{Z = [(T_s - 12)/1.654]\}$$

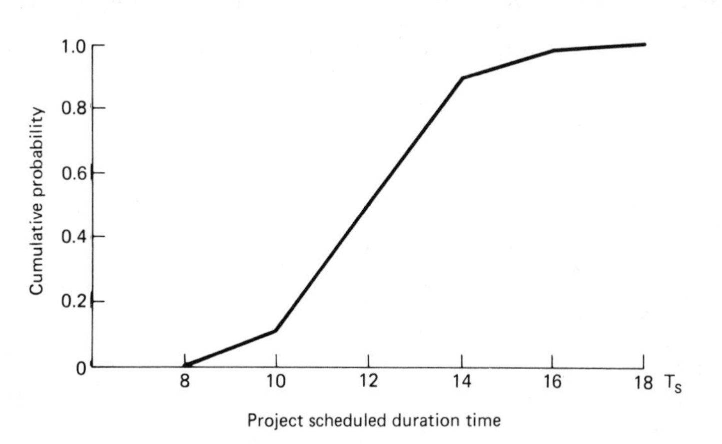


**Figure 9-11** Comulative probability of completing the project on or before the scheduled time $T_S$.

Now, the probability of meeting any possible schedule can be read directly from Figure 9-11. For example, the probability of meeting a schedule of $T_s = 13$ days is approximately 0.7.

## ILLUSTRATIVE EXAMPLE USING PERCENTILE ESTIMATES

To illustrate the use of Eqs. (3) and (4), an example will be given dealing with the overhaul of jet aircraft. The data given in Table 9-2 lists the nine activities that form the critical (longest) path through the network of overhaul activities. Because of physical space limitations, the number of men that can be assigned to an overhaul task is almost fixed. Thus, an activity cannot be accelerated by adding manpower when an overhaul project is behind schedule, and for this reason, it is reasonable to assume that the actual activity duration times are statistically independent. Also, *the* critical path is considerably longer than the next most critical path, so that the usual PERT application of the Central Limit Theorem is justified. The results given in Table 9-2 indicate that the expected duration of this series of activities is 169.0 hours (7.05 days), with a standard deviation of the actual performance times equal to 11.3 hours (0.47 days). The overhaul duration that will not be exceeded more often than say one time in ten can be estimated by substituting these statistics in eq. (5) and solving for $T_S$, which is found to be 7.65 days.

In this example, the objective was to accomplish the overhaul project in seven days. Thus, technological changes in the techniques of overhaul or in the extent of the work to be accomplished must be made to reduce the overall duration of

**Table 9-2. PERT Probability Analysis of an Aircraft Overhaul Project**

| | *Estimated Activity Durations (Hours)* | | | | |
|---|---|---|---|---|---|
| *Activity Description* | *Optimistic* | *Most Likely* | *Pessimistic* | *Mean Time** | *Variance*** |
| Open Pylons | 4 | 4.5 | 6 | 4.7 | 0.39 |
| Open Engines | 1 | 1.5 | 3 | 1.7 | 0.39 |
| Cable Checking | 10 | 12 | 16 | 12.3 | 3.53 |
| Remove Engines | 1 | 1 | 1.5 | 1.1 | 0.02 |
| Pylon Rework | 96 | 110 | 126 | 110.3 | 88.23 |
| Reassemble Pylon, Etc. | 12 | 16 | 20 | 16.0 | 6.27 |
| Fuel Aircraft | 1 | 2 | 6 | 2.5 | 2.45 |
| Check Fuel Tanks, Etc. | 4 | 8 | 12 | 8.0 | 6.27 |
| Wing Closures | 2 | 2.5 | 4 | 2.7 | 0.39 |
| Final Checkout | 6 | 8 | 20 | 9.7 | 19.03 |
| TOTALS: | | | | 169.0 | 126.97 |
| Standard Deviation = 11.3 Hours | | | | | |

* Based on Eq. (3)
** Based on Eq. (4)

the project by at least 0.65 days, or about two work shifts, or a reduction in the standard deviation of the project duration, or some combination of both.

## MONITORING ACTIVITY TIME ESTIMATES AND PERFORMANCE

One of the frequent criticisms of the PERT statistical approach is that the persons supplying values of $a$, $m$, and $b$ do not have the experience to furnish accurate data. This criticism is largely due to the nature of the work being planned. Also, the estimators are not consistent, some being conservative while others are liberal in making their estimates. In addition, there is the real possibility that the estimates are biased by a knowledge of what some higher authority would like the times to be for aribtrary reasons.

This problem was studied by MacCrimmon and Ryavec.[8] They studied the effects of various sources of errors on the estimates of the mean, $t_e$, and the standard deviation $(V_t)^{1/2}$. They considered errors introduced by (1) assuming the activity time distribution was a beta distribution; (2) by using the PERT approximate formulas given by equations (3)* and (4); and (3) by using estimates of $a$, $m$, and $b$ in place of the true values. They concluded that these sources of error could cause absolute errors in estimates of $t_e$ and $(V_t)^{1/2}$ of 30 and 15 percent of the range $(b - a)$, respectively. Since these errors are both positive and negative, however, they will tend to cancel each other.

This problem was also considered by King and Wilson,[7] who studied actual data obtained from a large scale development project involving a prime contractor and a number of subcontractors. They examined the hypothesis that there is a general increase in the accuracy of pre-activity time estimates as the beginning of the activity approaches. They rejected this hypothesis. Their data also indicated that most of the time estimates were optimistic, some being as low as 13 percent of the actual activity duration time. The same conclusion was reached in regard to ability to improve the estimates of remaining life after the activity was started. In this study, the estimate of remaining life was, on the average, 72 percent of the actual value. On the basis of these findings, the authors proposed for consideration, the upward adjustment of all time estimates. In the project studied, a multiplier of 1.39 would have reduced the average error in the activity duration times to zero.

Another serious aspect of this problem that has not received much attention is the effect of the "Peter Principle" applied to activity performance times. It might be stated as follows: *The actual time required to perform an activity will rise to at least fill the time allotted to it*. Referring to Fig. 9-9, the actual time to perform an activity would be expected to be less than the mean, $t_e$, *about* as

---

*They assumed 0 and 100 percentiles were used, and a corresponding divisor of 6 in equation (3).

often as it exceeds the mean. (*About* is used here because for a distribution skewed to the right (or left), the mean will be slightly greater (or less) than the median or 50% point.) Experience would undoubtedly not bear this out because of the ever present human temptation to take a situation where the performance time could be less than the mean, and drag it out to refine or improve the results beyond the initial product specification, or by the diversion of efforts away from the activity to complete "other things." This presents a real challenge to management to create a climate where performance times less than the expected values ($t_e$) will occur. The control chart procedure described below may be of some help in alleviating this problem.

Another approach to this problem, which these authors feel has more promise, is to work with the individuals making the time estimates to improve their future estimates by supplying them with positive feedback information. It is suggested that records be kept for each person supplying activity time estimates. These records should give the deviation of the estimated and actual activity performance times in units of standard deviations as shown in equation (6).

$$Z = (t_e - t)/\sqrt{V_t} \tag{6}$$

$Z$ = difference between estimated and actual duration time divided by the estimated standard deviation of the duration time

One problem which complicates this analysis is that the specifications for the work comprising the activity, or the level of effort applied to the activity, may be changed before the activity is completed. In these cases, the only valid procedure is to use the estimated mean time, $t_e$, which was made immediately after the final deviation in the activity specification or level of effort occurred.

The $Z$ values computed for a particular estimator can then be tabulated and studied, as additional activities with which he is associated are completed. Theoretically, these $Z$ values should vary randomly about zero from about $-3$ to $+3$, with the majority of the values near zero. Deviations from this pattern have a very logical interpretation; they should be studied by the estimator who supplied the data so that improvements in the estimates can be made. If the points vary about a mean less than the zero centerline, then the estimator has the common problem of underestimating the time required to perform assigned tasks. Similarly, variation about a mean greater than zero would indicate a consistent overestimation of performance times. Also, the points should vary from $-3$ to $+3$, or possibly $-4$ to $+4$ if the performance time distributions are skewed. If they vary over too wide a range, say $\pm 5$, then generally, the difference between the pessimistic and optimistic estimates $(b - a)$ must be increased. The other possibility is not too likely, but consistent variation over too narrow a range would call for a reduction in the difference $(b - a)$.

## SUMMARY OF HAND PROBABILITY COMPUTATIONS

PERT probability computations can be handled in much the same way as time-cost trade-offs were handled in Chapter 8. If a computer is being used, then estimates of $a$, $m$, and $b$ must be obtained and used on all activities. In this case, probabilities of meeting specified scheduled times are given in the computer output, such as are shown in Figure 9-13, for the network shown in Figure 9-12. However, if the computations are being made by hand, one could start off with a single time estimate for the *mean* performance time for each activity, and then obtain $a$, $m$, and $b$ values only for those activities on the critical path, or the longest path leading to the scheduled event in question. This procedure is summarized below.

1. Make the usual forward and backward pass computations based on a single-time estimate, $D$, for each activity.
2. Suppose one wishes to compute the probability of meeting a specified scheduled time for event $X$. Then obtain estimates of $a$, $m$ and $b$ for only those activities that comprise the "longest path" from the initial event to the event $X$. If necessary, adjust the length of this path as dictated by the new $t_e$ values based on $a$, $m$ and $b$.
3. Compute the variance for event $X$, ($V_T$), by summing the variances for the activities listed in Step 2. $V_T$ = sum of values of $[(b - a)/3.2]^2$, for each activity on the "longest path" leading to event $X$.
4. Compute $z$ using equation (5) and look up the corresponding probability in the normal curve table given in Appendix 9-1.

To use the PNET method described below, values of $a$ and $b$ must also be collected on the "near" critical paths, to be defined.

Referring again to Figure 9-12, the practice of labeling the events on the network, rather than the activities, should be noted. Event labeling in this manner is not recommended because it confuses the reader; one cannot be sure which arrows and time estimates apply to which events. This procedure was adopted to be compatible with the early government practice of controlling this type of project by monitoring milestone events. Milestones are, of course, key points in time in the life of a project and are, therefore, events which occur at the completion of one or more activities. Upper management keys on the status of milestone events, while lower levels of management concentrate on the performance of individual activities as shown in Figure 9-13. To facilitate this reporting at several levels of detail, computer programs are utilized to summarize networks by retaining only the specified set of milestone events. (This concept was discussed under Network Condensation in Chapter 3.)

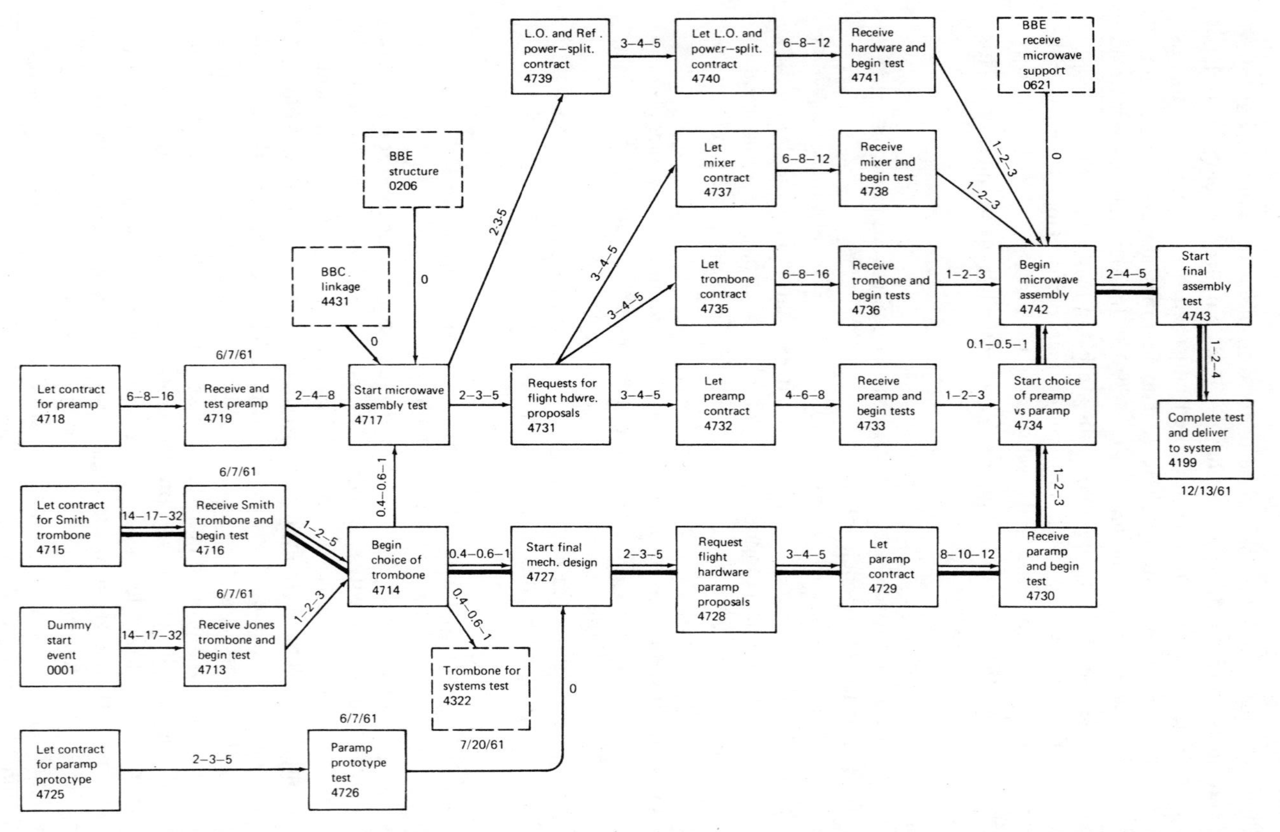


**Figure 9-12** Typical PERT network of an electronic module development project. (*Courtesy of the Applied Physics Laboratory, John Hopkins University.*)

PAGE 1

PERT SYSTEM
PERT SYSTEM

RUN 1 ENDING EVENT
BY PATHS OF CRITICALITY DATE 06-07-61
CHART AJ LR SN 9 ELECTRONIC MODULE (ILLUSTRATIVE NETWORK) SYSTEM W 034

| EVENT PREDECESSOR | SUCCESSOR | NOMENCLATURE | DEP | DATE EXPECTED | ALLOWED | DATE SCHD/ACT | PROB | SLACK | EXP TIME | EXP VAR |
|---|---|---|---|---|---|---|---|---|---|---|
| 4004-715 | 4004-716 | REV DATE (SMITH TROMBONE RECD-BEG TEST) | 98 | | 05-26-61 | A06-07-61 | | -1.6 | + | |
| 4004-716 | 4004-714 | SMITH TROMBONE TESTED | 0146 | 06-23-61 | 06-12-61 | | | -1.6 | + 2.3 | .4 |
| 4004-714 | 4004-727 | TROMBONE CHOSEN-BEGIN MECH DESIGN | 0146 | 06-28-61 | 06-16-61 | | | -1.6 | + 3.0 | .5 |
| 4004-727 | 4004-728 | RFP PARAMP FLIGHT HARDWARE | 0146 | 07-20-61 | 07-08-61 | | | -1.6 | + 6.1 | .7 |
| 4004-728 | 4004-729 | PARAMP CONTRACT LET | 0146 | 08-17-61 | 08-05-61 | | | -1.6 | +10.1 | .8 |
| 4004-729 | 4004-730 | PARAMP RECEIVED | | 10-26-61 | 10-14-61 | | | -1.6 | +20.1 | 1.3 |
| 4004-730 | 4004-734 | PARAMP TESTED | 0146 | 11-09-61 | 10-28-61 | | | -1.6 | +22.1 | 1.4 |
| 4004-734 | 4004-742 | CHOICE BETWEEN PREAMP-PARAMP | 0146 | 11-13-61 | 11-01-61 | | | -1.6 | +22.6 | 1.4 |
| 4004-742 | 4004-743 | COMPL MICROWAVE ASSY | 0146 | 12-09-61 | 11-28-61 | | | -1.6 | +26.5 | 1.6 |
| 4004-743 | 4004-199 | COMPL FINAL TEST MICWAVE ASSY-DELIVERED | 0146 | 12-25-61 | 12-13-61 | 12-13-61 | .12 | -1.6 | +28.6 | 1.9 |
| | | | | | | | | | | |
| 4000-001 | 4004-713 | REV DATE (JONES TROMBONE RECD-BEG TEST) | 99 | | 05-29-61 | A06-07-61 | | -1.3 | + | |
| 4004-713 | 4004-714 | JONES TROMBONE TESTED | 0146 | 06-21-61 | 06-12-61 | | | -1.3 | + 2.0 | .1 |
| | | | | | | | | | | |
| 4004-714 | 4004-717 | TROMBONE CHOSEN-BEGIN MICWAVE ASSY TEST | 0146 | 06-28-61 | 07-02-61 | | | + .5 | + 3.0 | .5 |
| 4004-717 | 4004-731 | RFP FOR FLIGHT HDW-MIXER-TROMB-PREAMP | 0146 | 07-20-61 | 07-24-61 | | | + .5 | + 6.1 | .7 |
| 4004-717 | 4004-739 | COMPL MICWAVE ASSY TEST-RFP LOC OSCIL | 0146 | 07-20-61 | 07-24-61 | | | + .5 | + 6.1 | .7 |
| 4004-731 | 4004-735 | TROMBONE CONTRACT LET | 0146 | 08-17-61 | 08-21-61 | | | + .5 | +10.1 | .8 |
| 4004-731 | 4004-737 | MIXER CONTRACT LET | 0146 | 08-17-61 | 08-21-61 | | | + .5 | +10.1 | .8 |
| 4004-739 | 4004-740 | CONTRACT LET FOR LOC OSCIL AND PWR SPLT | 0146 | 08-17-61 | 08-21-61 | | | + .5 | +10.1 | .8 |
| 4004-735 | 4004-736 | TROMBONE RECEIVED | | 10-14-61 | 10-18-61 | | | + .5 | +18.5 | 1.8 |
| 4004-737 | 4004-738 | MIXER RECEIVED | | 10-14-61 | 10-18-61 | | | + .5 | +18.5 | 1.8 |
| 4004-740 | 4004-741 | LOC OSC-PWR SPLITTER RECEIVED | | 10-14-61 | 10-18-61 | | | + .5 | +18.5 | 1.8 |
| 4004-736 | 4004-742 | TROMBONE TESTED | 0146 | 10-28-61 | 11-01-61 | | | + .5 | +20.5 | 1.9 |
| 4004-738 | 4004-742 | MIXER TESTED | 0146 | 10-28-61 | 11-01-61 | | | + .5 | +20.5 | 1.9 |
| 4004-741 | 4004-742 | LOC OSC-PWR SPLITTER TESTED | 0146 | 10-28-61 | 11-01-61 | | | + .5 | +20.5 | 1.9 |

**Figure 9-13** Typical PERT computer output. (First three-paths shown here.)

The critical path through the project network in Figure 9-12 is marked by the heavy line. The computer output in Figure 9-13 is based on a primary sort of slack, and a secondary sort of early start time. This places the critical path as the first set of 10 activities, the next most critical path as the next set of 2 activities, and the third most critical paths as the final set of 12 activities. Notice that the slack on the critical path is -1.6 weeks, because the scheduled completion time (12-13-61) is earlier than the expected time (12-25-61). Because of this, the probabiliy of meeting this schedule is a low value of 0.12. Clearly, this would be a red light signal, calling for some immediate application of time-cost trade-off procedures to raise this probability to an acceptable level. Furthermore, it would be unwise to wait further to consider this problem, because as time goes by, the most economical way to buy time may have passed by.

## INVESTIGATION OF THE MERGE EVENT BIAS PROBLEM

As pointed out in the introduction to this chapter, the conventional PERT procedure described above always leads to an optimistically biased estimate of the earliest (expected) occurrence time for the network events. This bias arises because all subcritical paths are ignored in making the forward pass computations. If the longer path leading to a merge event is much longer than the second longest path, and/or the variance of the activities on the longest path is small, this bias will be insignificant. The first part of this section will be devoted to a series of examples designed to give the reader an appreciation for the significance of this bias problem. Finally, a solution to this problem using the PNET procedure and Monte Carlo simulation will be described in some detail.

### Magnitude of Bias

A study of this problem was made by MacCrimmon and Ryavec,[8] who considered two of the more important factors affecting the magnitude of the merge event bias. First, one would intuitively expect the bias to increase as the number of parallel paths to the network end event increases. This is studied in Figure 9-14 below. Second, one would also expect the bias to increase as the expected length of the parallel paths become equal. This is studied in Figure 9-15.

Consider the four activity network in Figure 9-14a. The particular discrete distribution assumed for each of these activities can be identified in Table 9-3 by the corresponding mean shown on the network activities. There are two paths, *ABD* and *ACD*, both having a mean length of 6. The mean of the maximum time distribution, or the earliest expected time for event $D$ is 6.89. Thus, the error in the PERT calculated mean is 12.9 percent of the actual mean.

There are two possible ways a third path, with a mean length of 6, may be created by adding one more activity. In one case the path may be completely independent of the other two paths, thus resulting in a third parallel element,

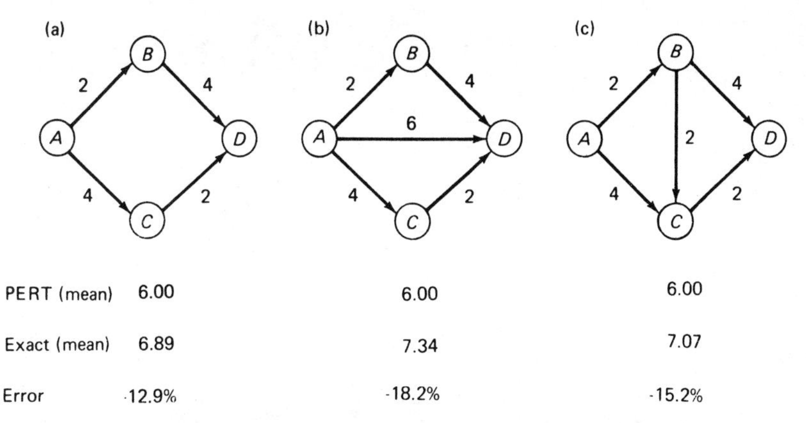


**Figure 9-14** Effect of parallel paths, with and without correlation, on the merge event bias.

*AD*, as depicted in Figure 9-14b. Alternatively, an activity *BC* can be added with a mean time of 2, thus creating path *ABCD*, shown in Figure 9-14c. In both cases there are three paths, all of mean length 6, and the network has four events and five activities.

The addition of the third path in parallel leads to an increase in the deviation of the PERT-calculated mean (still 6) from the actual mean, which in this case is 7.336. Thus, the error has increased to 18.2 percent. Figure 9-14c, on the other hand, is a network configuration, where there is a cross connection between two parallel paths. Since there are three paths, one would expect a larger error than in a similar network with only two paths (such as Figure 9-14a), although not as large an error as in Figure 9-14b, where the three paths are in parallel. The correlation (resulting from the common activities) in the network of Figure 9-14c does indeed have the effect discussed, and the mean of the maximum time distribution lies between these two bounds, being 7.074. The error as a percent of the actual mean is 15.2 percent.

The examples given above are extreme cases, since all the paths have the same expected duration—hence, they are all critical paths. If the durations of some paths are shorter than the duration of the longest path, their effect on the project mean and standard deviation would not be as great. However, if they have a mean duration very close to the mean duration of the critical path, they would not be critical but they would have an effect almost as significant as the examples of the previous sections. The following examples shown in Figure 9-15 indicate the effect of slack in a path length.

The simple network has only two paths, *ABC* and *AC*. All activities are assumed to be normally distributed, with standard deviation equal to 1, and the appropriate mean given on the diagram. It may be noted from the diagrams that various lengths are assumed for paths *ABC* and *AC*, ranging from both of them being of equal length, to path *AC* being only $\frac{1}{4}$ the length of path *ABC*.

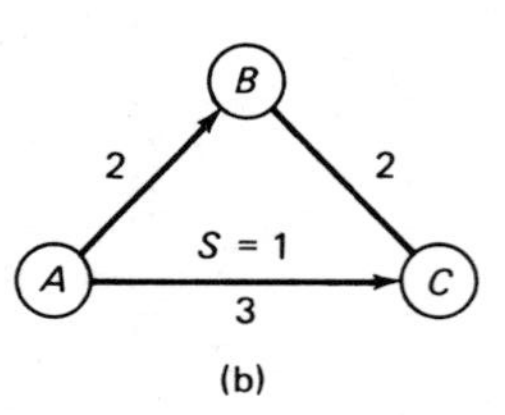


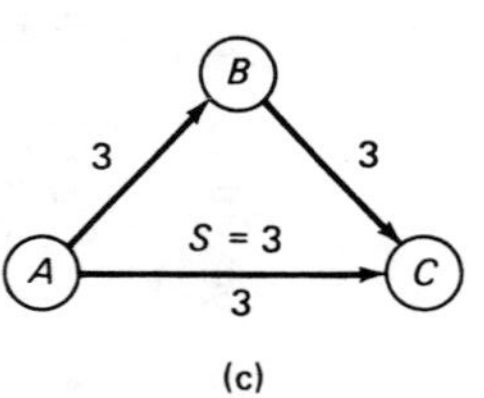


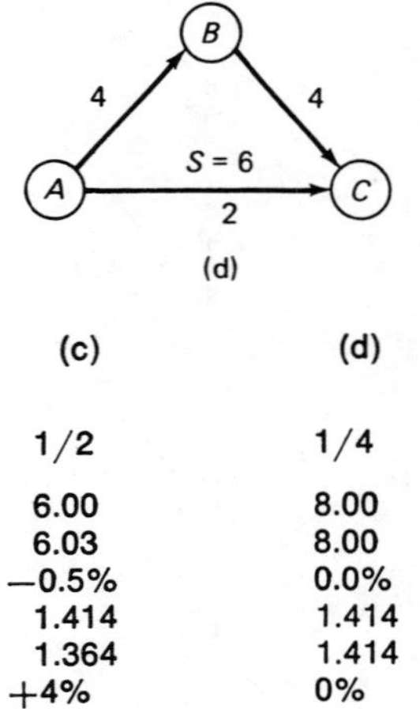


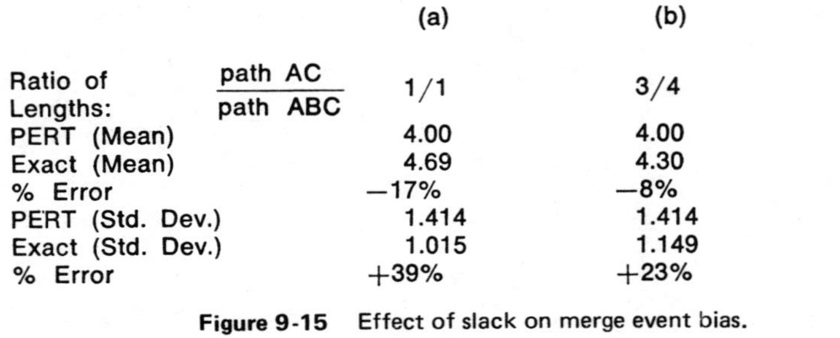

| | (a) | (b) | (c) | (d) |
|---|---|---|---|---|
| Ratio of Lengths: path AC / path ABC | 1/1 | 3/4 | 1/2 | 1/4 |
| PERT (Mean) | 4.00 | 4.00 | 6.00 | 8.00 |
| Exact (Mean) | 4.69 | 4.30 | 6.03 | 8.00 |
| % Error | −17% | −8% | −0.5% | 0.0% |
| PERT (Std. Dev.) | 1.414 | 1.414 | 1.414 | 1.414 |
| Exact (Std. Dev.) | 1.015 | 1.149 | 1.364 | 1.414 |
| % Error | +39% | +23% | +4% | 0% |

**Figure 9-15** Effect of slack on merge event bias.

**Table 9-3. Discrete Distribution for Activities in Figure 9-14**

| $t$ | *Probability* | $t$ | *Probability* |
|---|---|---|---|
| 1 | ¼ | 2 | ¼ |
| 2 | ½ | 4 | ½ |
| 3 | ¼ | 6 | ¼ |
| Mean = $t_e$ | = 2 | Mean = $t_e$ | = 4 |
| Std. Dev. = $(V_t)^{1/2}$ | = 0.707 | Std. Dev. = $(V_t)^{1/2}$ | = 1.414 |
| Coef. of Var. = $(V_t)^{1/2}/t_e$ | = 35% | Coef. of Var. = $(V_t)^{1/2}/t_e$ | = 35% |

This example indicates that the deviation of the PERT-calculated mean and standard deviation, from the actual mean and standard deviation, may be quite large when the paths are about equal in length, but the difference decreases substantially as the path lengths become farther apart.

### Rules of Thumb on Merge Point Bias

To summarize qualitatively the above results on merge event bias, it can be noted that the magnitude of the bias correction at a given merge event increases as

1. the number of merging activity increases,
2. the expected complete times of the merging activities get closer together,
3. the variances of the merging activities increase, and
4. the correlation among the merging activity complete times approaches zero.

Because of point 2 above, the correction at most merge events will be negligible and thus can be ignored. From a study of tables derived by Clark,[3] giving the expected value of the greatest of a finite set of random variables, this can be stated as a useful rule of thumb as follows.

*Rule:*

If the difference between the expected complete times of the two merging activities being considered is greater than the larger of their respective standard deviations, then the bias correction will be small; if the difference is greater than two standard deviations, the bias will be less than a few percent and can be ignored. (The difference referred to here is what has been defined in Chapter 4 as activity free slack.) If there are more than two merging activities, this rule should be applied to the two with the latest expected finish times.

The validity of this rule is illustrated in Figure 9-15. The difference (slack) in the expected time of the two merging activities is less than one standard deviation in Figure 9-15b and is greater than two in Figure 9-15c. The corresponding biases of 8 percent and 0.5 percent are appreciable and insignificant, respectively, as suggested by this rule.

If the above does not rule out the need for a bias correction, then it should be made by one of the procedures described below.

### Analytical Merge Event Bias Correction Procedures

The merge event bias correction problem is essentially a statistical problem, dealing with a random variable defined as the maximum value of a set of random variables, not necessarily statistically independent. The latter condition complicates the problem greatly. The maximum value is the earliest expected occurrence time of the (merge) event in question, and the set of random variables is the actual complete times of the activities merging to the event in question. These latter times are not always statistically independent, because of the network crossover condition previously illustrated in Figure 9-14c.

This is an intriguing statistical problem that has caught the fancy of many researchers, and dozens of papers have been written about it. Only the relatively recent work of Ang, Abdelnour and Chaker[1] (1974) will be presented here because it is the simplest of the procedures that produce a satisfactory solution. Their procedure is called PNET.

To introduce this procedure, consider again the network shown earlier in Figure 9-1. The conventional critical path is 0-3-7-8, having an expected time of 15, with a standard deviation 1.68 days. Suppose the three time estimates for the additional activities 3-4, 4-5, and 5-8 are as shown in Table 9-4. These data

**Table 9-4. Mean and Standard Deviation of the Critical and Near Critical Paths for the Network in Figure 9-1**

| | *Time Estimates* | | | *Path* 0-3-7-8 | | *Path* 0-3-4-5-8 | |
|---|---|---|---|---|---|---|---|
| *Activity* | *a* | *m* | *b* | *Mean* | *Variance* | *Mean* | *Variance* |
| 0-3 | 1 | 2 | 3 | 2 | 0.39 | 2 | 0.39 |
| 3-7 | 6 | 8 | 10 | 8 | 1.56 | — | — |
| 7-8 | 3.5 | 5 | 6.5 | 5 | 0.88 | — | — |
| 3-4 | 1 | 4 | 13 | — | — | 5 | 14.06 |
| 4-5 | 2 | 4 | 6 | — | — | 4 | 1.56 |
| 5-8 | 2 | 3 | 4 | — | — | 3 | 0.39 |
| | | | Totals* | 15.0 | 2.83 | 14.0 | 16.40 |
| | | | Standard Deviation | — | 1.68 | — | 4.05 |

*The mean and variance of the duration of a path is merely the sum of the means and variances of the activities along the path in question; the standard deviation of the path duration is then obtained as the square root of its variance.

lead to the mean, variance, and standard deviation statistics for the critical path, 0-3-7-8, and the near critical path, 0-3-4-5-8, as shown at the bottom of the table. Using eq. (5), the cumulative probability curves for each of these paths are then computed and are shown in Figure 9-16. They appear as straight lines because the graph paper used here, called Normal Probability paper, has a vertical scale adjusted to produce a straight line for any normally distributed random variable. It is obvious from these two graphs that the PERT probability procedure, which considers *only* path 0-3-7-8, is biased on the high side. In fact, for any scheduled time greater than the cross over point (about 15.7) the cumulative probability for the near critical path is considerably *less* than for the critical path itself. This anomoly occurs because of the very large variance for the near critical path; a situation which could occur in practice. It should also be noted that although the individual path lengths have Normal distributions (by virtue of the Central Limit Theorem) the distribution of the project duration, considering all paths, is not Normal. This distribution, to be derived below, is shown by the dashed line in Figure 9-16. This censoring of the lower tail of the distribution is a characteristic result which has been verified repeatedly by simulation. It can be explained by recalling that the actual project duration is the *maximum* of a set of random variables. The lower tail of this distribution of the *maximums* is

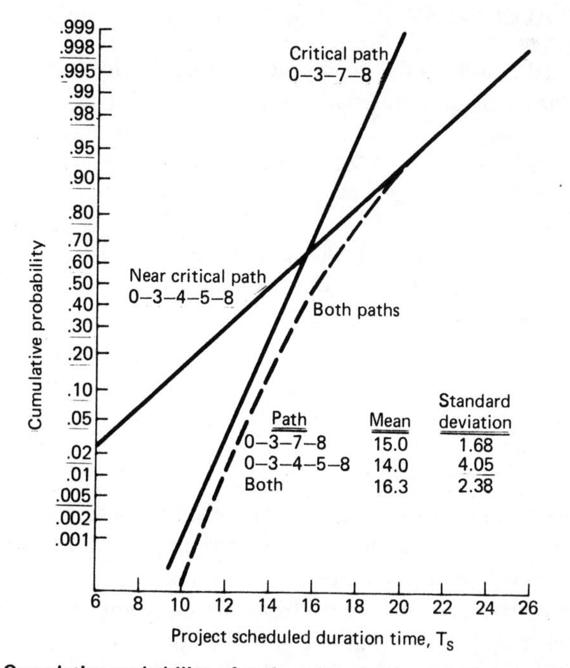


**Figure 9-16** Cumulative probability of path and project completion on or before time $T_S$.

truncated because a low maximum requires *all* of the paths to be low, and hence it occurs infrequently. However, a high maximum will result when only *one or more* high path times occur, and hence it occurs more frequently. At this point it is appropriate to ask, "What is the correct cumulative probability?"

If we are justified in ignoring all but these two paths, we could proceed to answer this question by first assuming they are statistically independent. This is a very good assumption because these paths have only one element in common, activity 0-3, and its variance is negligible in comparison to the total path variances. Thus, from basic probability theory of independent events, the probability that the project will be completed on or before time $T_s$ is merely the *product* of the cumulative probabilities that each path will be completed on or before time $T_s$. For example, for $T_s = 14$, this probability (reading from Figure 9-16) is $0.50 \times 0.27 = 0.135$. The latter value is shown by the dashed curve, labeled Both Paths. This is essentially the procedure developed by Ang,[1] called PNET. It amounts to finding the set of assumed independent paths which "represent" the network, and then set the cumulative probability for the project equal to the product of the cumulative probabilities for each of the "representative" paths. It can be shown that the dashed PNET curve in Figure 9-16 is almost identical to the correct curve.

## ANG'S PNET ALGORITHM

The PNET algorithm utilizes the important fact that the linear correlation coefficient between any pair of path lengths, say $T_i$ and $T_j$ for paths $i$ and $j$, is given by

$$\rho_{ij} = \sum_k v_k / \sqrt{V_i} \sqrt{V_j} \tag{7}$$

where the sum in the numerator is the sum of the variances of the activities that are common to both paths $i$ and $j$, and the denominator is merely the product of the standard deviations of the total durations of paths $i$ and $j$. The value of $\rho_{ij}$ varies from 0 for completely independent paths, to 1 for identical paths. For the two paths in the above example, the data in Table 9-4 shows that $\rho_{ij}$ is only 0.057, indicating almost complete independence of these two paths.

$$\rho_{0\text{-}3\text{-}7\text{-}8,\ 0\text{-}3\text{-}4\text{-}5\text{-}8} = 0.39/1.68 \times 4.05 = 0.057$$

In the PNET algorithm, this correlation coefficient is used to determine the set of representative paths for the network. If two paths have a correlation coefficient greater than 0.5, then the longer of the two paths is said to represent both paths, and the shorter path is dropped from consideration. Similarly, if the correlation coefficient is less than 0.5, the two paths are assumed independent, and both remain under consideration. Then, the probability that a project will meet

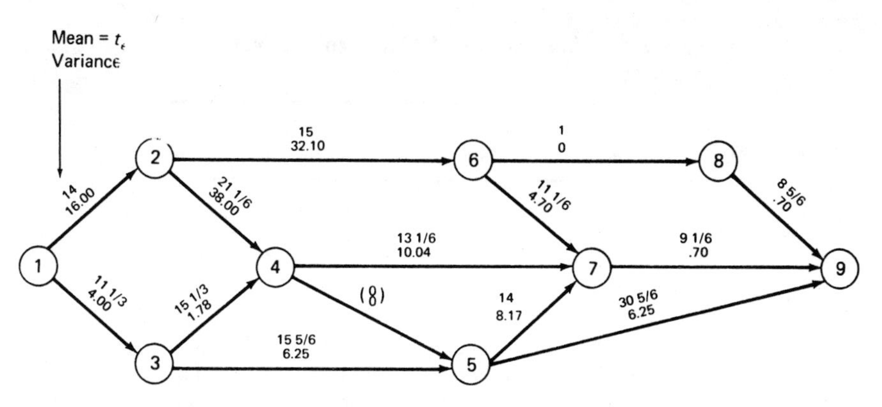


**Figure 9-17** Illustrative network for Monte Carlo simulation and PNET.

a specified completion date is taken as the product of the probabilities that each of the representative paths will meet the specified completion date.

An application of this procedure, given in the next section, shows excellent agreement with the correct results obtained by simulation. Ang gives three applications of this procedure, and each shows excellent agreement with the results obtained by simulation. One explanation of why this procedure seems to work better than expected, based on intuition, is that a form of the Central Limit Theorem is in operation. That is, dropping paths with correlations greater than 0.5, tends to cancel the error effects of assuming paths with correlations less than 0.5 are independent. A modification of Ang's original PNET procedure is given below, along with its application to the network shown above in Figure 9-17.

## Modified PNET Algorithm

1. Generate the list of major network paths* sequentially with decreasing mean path durations, starting with the critical path(s). In case of ties, arrange the paths with decreasing standard deviations. The major paths are defined as those whose mean path durations are at least a certain percentage of the expected critical path duration. (The rule developed in the previous section and adopted here, suggests the retention of all paths whose lengths differ from the critical path by less than twice the larger of the standard deviations of the two paths.) Number the major paths from 1 to $N$. This step is illustrated for the network shown in Figure 9-17, and the results are given in Table 9-5. All

*The listing of paths in order of criticality could be accomplished from a slack sort of the project activities, modified so that each slack path extends from the initial to the terminal network event.

**Table 9-5. Application of PNET to the Network in Figure 9-17.**

| *No.* | *Path* | *Mean* | *Variance* | *Standard Deviation* |
|---|---|---|---|---|
| 1 | 12459 | 66 | 60.3 | 7.8 |
| 2 | 124579 | 58.7 | 62.9 | 7.9 |
| 3 | 1359 | 58 | 16.5 | 4.1 |
| 4 | 13459 | 57.5 | 12.0 | 3.5 |
| 5 | 12479 | 57.5 | 64.7 | 8.0 |
| 6 | 13579 | 50.3 | 19.1 | 4.4 |
| 7 | 134579 | 49.8 | 14.7 | 3.8 |
| 8 | 12679 | 49.3 | 53.5 | 7.3 |
| 9 | 13479 | 49 | 16.5 | 4.1 |
| 10 | 12689 | 38.8 | 48.8 | 7.0 |

10 network paths are shown here for illustrative purposes; the listing could have stopped after generating path 6 and showing that its mean differs from the first (critical) path by 66 - 50.3 = 15.7, which exceeds twice the larger of the path standard deviations, i.e., 2 X 7.8 = 15.6.

2. Compute the correlation coefficient for each pair of major paths using equation (7), and arrange them in matrix form. For the above example, this matrix is shown in Table 9-6.

3. Determine the set of representative paths for the network. This procedure is given for the correlation matrix shown in Table 9-6.

**Table 9-6. Correlation Matrix for the Five Major Paths in the Network in Figure 9-17**

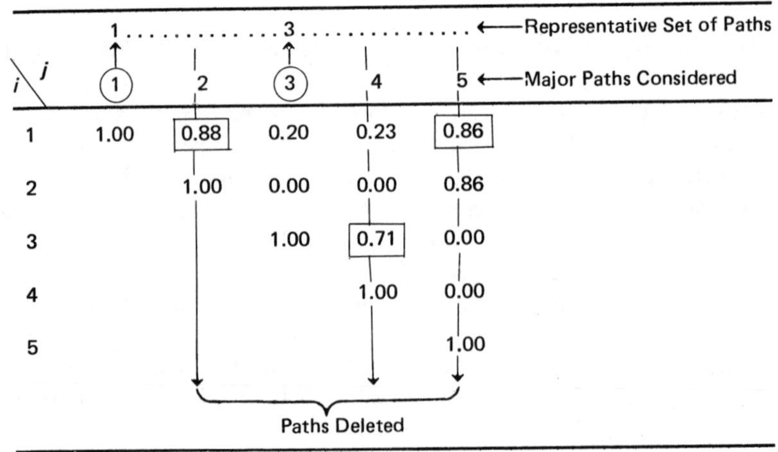

1 . . . 3 . . . ←Representative Set of Paths

| $i$ \ $j$ | (1) | 2 | (3) | 4 | 5 ←Major Paths Considered |
|---|---|---|---|---|---|
| 1 | 1.00 | 0.88 | 0.20 | 0.23 | 0.86 |
| 2 | | 1.00 | 0.00 | 0.00 | 0.86 |
| 3 | | | 1.00 | 0.71 | 0.00 |
| 4 | | | | 1.00 | 0.00 |
| 5 | | | | | 1.00 |

Paths Deleted

3.1. Start with path 1, enter it into the set of representative paths, and move across row 1 of the correlation matrix deleting all paths with correlations greater than 0.5. In this example, delete paths 2 and 5 because $r_{12} = 0.88$ and $r_{15} = 0.86$ are both greater than 0.5.

3.2. Move to the next remaining path, enter it into the set of representative paths, and move across this row of the correlation matrix again deleting all paths with correlations greater than 0.5. In this example, enter path 3 into the set of representative paths, and delete path 4 because $r_{34} = 0.71$ is greater than 0.5.

3.3. Continue Step 3.2 until there are no remaining major paths to be considered; at this point stop, list the set of representative paths and number them $1, 2, \ldots, K$. In this example the procedure stops after Step 3.2 because no major paths remain to be considered, and the set of representative paths consists of paths 1 and 3.

4. Calculate the probability that the duration of each representative path $(T_1, \ldots, T_K)$ will not exceed some scheduled project duration $(T_s)$, assuming that each path duration has a normal distribution with mean and standard deviation computed in Step 1, i.e., calculate $P(T_1 \leqq T_s), \cdots, P(T_K \leqq T_s)$. These probabilities are shown on the left side of Table 9-7, for the example problem.

5. The probability that the project completion time, $T$, will meet the schedule time, $T_s$, is denoted by $P(T \leqq T_s)$, and is then given approximately by the product

$$P(T \leqq T_s) \cong P(T_1 \leqq T_s) \times \cdots \times P(T_K \leqq T_s).$$

For the example, there are two representative paths, and this probability is given as follows:

$$P\{T \leqq T_s\} \cong P\{Z \leqq (T_s - 66)/7.8\} \times P\{Z \leqq (T_s - 58)/4.1\}$$

The results of these computations are given in Table 9-7. It will be shown below, that these PNET probability estimates are not significantly different from the results obtained by Monte Carlo simulation, where the latter are taken as the "correct" probabilities. The computation of the estimates of the PNET mean and standard deviation are also shown at the bottom of Table 9-7.

### Monte Carlo Simulation Approach to Pert Probabilities

The Monte Carlo simulation approach to the solution of this problem was used by Van Slyke[12] in 1963. He recognized this problem as one of simply solving the stochastic network model to find something that corresponds, in some sense, to the project duration and critical path in the deterministic case. The difficulty here was avoided to some extent by approximating the random problem by a series of problems of the deterministic form. To accomplish this, Monte Carlo simulation was used. A bonus from this approach was that it not only gave

**Table 9-7. Computation of PNET Project Probabilities from Representative Path Probabilities**

| | *Path* 1 = 1-2-4-5-9 | | *Path* 3 = 1-3-5-9 | | |
|---|---|---|---|---|---|
| $T_s$ | $z_1 = (T - 66)/7.8$ | $P(Z \leqq z_1)$ | $z_3 = (T_s - 58)/4.1$ | $P(Z \leqq z_3)$ | PNET = $P(Z \leqq z_1) \cdot P(Z \leqq z_3)$ |
| 55 | −1.41 | .079 | −0.73 | .233 | .018 |
| 60 | −0.77 | .221 | 0.49 | .688 | .152 |
| 65 | −0.13 | .448 | 1.71 | .956 | .428 |
| 70 | 0.51 | .695 | 2.93 | .998 | .694 |
| 75 | 1.15 | .875 | 4.14 | 1.000 | .875 |
| 80 | 1.79 | .963 | 5.37 | 1.000 | .963 |
| 85 | 2.44 | .993 | 6.59 | 1.000 | .993 |
| MEAN | 66 | | 58 | | 66.9* |
| STANDARD DEVIATION | 7.8 | | 4.1 | | 6.4 |

*This mean value was computed from the PNET cumulative probability distribution using the following formula:

$$\text{Mean} = \sum_{i=1}^{K} t_i \times p(t_i)$$

where $t_i$ is the class mark (mid-point) of the $i$th class interval ($i = 1, 2, \ldots, K$; where $K = 7$ in this example), and $p(t_i)$ is the probability that the (random) project duration will fall in the $i$th class interval.

$$\text{Mean} = (.018 - 0)\ 52.5 + (.152 - .018)\ 57.5 + \cdots + (.993 - .963)\ 82.5 + (1.000 - .993)\ 87.5 = 66.9$$

The standard deviation, or variance, was computed in a similar fashion, replacing the class mark (e.g., 52.5), by the (class mark-mean)$^2$ = $(52.5 - 66.9)^2$.

unbiased estimates of the mean and variance of the project duration, along with the distribution of total project time, but it also gave estimates for quantities not obtainable from the standard PERT approach. In particular, the 'criticality' of an activity, i.e., the probability of an activity being on the critical path, can be calculated. One of the more misleading aspects of conventional PERT methods is the implication that there is a unique critical path. In general, any of a number of paths could be critical, depending on the particular realization of the random activity durations that actually occur. Thus, it makes sense to talk about a criticality index. This appears to be an exceedingly useful measure of the degree of attention on activity should receive by management, and is not as misleading as the critical path concept used in PERT. It should be added that the probability of an activity being on the critical path is not correlated too well with slack, as computed by the conventional PERT procedure, which is the factor that usually determines the degree of attention that a particular activity receives.

The Monte Carlo simulation procedure was applied by Van Slyke to the network given in Figure 9-17. Each activity was assumed to have a beta distribution with mean, $t_e$, and a variance, $V_t$, as noted on each activity. As generally recommended by Van Slyke for this purpose, 10,000 sets of random times were generated for each activity in the network. For each of these sets, the longest path through the network was determined: its duration was noted, as well as a count for each activity on the critical path. The results of these 10,000 simulations are given in Figure 9-18, where the probability that an activity was on the critical path is noted on each activity. For example, 0.737 on activity 1-2 means that in 7370 of the 10,000 simulations, this activity was on the longest path in the network. Also, given at the bottom of Figure 9-18 are the statistics pertaining to the total project duration. We note here that the PERT estimate of the project mean was low (optimistic) by only 1.5 percent; however, the variance was estimated too high by 42 percent, and the results previously obtained by PNET are well inside the 95% confidence intervals based on the simulation estimates. The PERT result is about as expected according to the rule given in the previous section, because there is a considerable amount of slack along the subcritical path(s) at each merge event. It is interesting to note, however, that activities 1-3 and 3-5, which are not on the conventional PERT critical path, have appreciable probability of ending up on the actual critical path.

Another output of the Monte Carlo simulation study is given in Figure 9-19, where the cumulative probability of a specified project duration is given for simulation and PNET, which were almost identical, and is compared with the results given by the conventional PERT procedure.

## SIMULATION SAMPLE SIZE

A final note on the simulation procedure regards the cost of obtaining these results. A sample size of 10,000 simulations was used in the example in Figures 9-17 and 9-18, however, this is considerably larger than justified for this problem. A recent empirical study by Crandall[5] indicates that this sample size can be reduced to 1000 or less and still obtain an adequate level of confidence in the final estimates. Other practitioners of simulation suggest samples as low as 400. This is the generally recommended range of sample sizes for this problem.

A statistical analysis for the criticality index can be based on the binomial distribution, since for each simulation trial, any particular activity is either on, or not on, the critical path. The (1-$\alpha$) level confidence interval could then be constructed on the true criticality index, $p$, from the equation

$$\text{Probability}\left\{\hat{p} - Z_{\alpha/2}\sqrt{\frac{\hat{p}(1-\hat{p})}{n}} \leqq p \leqq \hat{p} + Z_{\alpha/2}\sqrt{\frac{\hat{p}(1-\hat{p})}{n}}\right\} = (1-\alpha)$$

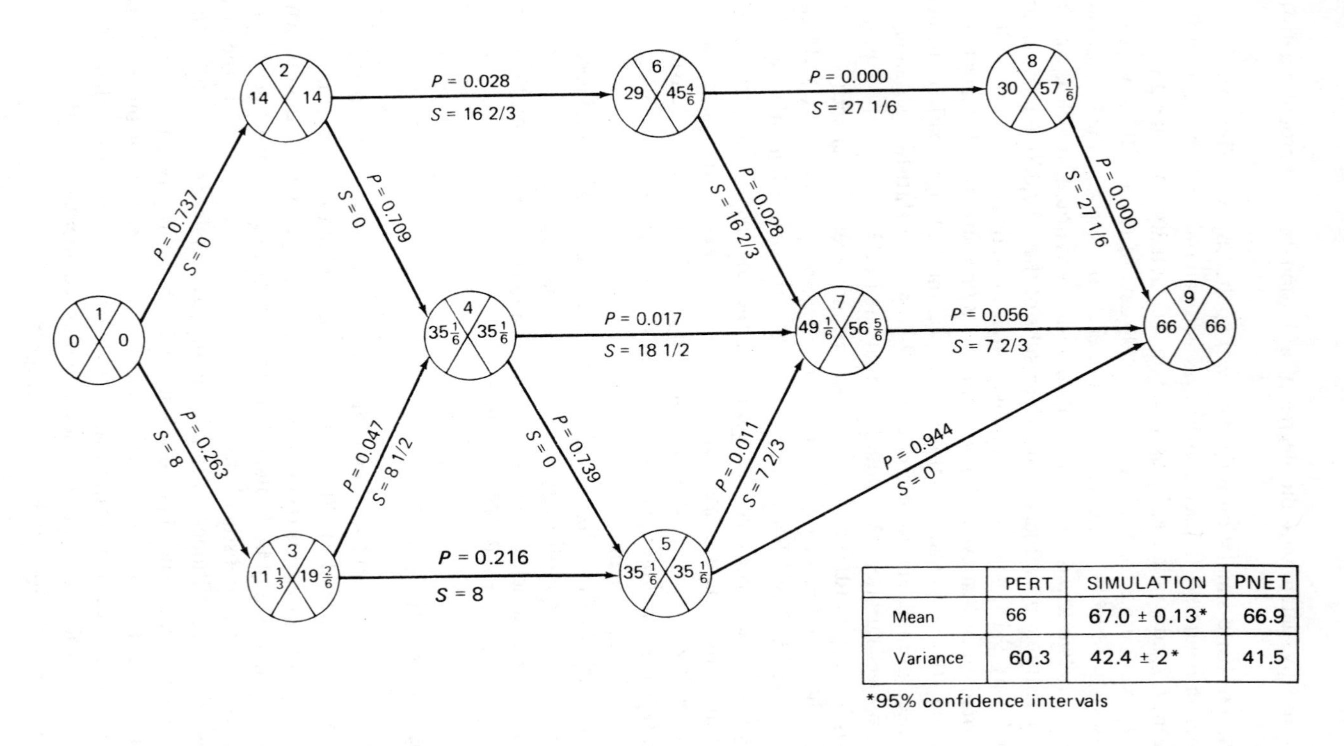


| | PERT | SIMULATION | PNET |
|---|---|---|---|
| Mean | 66 | 67.0 ± 0.13* | 66.9 |
| Variance | 60.3 | 42.4 ± 2* | 41.5 |

*95% confidence intervals

**Figure 9-18** Activity critically and project duration statistics determined by Monte Carlo simulation, conventional PERT, and PNET.

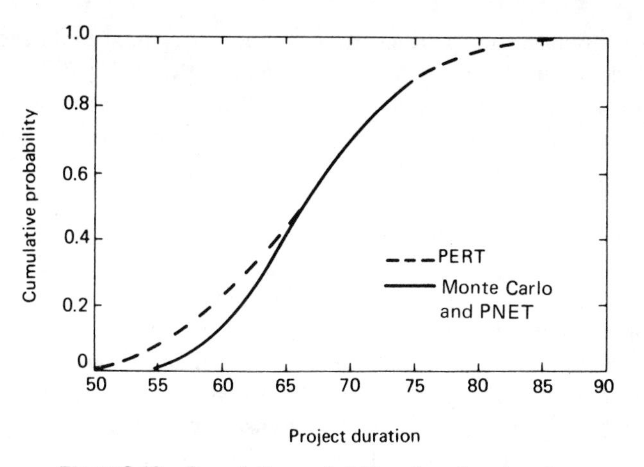


**Figure 9-19** Cumulative probability of project duration.

where $\hat{p}$ is the *estimated* criticality index, $n$ = sample size, and $Z_{\alpha/2}$ is the standard normal deviate. Application of this formula at the 95% confidence level ($Z_{.05/2} = 1.96$) would indicate that if $p$ is in the 10% to 90% range, and sample sizes in the 400 to 1000 range, then the estimates would fall in the ranges $(100\,p \pm 2)\%$ to $(100\,p \pm 5)\%$. Estimates within 2% to 5% certainly represent an acceptable level of accuracy.

Regarding the estimation of the mean project duration, some assumption must be made regarding the coefficient of variation ($CV$) of the project duration, where $CV$ = (Standard Deviation/Mean) × 100%. From the real examples reported by Ang[1], and many others reported in the literature, the CV was found to vary from 5% to 15%. In this case, the width of the 95% confidence interval on the estimate of the mean will vary from ±0.5 to ±1.5% of the mean for sample size 400, and ±0.3 to ±1.0% for sample size 1000. This can be computed from the formula:

$$\pm \text{Error}\% \cong Z_{\alpha/2}(CV\%)/\sqrt{N}$$

The estimation of the corresponding standard deviation of project duration would not be quite as good, ranging from ±4 to ±7% of the true value. Similarly, this can be computed from the following formula, which again indicates that the error levels are quite acceptable.

$$\pm \text{Error}\% \cong 100\,Z_{\alpha/2}/\sqrt{2N}$$

Finally, a confidence interval could be placed on the cumulative probability curve by the use of the Kolmogorov-Smirnov[6] test. Essentially, this method states that the observed cumulative distribution curve will not deviate from the true curve by more than $1.22/\sqrt{N}$, $1.36/\sqrt{N}$, or $1.63/\sqrt{N}$, at the 90%, 95%, and 99% confidence levels, respectively. For example, a confidence interval statement can be made about the observed cumulative distribution curve (solid line) shown in Figure 9-19, which was based on a sample size of $N = 10{,}000$. That is, we can state with 95% confidence, that the entire observed cumulative distribution does not deviate from the true curve of more than $1.36/\sqrt{10{,}000} = 0.0136$, or about 1.4%. If a sample size of only 400 or 200 had been used, this maximum deviation would increase to 6.8% and 9.6%, respectively. The latter is about as large as one could consider tolerable and thus $N = 200$ represents about the smallest sample size that should be used.

With these recommended sample sizes of 400 to 1000, a number of simulation computer programs are available that could handle most networks at an affordable cost. Where cost is a severe limitation, a sample size as small as 200 could even be used, however, values below this would not give reliable estimates of the shape of the distribution of project durations. The use of GERT III for this type of simulation will be illustrated in Chapter 10, Figures 10-4 through 10-7.

The application of PNET to precedence diagramming is treated in exercise 11.

## SUMMARY

In this chapter, the PERT statistical approach to project planning and control was given, which leads to a probability that a given scheduled event occurrence time will be met, *without having to expedite the project.* The conventional PERT procedure derives its measure of uncertainty in the event occurrence times from the three performance time estimates, optimistic, pessimistic, and most likely, for each network activity. This procedure was modified by defining the optimistic and pessimistic times as 5 and 95 percentiles, respectively, of the hypothetical activity performance time distribution, rather than the end points of the distribution. Based on the Central Limit Theorem, estimates of the mean and variance in activity performance times were then used to compute a probability of meeting arbitrary scheduled times for special network events. It was recognized that it is difficult to obtain accurate estimates of the activity performance times, and procedures for improving the estimation by feedback of past estimation performance plotted on a standardized control chart was outlined.

The merge event bias, introduced in conventional PERT by ignoring all but the critical path, was then discussed with examples illustrating the magnitude of this problem. A simple rule of thumb was given to determine, from the completed conventional PERT analysis, whether this bias will be serious or not. This rule simply states that at merge events, one can ignore the path with free slack in excess of twice the larger of the two paths standard deviations.

Two practical procedures to overcome the statistical errors in conventional PERT were described, that is, PNET and Monte Carlo simulation. Both methods will give accurate estimates of the expected project duration and the probability of meeting a range of scheduled duration times. It is, however, easier to collect this information on intermediate network events with simulation than with PNET. Also, simulation alone will give the criticality index on each project activity. Something approaching this might be achieved with PNET if it were programmed to give the cumulative probability curves for the critical and near critical paths, but this would not be as easy to interpret as the criticality index. Finally, it has been estimated by Ang[1] that the computer cost of simulation is about an order of magnitude greater than running PNET.

From the above comparison, it appears that PNET should become a standard procedure, expected of all PERT computer routines. If the importance of the project warrants the additional expenditure, then simulation should be utilized, primarily to give information on all milestone network events, and the criticality index on all activities. With these improvements in PERT methodology in place, interest in this technique just may be rekindled.

## REFERENCES

1. Ang, A. H-S, J. Abdelnour, and A. A. Chaker, "Analysis of Activity Networks Under Uncertainty," *Journal of the Engineering Mechanics Division, Proceedings of American Society of Civil Engineers*, Vol. 101, No. EM4, August 1975, pp. 373–387.
2. Clark, C. E., "The PERT Model for the Distribution of an Activity Time," *Operations Research*, Vol. 10, No. 3, May–June 1962, pp. 405 and 406.
3. Clark, C. E., "The Greatest of a Finite Set of Random Variables," *Operations Research*, Vol. 9, No. 2, March–April 1961, pp. 145–162.
4. Crandall, K. C., "Probabilistic Time Scheduling," *Journal of the Construction Division, Proceedings of American Society of Civil Engineers*, Vol. 102, No. C03, Sept. 1976, pp. 415–423.
5. Crandall, K. C., "Analysis of Schedule Simulations," *Journal of the Construction Division, Proceedings of American Society of Civil Engineers*, Vol. 103, No. C03, Sept. 1977, pp. 387–394.
6. Kendall, M. G. and A. S. Stuart, "The Advanced Theory of Statistics," *Griffin*, London, Vol. 2, 1973, p. 473.
7. King, W. R. and T. A. Wilson, "Subjective Time Estimates in Critical Path Planning—A Preliminary Analysis," *Management Science*, Vol. 13, No. 5, January 1967, pp. 307–320.
8. MacCrimmon, K. R. and C. A. Ryavec, "An Analytical Study of the PERT Assumptions," *Operations Research*, Vol. 12, No. 1, January–February 1964, pp. 16–37.
9. Malcolm, D. G., J. H. Roseboom, C. E. Clark, and W. Fazar, "Applications of a Technique for R and D Program Evaluation," (PERT) *Operations Research*, Vol. VII, No. 5, September–October 1959, pp. 646–669.

10. Moder, J. J. and E. G. Rodgers, "Judgment Estimates of the Moments of PERT Type Distributions," *Management Science*, Vol. 15, No. 2, October 1968, pp. B-76, B-83.
11. PERT, *Program Evaluation Research Task, Phase 1 Summary Report*, Special Projects Office, Bureau of Ordnance, Department of the Navy, Washington, July, 1958.
12. Van Slyke, R. M., "Monte Carlo Methods and the PERT Problem," *Operations Research*, Vol. 11, No. 5, September–October 1963, pp. 839–860.

## EXERCISES

1. Verify the Central Limit Theorem by sampling, that is, by tossing three dice and recording the results on graphs, such as those shown in Figure 9-6. For convenience, use a white, a red, and a green die. Call the number of spots on the white die, $X$, and plot on the first figure. Call the number of spots on the white plus the red die, $Y$, and plot on the second figure. Finally, call the number of spots on all three dice, $Z$, and plot on the third figure. Compare the results of your experiment with the theoretical values given in Figure 9-6.
2. Verify the expected time and variance for event (4004-199), and verify the the probability of 0.12 given for activity (4004-743)–(4004-199) in Figure 9-13b. Note: the values of $a$ and $b$ given in this figure are the end points of the distribution of activity performance time as used in conventional PERT Hence $(V_t)^{1/2} = (b - a)/6$ should be used in place of equation (3) given in this text. Also, note that the time interval 12-13-61 to 12-25-61 is equivalent to 1.6 working weeks.
3. Consider the oversimplified network given in Figure 9-20 which might be only a portion of a larger network, a portion which is subject to considerable chance variation in the performance times. In Figure 9-20 $a$ and $b$ are 5 and 95 percentiles, respectively.
   a. Compute $t_e$ and $V_t$ for each of the four activities.
   b. What is the earliest expected time of event 3?

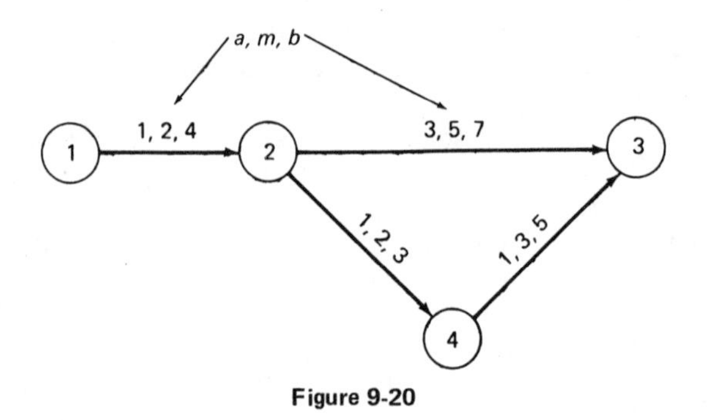


Figure 9-20

c. What is the variance, $V_T$, for the actual occurrence time for event 3?
d. What is the probability that the project will be completed by time 8? By time 9?
e. What time are you fairly sure (say 95 percent confident) of meeting for the completion of the project, i.e., the occurrence of event 3?

4. With reference to Figure 9-1, what is the probability that event 8 will be completed on or before the end of the 12th day, assuming all activities are started as early as possible? What scheduled time for the entire project would you feel 95 percent confident of meeting without having to expedite the project?

5. With reference to Figure 9-1, what is the (conditional) probability that event 8 will be completed on or before the end of the 16th day, assuming that a scheduled time of 10 is met on event 7?

6. Repeat exercise 5 on the assumption that event 7 has occurred 2 days late, i.e., at the end of the 12th day.

7. Solve exercise 3 using the Monte Carlo approach described in the text. For simplicity, assume that the distribution of performance times for each activity is rectangular between the limits given by $a$ and $b$. Using a table of random numbers that are uniformly distributed on the range from 0 to 1, given in most standard statistics texts, transform them to the distribution required for each of the activities in Figure 9-20. For example, activity 2-3 has a possible range of 3 to 7, or 4 time units, and a mean of 5. Letting $r$ denote a random number uniformly distributed on the interval 0 to 1, then, $t = 3 + 4r$, will be uniformly distributed on the range 3 to 7 as desired. Decide for yourself how many simulations of the network should be made, and express the results in a form such as given in Figures 9-18 and 9-19. Discuss your results.

8. Repeat exercise 7 for the network given in Figure 9-1 making the same assumptions as suggested in exercise 7 about the distributions of activity performance times. To carry out this exercise, use the following values of $a$, $m$, and $b$.

| *Activity* | *a* | *m* | *b* |
|---|---|---|---|
| 0–1 | 1.5 | 2 | 2.5 |
| 0–3 | 1 | 2 | 3 |
| 0–6 | 1 | 1 | 1 |
| 1–2 | 3 | 4 | 5 |
| 2–5 | 1 | 1 | 1 |
| 3–4 | 2 | 5 | 8 |
| 3–7 | 6 | 8 | 10 |
| 4–5 | 3 | 4 | 5 |
| 5–8 | 2 | 3 | 4 |
| 6–7 | 2 | 3 | 4 |
| 7–8 | 3.5 | 5 | 6.5 |

a. Applying the rule of thumb given in the text on merge event bias, will the latter be significant in this problem? Why? At what merge events?
b. If we wanted to simplify the Monte Carlo simulation, would the elimination of slack paths 0-6-7 and 0-1-2-5 be justified?
c. Perform the Monte Carlo simulation and comment on the results obtained.

9. Referring to exercise 2 in Chapter 8 assume the times given under $a$ and $b$ are 0 and 100 percentiles, so that the standard deviation, $(V_t)^{1/2} = (b - a)/6$, and the times under the $m$ column are single time estimates of the mean, i.e., $t_e = m$, in this case. Without making a *complete* PERT statistical analysis, answer the following questions.
   a. Is there a significant merge event bias problem for this network?
   b. What is the mean and *approximate* standard deviation of the total time for this project?
   c. What are the *approximate* chances of completing this project in 400 hours?

10. Apply the PNET procedure to the network shown in Figures 9-12 and 9-13.
   a. Compute the corrected probability of meeting the schedule of 12/13/61 for event 4199. (This is equivalent to a project duration of 27.0 weeks).
   b. Compute the cumulative probability curve for a range of scheduled times from 26 to 36 weeks.
   c. What project scheduled duration currently has a probability of 0.90 of being met?

11. How would PERT be applied to precedence diagramming? Answer this question by applying the Central Limit Theorem.
   a. Describe a procedure that could be used assuming the critical path is specified as suggested in Chapter 4. For example, the network in Figure 4-11f has the following critical path. Framing(NC)–FF1–Electrical(RC)–SS1–Finishing(NC). Assuming estimates of the mean and variance for each activity duration are available, give equations for the mean and variance for the project duration, T. What assumptions are required for this procedure?
   b. Does the Monte Carlo simulation approach require any modification for precedence diagramming?
   c. Does the PNET procedure require any modification for precedence diagramming?

# APPENDIX 9-1a
# THE CUMULATIVE NORMAL DISTRIBUTION FUNCTION†

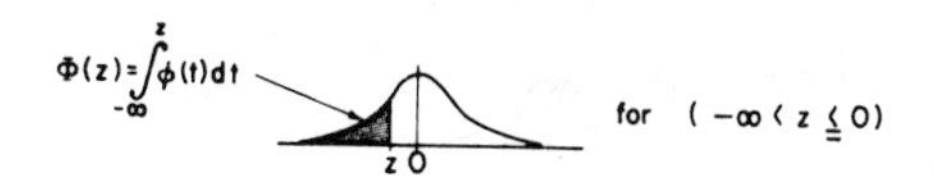


| z | .00 | .01 | .02 | .03 | .04 | .05 | .06 | .07 | .08 | .09 |
|---|---|---|---|---|---|---|---|---|---|---|
| −.0 | .5000 | .4960 | .4920 | .4880 | .4840 | .4801 | .4761 | .4721 | .4681 | .4641 |
| −.1 | .4602 | .4562 | .4522 | .4483 | .4443 | .4404 | .4364 | .4325 | .4286 | .4247 |
| −.2 | .4207 | .4168 | .4129 | .4090 | .4052 | .4013 | .3974 | .3936 | .3897 | .3859 |
| −.3 | .3821 | .3783 | .3745 | .3707 | .3669 | .3632 | .3594 | .3557 | .3520 | .3483 |
| −.4 | .3446 | .3409 | .3372 | .3336 | .3300 | .3264 | .3228 | .3192 | .3156 | .3121 |
| −.5 | .3085 | .3050 | .3015 | .2981 | .2946 | .2912 | .2877 | .2843 | .2810 | .2776 |
| −.6 | .2743 | .2709 | .2676 | .2643 | .2611 | .2578 | .2546 | .2514 | .2483 | .2451 |
| −.7 | .2420 | .2389 | .2358 | .2327 | .2297 | .2266 | .2236 | .2206 | .2177 | .2148 |
| −.8 | .2119 | .2090 | .2061 | .2033 | .2005 | .1977 | .1949 | .1922 | .1894 | .1867 |
| −.9 | .1841 | .1814 | .1788 | .1762 | .1736 | .1711 | .1685 | .1660 | .1635 | .1611 |
| −1.0 | .1587 | .1562 | .1539 | .1515 | .1492 | .1469 | .1446 | .1423 | .1401 | .1379 |
| −1.1 | .1357 | .1335 | .1314 | .1292 | .1271 | .1251 | .1230 | .1210 | .1190 | .1170 |
| −1.2 | .1151 | .1131 | .1112 | .1093 | .1075 | .1056 | .1038 | .1020 | .1003 | .09853 |
| −1.3 | .09680 | .09510 | .09342 | .09176 | .09012 | .08851 | .08691 | .08534 | .08379 | .08226 |
| −1.4 | .08076 | .07927 | .07780 | .07636 | .07493 | .07353 | .07215 | .07078 | .06944 | .06811 |
| −1.5 | .06681 | .06552 | .06426 | .06301 | .06178 | .06057 | .05938 | .05821 | .05705 | .05592 |
| −1.6 | .05480 | .05370 | .05262 | .05155 | .05050 | .04947 | .04846 | .04746 | .04648 | .04551 |
| −1.7 | .04457 | .04363 | .04272 | .04182 | .04093 | .04006 | .03920 | .03836 | .03754 | .03673 |
| −1.8 | .03593 | .03515 | .03438 | .03362 | .03288 | .03216 | .03144 | .03074 | .03005 | .02938 |
| −1.9 | .02872 | .02807 | .02743 | .02680 | .02619 | .02559 | .02500 | .02442 | 0.2385 | .02330 |
| −2.0 | .02275 | .02222 | .02169 | .02118 | .02068 | .02018 | .01970 | .01923 | .01876 | .01831 |
| −2.1 | .01786 | .01743 | .01700 | .01659 | .01618 | .01578 | .01539 | .01500 | .01463 | .01426 |
| −2.2 | .01390 | .01355 | .01321 | .01287 | .01255 | .01222 | .01191 | .01160 | .01130 | .01101 |
| −2.3 | .01072 | .01044 | .01017 | $.0^{2}9903$ | $.0^{2}9642$ | $.0^{2}9387$ | $.0^{2}9137$ | $.0^{2}8894$ | $.0^{2}8656$ | $.0^{2}8424$ |
| −2.4 | $.0^{2}8198$ | $.0^{2}7976$ | $.0^{2}7760$ | $.0^{2}7549$ | $.0^{2}7344$ | $.0^{2}7143$ | $.0^{2}6947$ | $.0^{2}6756$ | $.0^{2}6569$ | $.0^{2}6387$ |
| −2.5 | $.0^{2}6210$ | $.0^{2}6037$ | $.0^{2}5868$ | $.0^{2}5703$ | $.0^{2}5543$ | $.0^{2}5386$ | $.0^{2}5234$ | $.0^{2}5085$ | $.0^{2}4940$ | $.0^{2}4799$ |
| −2.6 | $.0^{2}4661$ | $.0^{2}4527$ | $.0^{2}4396$ | $.0^{2}4269$ | $.0^{2}4145$ | $.0^{2}4025$ | $.0^{2}3907$ | $.0^{2}3793$ | $.0^{2}3681$ | $.0^{2}3573$ |
| −2.7 | $.0^{2}3467$ | $.0^{2}3364$ | $.0^{2}3264$ | $.0^{2}3167$ | $.0^{2}3072$ | $.0^{2}2980$ | $.0^{2}2890$ | $.0^{2}2803$ | $.0^{2}2718$ | $.0^{2}2635$ |
| −2.8 | $.0^{2}2555$ | $.0^{2}2477$ | $.0^{2}2401$ | $.0^{2}2327$ | $.0^{2}2256$ | $.0^{2}2186$ | $.0^{2}2118$ | $.0^{2}2052$ | $.0^{2}1988$ | $.0^{2}1926$ |
| −2.9 | $.0^{2}1866$ | $.0^{2}1807$ | $.0^{2}1750$ | $.0^{2}1695$ | $.0^{2}1641$ | $.0^{2}1589$ | $.0^{2}1538$ | $.0^{2}1489$ | $.0^{2}1441$ | $.0^{2}1395$ |
| −3.0 | $.0^{2}1350$ | $.0^{2}1306$ | $.0^{2}1264$ | $.0^{2}1223$ | $.0^{2}1183$ | $.0^{2}1144$ | $.0^{2}1107$ | $.0^{2}1070$ | $.0^{2}1035$ | $.0^{2}1001$ |
| −3.1 | $.0^{3}9676$ | $.0^{3}9354$ | $.0^{3}9043$ | $.0^{3}8740$ | $.0^{3}8447$ | $.0^{3}8164$ | $.0^{3}7888$ | $.0^{3}7622$ | $.0^{3}7364$ | $.0^{3}7114$ |
| −3.2 | $.0^{3}6871$ | $.0^{3}6637$ | $.0^{3}6410$ | $.0^{3}6190$ | $.0^{3}5976$ | $.0^{3}5770$ | $.0^{3}5571$ | $.0^{3}5377$ | $.0^{3}5190$ | $.0^{3}5009$ |
| −3.3 | $.0^{3}4834$ | $.0^{3}4665$ | $.0^{3}4501$ | $.0^{3}4342$ | $.0^{3}4189$ | $.0^{3}4041$ | $.0^{3}3897$ | $.0^{3}3758$ | $.0^{3}3624$ | $.0^{3}3495$ |
| −3.4 | $.0^{3}3369$ | $.0^{3}3248$ | $.0^{3}3131$ | $.0^{3}3018$ | $.0^{3}2909$ | $.0^{3}2803$ | $.0^{3}2701$ | $.0^{3}2602$ | $.0^{3}2507$ | $.0^{3}2415$ |
| −3.5 | $.0^{3}2326$ | $.0^{3}2241$ | $.0^{3}2158$ | $.0^{3}2078$ | $.0^{3}2001$ | $.0^{3}1926$ | $.0^{3}1854$ | $.0^{3}1785$ | $.0^{3}1718$ | $.0^{3}1653$ |
| −3.6 | $.0^{3}1591$ | $.0^{3}1531$ | $.0^{3}1473$ | $.0^{3}1417$ | $.0^{3}1363$ | $.0^{3}1311$ | $.0^{3}1261$ | $.0^{3}1213$ | $.0^{3}1166$ | $.0^{3}1121$ |
| −3.7 | $.0^{3}1078$ | $.0^{3}1036$ | $.0^{4}9961$ | $.0^{4}9574$ | $.0^{4}9201$ | $.0^{4}8842$ | $.0^{4}8496$ | $.0^{4}8162$ | $.0^{4}7841$ | $.0^{4}7532$ |
| −3.8 | $.0^{4}7235$ | $.0^{4}6948$ | $.0^{4}6673$ | $.0^{4}6407$ | $.0^{4}6152$ | $.0^{4}5906$ | $.0^{4}5669$ | $.0^{4}5442$ | $.0^{4}5223$ | $.0^{4}5012$ |
| −3.9 | $.0^{4}4810$ | $.0^{4}4615$ | $.0^{4}4427$ | $.0^{4}4247$ | $.0^{4}4074$ | $.0^{4}3908$ | $.0^{4}3747$ | $.0^{4}2594$ | $.0^{4}3446$ | $.0^{4}3304$ |
| −4.0 | $.0^{4}3167$ | $.0^{4}3036$ | $.0^{4}2910$ | $.0^{4}2789$ | $.0^{4}2673$ | $.0^{4}2561$ | $.0^{4}2454$ | $.0^{4}2351$ | $.0^{4}2252$ | $.0^{4}2157$ |
| −4.1 | $.0^{4}2066$ | $.0^{4}1987$ | $.0^{4}1894$ | $.0^{4}1814$ | $.0^{4}1737$ | $.0^{4}1662$ | $.0^{4}1591$ | $.0^{4}1523$ | $.0^{4}1458$ | $.0^{4}1395$ |
| −4.2 | $.0^{4}1335$ | $.0^{4}1277$ | $.0^{4}1222$ | $.0^{4}1168$ | $.0^{4}1118$ | $.0^{4}1069$ | $.0^{4}1022$ | $.0^{5}9774$ | $.0^{5}9345$ | $.0^{5}8934$ |
| −4.3 | $.0^{5}8540$ | $.0^{5}8163$ | $.0^{5}7801$ | $.0^{5}7455$ | $.0^{5}7124$ | $.0^{5}6807$ | $.0^{5}6503$ | $.0^{5}6212$ | $.0^{5}5934$ | $.0^{5}5668$ |
| −4.4 | $.0^{5}5413$ | $.0^{5}5169$ | $.0^{5}4935$ | $.0^{5}4712$ | $.0^{5}4498$ | $.0^{5}4294$ | $.0^{5}4098$ | $.0^{5}3911$ | $.0^{5}3732$ | $.0^{5}3561$ |
| −4.5 | $.0^{5}3398$ | $.0^{5}3241$ | $.0^{5}3092$ | $.0^{5}2949$ | $.0^{5}2813$ | $.0^{5}2682$ | $.0^{5}2558$ | $.0^{5}2439$ | $.0^{5}2325$ | $.0^{5}2216$ |
| −4.6 | $.0^{5}2112$ | $.0^{5}2013$ | $.0^{5}1919$ | $.0^{5}1828$ | $.0^{5}1742$ | $.0^{5}1660$ | $.0^{5}1581$ | $.0^{5}1506$ | $.0^{5}1434$ | $.0^{5}1366$ |
| −4.7 | $.0^{5}1301$ | $.0^{5}1239$ | $.0^{5}1179$ | $.0^{5}1123$ | $.0^{5}1069$ | $.0^{5}1017$ | $.0^{6}9680$ | $.0^{6}9211$ | $.0^{6}8765$ | $.0^{6}8339$ |
| −4.8 | $.0^{6}7933$ | $.0^{6}7547$ | $.0^{6}7178$ | $.0^{6}6827$ | $.0^{6}6492$ | $.0^{6}6173$ | $.0^{6}5869$ | $.0^{6}5580$ | $.0^{6}5304$ | $.0^{6}5042$ |
| −4.9 | $.0^{6}4792$ | $.0^{6}4554$ | $.0^{6}4327$ | $.0^{6}4111$ | $.0^{6}3906$ | $.0^{6}3711$ | $.0^{6}3525$ | $.0^{6}3348$ | $.0^{6}3179$ | $.0^{6}3019$ |

**Example: $\Phi(-3.57) = .0^{3}1785 = 0.0001785$.**

†By permission from A. Hald, *Statistical Tables, and Formulas,* John Wiley & Sons. Inc., New York, 1952.

# APPENDIX 9-1b
# THE CUMULATIVE NORMAL DISTRIBUTION FUNCTION†

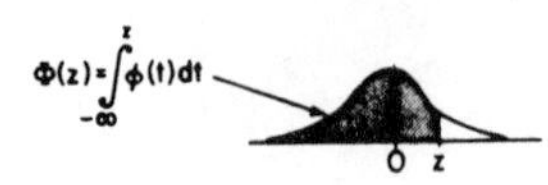


for $(0 \leq z < \infty)$

| z | .00 | .01 | .02 | .03 | .04 | .05 | .06 | .07 | .08 | .09 |
|---|---|---|---|---|---|---|---|---|---|---|
| .0 | .5000 | .5040 | .5080 | .5120 | .5160 | .5199 | .5239 | .5279 | .5319 | .5359 |
| .1 | .5398 | .5438 | .5478 | .5517 | .5557 | .5596 | .5636 | .5675 | .5714 | .5753 |
| .2 | .5793 | .5832 | .5871 | .5910 | .5948 | .5987 | .6026 | .6064 | .6103 | .6141 |
| .3 | .6179 | .6217 | .6255 | .6293 | .6331 | .6368 | .6406 | .6443 | .6480 | .6517 |
| .4 | .6554 | .6591 | .6628 | .6664 | .6700 | .6736 | .6772 | .6808 | .6844 | .6879 |
| .5 | .6915 | .6950 | .6985 | .7019 | .7054 | .7088 | .7123 | .7157 | .7190 | .7224 |
| .6 | .7257 | .7291 | .7324 | .7357 | .7389 | .7422 | .7454 | .7486 | .7517 | .7549 |
| .7 | .7580 | .7611 | .7642 | .7673 | .7703 | .7734 | .7764 | .7794 | .7823 | .7852 |
| .8 | .7881 | .7910 | .7939 | .7967 | .7995 | .8023 | .8051 | .8078 | .8106 | .8133 |
| .9 | .8159 | .8186 | .8212 | .8238 | .8264 | .8289 | .8315 | .8340 | .8365 | .8389 |
| 1.0 | .8413 | .8438 | .8461 | .8485 | .8508 | .8531 | .8554 | .8577 | .8599 | .8621 |
| 1.1 | .8643 | .8665 | .8686 | .8708 | .8729 | .8749 | .8770 | .8790 | .8810 | .8830 |
| 1.2 | .8849 | .8869 | .8888 | .8907 | .8925 | .8944 | .8962 | .8980 | .8997 | .90147 |
| 1.3 | .90320 | .90490 | .90658 | .90824 | .90988 | .91149 | .91309 | .91466 | .91621 | .91774 |
| 1.4 | .91924 | .92073 | .92220 | .92364 | .92507 | .92647 | .92785 | .92922 | .93056 | .93189 |
| 1.5 | .93319 | .93448 | .93574 | .93699 | .93822 | .93943 | .94062 | .94179 | .94295 | .94408 |
| 1.6 | .94520 | .94630 | .94738 | .94845 | .94950 | .95053 | .95154 | .95254 | .95352 | .95449 |
| 1.7 | .95543 | .95637 | .95728 | .95818 | .95907 | .95994 | .96080 | .96164 | .96246 | .96327 |
| 1.8 | .96407 | .96485 | .96562 | .96638 | .96712 | .96784 | .96856 | .96926 | .96995 | .97062 |
| 1.9 | .97128 | .97193 | .97257 | .97320 | .97381 | .97441 | .97500 | .97558 | .97615 | .97670 |
| 2.0 | .97725 | .97778 | .97831 | .97882 | .97932 | .97982 | .98030 | .98077 | .98124 | .98169 |
| 2.1 | .98214 | .98257 | .98300 | .98341 | .98382 | .98422 | .98461 | .98500 | .98537 | .98574 |
| 2.2 | .98610 | .98645 | .98679 | .98713 | .98745 | .98778 | .98809 | .98840 | .98870 | .98899 |
| 2.3 | .98928 | .98956 | .98983 | $.9^{2}0097$ | $.9^{2}0358$ | $.9^{2}0613$ | $.9^{2}0863$ | $.9^{2}1106$ | $.9^{2}1344$ | $.9^{2}1576$ |
| 2.4 | $.9^{2}1802$ | $.9^{2}2024$ | $.9^{2}2240$ | $.9^{2}2451$ | $.9^{2}2656$ | $.9^{2}2857$ | $.9^{2}3053$ | $.9^{2}3244$ | $.9^{2}3431$ | $.9^{2}3613$ |
| 2.5 | $.9^{2}3790$ | $.9^{2}3963$ | $.9^{2}4132$ | $.9^{2}4297$ | $.9^{2}4457$ | $.9^{2}4614$ | $.9^{2}4766$ | $.9^{2}4915$ | $.9^{2}5060$ | $.9^{2}5201$ |
| 2.6 | $.9^{2}5339$ | $.9^{2}5473$ | $.9^{2}5604$ | $.9^{2}5731$ | $.9^{2}5855$ | $.9^{2}5975$ | $.9^{2}6093$ | $.9^{2}6207$ | $.9^{2}6319$ | $.9^{2}6427$ |
| 2.7 | $.9^{2}6533$ | $.9^{2}6636$ | $.9^{2}6736$ | $.9^{2}6833$ | $.9^{2}6928$ | $.9^{2}7020$ | $.9^{2}7110$ | $.9^{2}7197$ | $.9^{2}7282$ | $.9^{2}7365$ |
| 2.8 | $.9^{2}7445$ | $.9^{2}7523$ | $.9^{2}7599$ | $.9^{2}7673$ | $.9^{2}7744$ | $.9^{2}7814$ | $.9^{2}7882$ | $.9^{2}7948$ | $.9^{2}8012$ | $.9^{2}8074$ |
| 2.9 | $.9^{2}8134$ | $.9^{2}8193$ | $.9^{2}8250$ | $.9^{2}8305$ | $.9^{2}8359$ | $.9^{2}8411$ | $.9^{2}8462$ | $.9^{2}8511$ | $.9^{2}8559$ | $.9^{2}8605$ |
| 3.0 | $.9^{2}8650$ | $.9^{2}8694$ | $.9^{2}8736$ | $.9^{2}8777$ | $.9^{2}8817$ | $.9^{2}8856$ | $.9^{2}8893$ | $.9^{2}8930$ | $.9^{2}8965$ | $.9^{2}8999$ |
| 3.1 | $.9^{3}0324$ | $.9^{3}0646$ | $.9^{3}0957$ | $.9^{3}1260$ | $.9^{3}1553$ | $.9^{3}1836$ | $.9^{3}2112$ | $.9^{3}2378$ | $.9^{3}2636$ | $.9^{3}2886$ |
| 3.2 | $.9^{3}3129$ | $.9^{3}3363$ | $.9^{3}3590$ | $.9^{3}3810$ | $.9^{3}4024$ | $.9^{3}4230$ | $.9^{3}4429$ | $.9^{3}4623$ | $.9^{3}4810$ | $.9^{3}4991$ |
| 3.3 | $.9^{3}5166$ | $.9^{3}5335$ | $.9^{3}5499$ | $.9^{3}5658$ | $.9^{3}5811$ | $.9^{3}5959$ | $.9^{3}6103$ | $.9^{3}6242$ | $.9^{3}6376$ | $.9^{3}6505$ |
| 3.4 | $.9^{3}6631$ | $.9^{3}6752$ | $.9^{3}6869$ | $.9^{3}6982$ | $.9^{3}7091$ | $.9^{3}7197$ | $.9^{3}7299$ | $.9^{3}7398$ | $.9^{3}7493$ | $.9^{3}7585$ |
| 3.5 | $.9^{3}7674$ | $.9^{3}7759$ | $.9^{3}7842$ | $.9^{3}7922$ | $.9^{3}7999$ | $.9^{3}8074$ | $.9^{3}8146$ | $.9^{3}8215$ | $.9^{3}8282$ | $.9^{3}8347$ |
| 3.6 | $.9^{3}8409$ | $.9^{3}8469$ | $.9^{3}8527$ | $.9^{3}8583$ | $.9^{3}8637$ | $.9^{3}8689$ | $.9^{3}8739$ | $.9^{3}8787$ | $.9^{3}8834$ | $.9^{3}8879$ |
| 3.7 | $.9^{3}8922$ | $.9^{3}8964$ | $.9^{4}0039$ | $.9^{4}0426$ | $.9^{4}0799$ | $.9^{4}1158$ | $.9^{4}1504$ | $.9^{4}1838$ | $.9^{4}2159$ | $.9^{4}2468$ |
| 3.8 | $.9^{4}2765$ | $.9^{4}3052$ | $.9^{4}3327$ | $.9^{4}3593$ | $.9^{4}3848$ | $.9^{4}4094$ | $.9^{4}4331$ | $.9^{4}4558$ | $.9^{4}4777$ | $.9^{4}4988$ |
| 3.9 | $.9^{4}5190$ | $.9^{4}5385$ | $.9^{4}5573$ | $.9^{4}5753$ | $.9^{4}5926$ | $.9^{4}6092$ | $.9^{4}6253$ | $.9^{4}6406$ | $.9^{4}6554$ | $.9^{4}6696$ |
| 4.0 | $.9^{4}6833$ | $.9^{4}6964$ | $.9^{4}7090$ | $.9^{4}7211$ | $.9^{4}7327$ | $.9^{4}7439$ | $.9^{4}7546$ | $.9^{4}7649$ | $.9^{4}7748$ | $.9^{4}7843$ |
| 4.1 | $.9^{4}7934$ | $.9^{4}8022$ | $.9^{4}8106$ | $.9^{4}8186$ | $.9^{4}8263$ | $.9^{4}8338$ | $.9^{4}8409$ | $.9^{4}8477$ | $.9^{4}8542$ | $.9^{4}8605$ |
| 4.2 | $.9^{4}8665$ | $.9^{4}8723$ | $.9^{4}8778$ | $.9^{4}8832$ | $.9^{4}8882$ | $.9^{4}8931$ | $.9^{4}8978$ | $.9^{5}0226$ | $.9^{5}0655$ | $.9^{5}1066$ |
| 4.3 | $.9^{5}1460$ | $.9^{5}1837$ | $.9^{5}2199$ | $.9^{5}2545$ | $.9^{5}2876$ | $.9^{5}3193$ | $.9^{5}3497$ | $.9^{5}3788$ | $.9^{5}4066$ | $.9^{5}4332$ |
| 4.4 | $.9^{5}4587$ | $.9^{5}4831$ | $.9^{5}5065$ | $.9^{5}5288$ | $.9^{5}5502$ | $.9^{5}5706$ | $.9^{5}5902$ | $.9^{5}6089$ | $.9^{5}6268$ | $.9^{5}6439$ |
| 4.5 | $.9^{5}6602$ | $.9^{5}6759$ | $.9^{5}6908$ | $.9^{5}7051$ | $.9^{5}7187$ | $.9^{5}7318$ | $.9^{5}7442$ | $.9^{5}7561$ | $.9^{5}7675$ | $.9^{5}7784$ |
| 4.6 | $.9^{5}7888$ | $.9^{5}7987$ | $.9^{5}8081$ | $.9^{5}8172$ | $.9^{5}8258$ | $.9^{5}8340$ | $.9^{5}8419$ | $.9^{5}8494$ | $.9^{5}8566$ | $.9^{5}8634$ |
| 4.7 | $.9^{5}8699$ | $.9^{5}8761$ | $.9^{5}8821$ | $.9^{5}8877$ | $.9^{5}8931$ | $.9^{5}8983$ | $.9^{6}0320$ | $.9^{6}0789$ | $.9^{6}1235$ | $.9^{6}1661$ |
| 4.8 | $.9^{6}2067$ | $.9^{6}2453$ | $.9^{6}2822$ | $.9^{6}3173$ | $.9^{6}3508$ | $.9^{6}3827$ | $.9^{6}4131$ | $.9^{6}4420$ | $.9^{6}4696$ | $.9^{6}4958$ |
| 4.9 | $.9^{6}5208$ | $.9^{6}5446$ | $.9^{6}5673$ | $.9^{6}5889$ | $.9^{6}6094$ | $.9^{6}6289$ | $.9^{6}6475$ | $.9^{6}6652$ | $.9^{6}6821$ | $.9^{6}6981$ |

**Example: $\Phi$ (3.57) = $.9^{3}8215$ = 0.9998215.**

† By permission from A. Hald, *Statistical Tables, and Formulas*, John Wiley & Sons, Inc., New York, 1952.

# APPENDIX 9-2 USE OF HISTORICAL DATA IN ESTIMATING *a*, *m*, AND *b*

Occasionally one may have historical (sample) activity duration data on which to base estimates of $t_e$ and $(V_t)^{1/2}$, or better, to estimate $a$, $m$, and $b$, which when processed in the usual manner, with equations (3) and (4), will give the desired estimates of $(V_t)^{1/2}$ and $t_e$. This procedure has merit, if the following conditions are satisfied.

1. The historical data are representative of the hypothetical population (Figures 9-2 and 9-8) to be "sampled" in the future for the activity in question; that is, the activity is precisely the same, and the conditions which prevailed during the collection of the historical data are representative of those expected to prevail in the future when the activity in question is to be performed.
2. The sample of historical data is of "sufficient" size. Quantitative specification of what is "sufficient" depends on the nature of the activity in question and the experience and abilities of the person supplying the estimates; however, a sample of less than four or five observations would generally not be considered "sufficient."

If the above assumptions are satisfied, estimates of $a$, $m$, and $b$ can be obtained from equations (8), (9), and (10) below, wherein

$R$ = range of sample data
= largest observation–smallest observation

**Table 9-8. Constant to Convert the Range to Estimates of the Standard Deviation**

| *Sample Size*† | *(Range/Std. Dev.)* $= d_2$†† | $k = 1.6/d_2$ |
|---|---|---|
| 2 | 1.13 | 1.416 |
| 3 | 1.69 | 0.947 |
| 4 | 2.06 | 0.777 |
| 5 | 2.33 | 0.687 |
| 6 | 2.53 | 0.632 |
| 7 | 2.70 | 0.593 |
| 8 | 2.85 | 0.561 |
| 9 | 2.97 | 0.539 |
| 10 | 3.08 | 0.519 |
| 12 | 3.26 | 0.491 |
| 15 | 3.47 | 0.461 |
| 20 | 3.74 | 0.428 |
| 25 | 3.93 | 0.407 |

† Although this table includes samples as small as two, one should not rely solely on the sample data unless the sample size is at least four.

†† The symbol $d_2$ used here is the universal designation of this ratio, which is widely used and tabled in statistical quality control literature; it assumes the random variable is normally distributed.

$k = 3/d_2$, where $d_2$ is the statistical quality control constant tabled as a function of the number, $n$, of acitivity times in the sample data. Actually, $d_2$ is the average of the ratio $R/(V_t)^{1/2}$. Values of $k$ are given in Table 9-8, and are used to compute the constants $a$ and $b$.

$\bar{t}$ = arithmetic average of the sample data

$$\text{Estimate of } m = \bar{t} \tag{8}$$

$$\text{Estimate of } a = \bar{t} - kR \tag{9}$$

$$\text{Estimate of } b = \bar{t} + kR \tag{10}$$

In situations where $kR$ is greater than $\bar{t}$, and hence $a$ as given by equation (9) is negative, it is suggested that the following be used.

$$a = 0 \tag{9a}$$

$$b = 2\bar{t} \tag{10a}$$

# 10

# OTHER NETWORKING SCHEMES

The graphical methods of drawing networks introduced in Chapter 2, and used throughout this text, were variations of two basic methods of expressing a project plan, that is, using activities-on-arrows (arrow diagrams) or activities-on-nodes (node or precedence diagrams). The original developers of PERT and CPM both utilized arrow diagrams based on the simple finish-start precedence relationship. It was also pointed out in Chapter 2 that for every such arrow diagram, an equivalent node diagram could be drawn by simply reversing the roles of the arrows and nodes. Using this scheme, which is growing in popularity, the nodes represent the activities and the arrows are merely connectors denoting finish-start precedence relationships. The principal advantage of this basic node scheme is that it eliminates the need for special dummies to correct false dependencies. For this reason, it is easier to learn to draw the networks, and computer processing is somewhat simplified since input/output lines devoted to dummy activities are unnecessary.

The final network development in Chapter 2 was the introduction of the precedence diagramming version of node diagrams. This scheme extended the simple logic of the basic node scheme. It permitted the start and/or finish of an activity to be tied into the start and/or finish of other predecessor or successor activities by the use of 4 types of precedence relationships. The addition of lead/lag times to these precedence relationships gave added flexibility to this method of network diagramming.

The basic arrow diagram and node diagram schemes have, indeed, served quite well in modeling project plans. However, at the cutting edge of project planning methodology, there are situations that still cannot be adequately modeled by these basic schemes. Arrow and node diagrams are based on deterministic network logic. That is, it is assumed that every path in the network is a necessary part of the project; there are no optional or alternative paths. However, we know that in some types of projects there is some uncertainty as to just which activities will be included. In research projects in particular, several different outcomes of the project may result, depending upon the outcomes of certain chains of activities. A network that shows only one plan with one possible outcome may not represent adequately the true nature of the project. The flexibility desired here is called *probabilistic branching*, where only one of several successor activities leaving a *burst event* are realized. Another important deficiency of basic network schemes is that they do not permit the incorporation of *loops* in the networks.

A simple example of an application of *looping* and *probabilistic branching* would be the activities following a "test" activity. If the test result is acceptable, then the branch to the next phase of the project would be followed. However, if the test result is unacceptable, then an alternate branch is followed which forms a loop to correct the unacceptable performance, and eventually leads back to repeat the same test activity.

The use of branching and looping in this situation could be avoided in simple arrow or node diagrams by incorporating testing, rework if needed, retesting, etc., all in one activity. These two alternatives are shown in Figure 10-1. If the PERT system of time estimation were utilized in this situation, the uncertainty in the Testing and Rework (10–20) activity could be accounted for by assigning

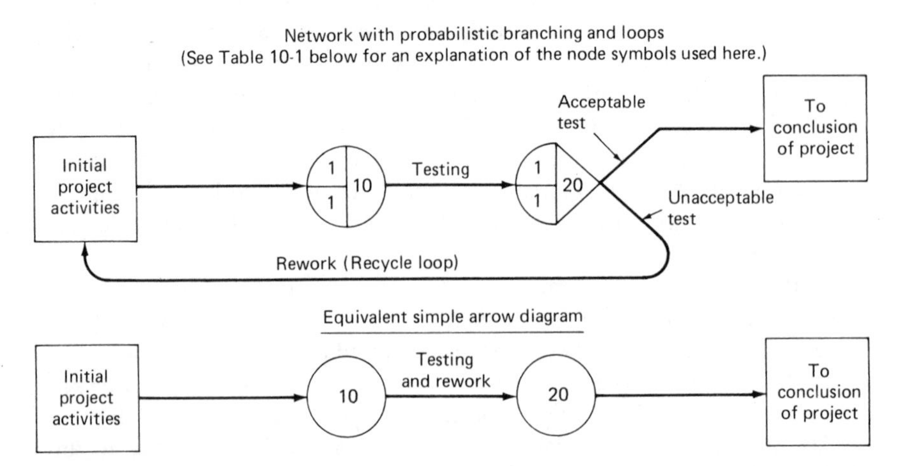


**Figure 10-1** Example of probabilistic branching and looping, with an equivalent single arrow diagram activity 10–20.

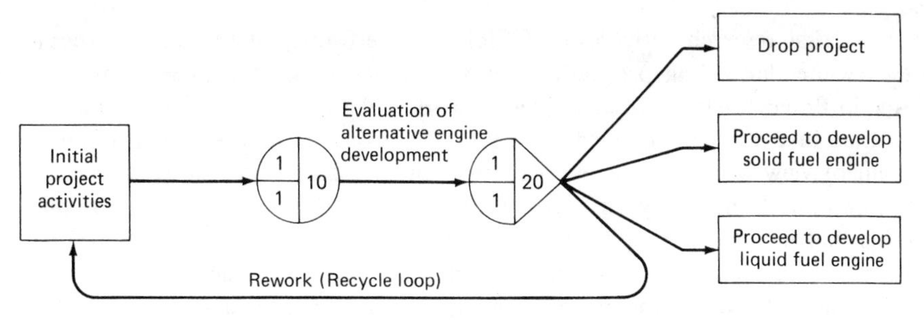


**Figure 10-2** Example of probabilistic branching and looping—no equivalent arrow diagram possible. (See Table 10-1 below for an explanation of the node symbols used here.)

an appropriate range of activity duration times, with the optimistic time corresponding to an acceptable test on the first try, and a pessimistic duration corresponding to the time required to traverse a maximum number of recycle loops that might reasonably be expected to occur in practice.

Now, consider a situation such as shown in Figure 10-2. Here, the testing or evaluation activity (10–20) has a probabilistic successor event (20) that branches to one of 4 possible successor activities, one of which forms a loop leading back to the same evaluation activity. Note also, in this case, that there are several different project end events, only one of which can ever be realized. The true essence of this project plan, and the evaluation of alternative project policies, could not be captured by basic arrow or node diagrams, such as was done in Figure 10-1. The generalized arrow diagram which incorporates the branching and looping shown in Figures 10-1 and 10-2 was developed by Alan Pritsker and is called GERT, the acronym for Graphical Evaluation and Review Technique (References 13–18).

## DEVELOPMENT OF GERT

One of the earliest extensions of the basic arrow and node diagrams was suggested by Freeman[6] and dealt with the portrayal of alternative networks to represent a project. Eisner[2] generalized this concept by including a new node type for branching events to graphically portray alternative paths in a network. The probabilities assigned to these paths add to one, and they are estimated by the project planners along with the activity duration times on the basis of judgment and experience with similar circumstances. A significant breakthrough in generalizing the structure of activity networks was then made by Elmaghraby[3,4] when he defined three types of nodes: AND nodes; EXCLUSIVE-OR nodes; and INCLUSIVE-OR nodes. The initial GERT developments were made by Pritsker[13,14,15,17] by building on Elmaghraby's node definitions.

The first research activities on GERT were performed at the Rand Corporation while developing procedures for automatic checkout equipment for the Apollo Program. Pritsker subsequently worked with many groups in leading up to the current program called Q-GERT. Notable from a project management point of view was the addition of costs to the usual time analysis of networks. Arisawa and Elmaghraby[1] modeled costs by considering setup costs and variable costs that increase in time for each project activity. P-GERT was developed by Pritsker[16] to handle node diagram notation. Finally, Hebert[8] developed the program called R-GERT which gives the basic outputs of a simulation of a PERT network (as discussed in Chapter 9) with the consideration of resource constraints. That is, the program gives the early/late start/finish times and criticality indices for each project activity, with the condition that an activity cannot be started until the stated resources required by the activity become available.

In the next section GERT nomenclature will be introduced and its use will be illustrated for a simple project of reviewing a paper for a technical journal. Then, a detailed example of Q-GERT will be presented which deals with an industrial sales negotiation process.

## GERT NOMENCLATURE

One form of basic GERT nomenclature is shown in Table 10-1. The basic arrow diagram logic of PERT/CPM is embodied in the circle node shown in the upper left-hand quadrant of this table. The other three node symbols in this table are GERT logic extensions which permit looping, branching, and multiple project end results.

An application of this notation is given in Figure 10-3, which treats the process of reviewing an article submitted for publication in a professional journal. The network contains one initial event and two terminal events; the latter have the symbol for infinity (∞) in the lower quadrant to denote that they can only be realized once. These events are *half nodes* because the initial event has no inputs, and the terminal events have no outputs. The network chart is essentially self-explanatory, which is one of the advantages of this nomenclature. This could be called a multiple reviewer policy which requires a favorable majority to accept the article for publication. The network involves branching at nodes 4, 5 and 8 to denote the appropriate reviewer decision. Also, there are two loops back to node 6 from nodes 8 and 13, and two end nodes denoting acceptance or rejection of the article. Of particular note is node 13 which is realized only if the first two reviewers do not agree (one accept and one reject); it triggers off a recycle loop to node 6 to select an alternate reviewer.

The diagram in Figure 10-3 could, by itself, be quite useful as a general tool of

**Table 10-1. A Form of Basic GERT Notation**

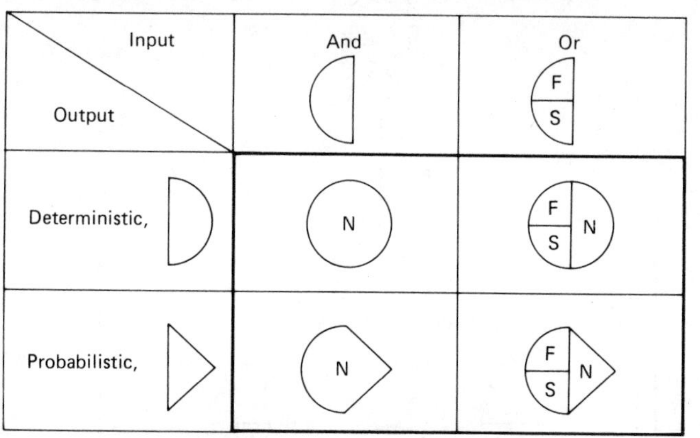


**N** Denotes the node identification number.

**AND** The AND node will be released only if *all* activities (arrows) leading into the node are realized. The time of realization is the largest of the completion times of the activities leading *into* the AND node (basic PERT/CPM input node logic).

**F** Denotes the number of predecessor activities that must be completed for the *F*irst realization of the node, and *S* denotes the number required for *S*ubsequent realizations.

**OR** The OR node will be realized the *first* time when any F of the total number of activities leading into the node are completed. If the node is contained in a loop, then the node can be realized the *second* and all *subsequent* times when any *S* of the total number of activities leading into the node are completed; usually $F \geqq S$.

**Deterministic** All activities emanating from the node are subsequently taken if the node is realized (basic PERT/CPM output node logic).

**Probabilistic** Only one activity emanating from the node is taken if the node is realized. The sum of the probabilities associated with each of the output activities equals one.

systems analysis, particularly where the problem is more complex than shown here. It could be used to show redundancies, inefficiencies, and inconsistencies in the policies or procedures in question. It could also be useful in speeding up action at each branching node in the network, since the requirements and implications of each alternative are clearly stated. If additional data is added to this network, the use of the Q-GERT simulation program could provide estimates of the probabilities of realization of each of the end nodes, and time and cost distributions for the realization of any selected nodes, considering resource constraints if desired. An example of this type will be presented in the next section.

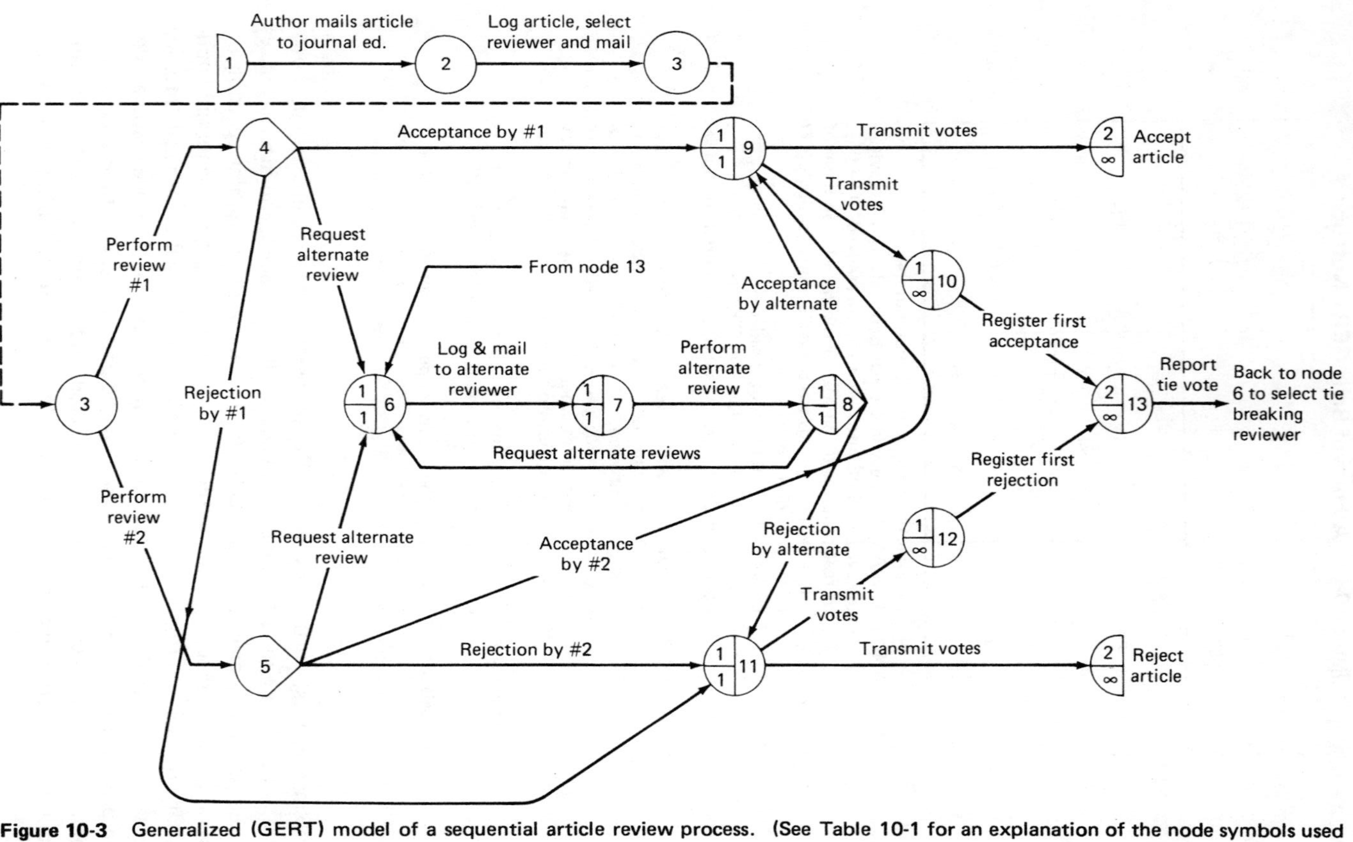


**Figure 10-3** Generalized (GERT) model of a sequential article review process. (See Table 10-1 for an explanation of the node symbols used here.)

## AN APPLICATION OF Q-GERT

Given below is an application of Q-GERT by Moore and Clayton[12]*, to model the costs associated with an industrial sales negotiation process. A major oil company wanted to buy a new gasoline plant as a completely installed package. Systematically, the purchasing agent and his buyers began to inform a select number of prospective vendors of the proposed gasoline plant and invited these vendors to negotiate for the job. Figure 10-4 is a Q-GERT network model which could have been used by one of the vendor firms. The activity numbers are referenced to Table 10-2 which contains a description of each activity along with the time distribution, cost estimates and branching probabilities for that activity. Fixed and variable costs are associated with the negotiation activities. A fixed cost is charged each time an activity occurs while variable costs accumulate linearly over the period of time required to complete the activity. Each activity will now be discussed with activity numbers given in brackets after the description.

The sales negotiation process involves the initial contact by company salesmen [1]. A report to the marketing vice president and to the president is then made [2, 3]. For this company, a decision to perform preliminary analyses is standard for the large type of project being considered. As will be seen, an evaluation of this policy could be cost effective. Next, on the Q-GERT network, five parallel studies are performed beginning at node 5. These involve engineering [4], production [5], financial [6], marketing [7] and purchasing [8]. Except for the production report, the evaluation of the reports is performed at the nodes following the activities, as represented by activities 9 and 10 and 13 through 18. If any of the reports is negative, the project is not negotiated further and the network terminates at node 22. Node 22 is repeated three times in Figure 10-4 for graphical convenience. The evaluation of the production schedule is slightly more complicated in that the possibility for production subcontracting is investigated [11]. An unfavorable production subcontract could also occur [19] which would lead to termination of the negotiation process. When all five reports are favorable, node 12 is released. Node 12 is a statistics node and the time at which node 12 is released will be collected. A user function is employed at node 12 to collect cost data with regard to the costs of the sales negotiation process up to the time at which five favorable reports are obtained. Based on the five reports, corporate level planning conferences are held [21]. Further information is then sought of a marketing and engineering nature [22, 23]. If either of these calls results in unfavorable information, the sales negotiation process is again terminated [24, 26]. If positive information is obtained from the calls [25, 27], marketing negotiation plans and engineering design plans are for-

*This example is taken from the article "Sales Negotiation Cost Planning for Corporate Level Sales," by M. M. Bird, E. R. Clayton, and L. J. Moore, *Journal of Marketing* (April), Vol. 37, pp. 7–13, by permission of the publisher, the American Marketing Association.

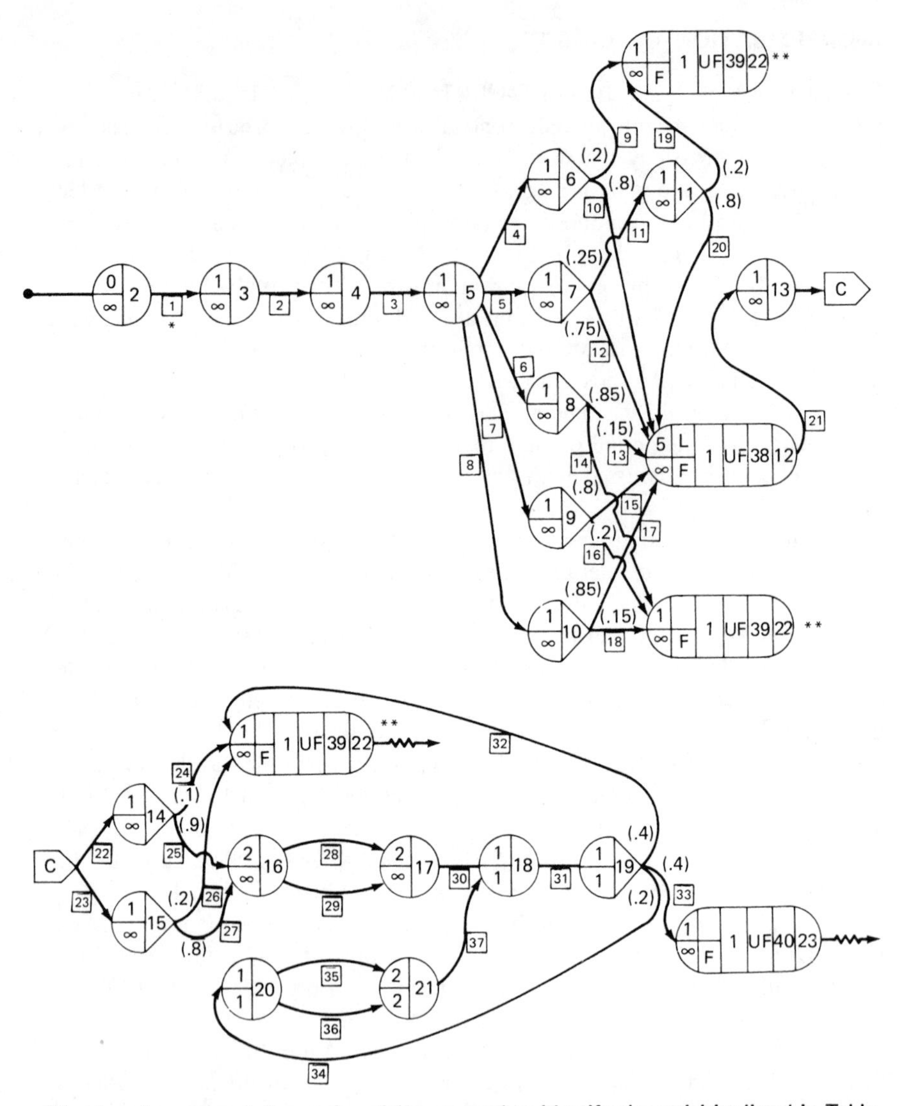


*Numbers in squares below each activity are used to identify the activities listed in Table 10-2; Activity 1, for example, could have alternatively been called 2-3, using the node numbers for identification.

**Node 22 appears three times in this network to facilitate drawing the network; they should be interpreted as the same node.

**Figure 10-4** Q-GERT diagram of industrial sales negotiation model.

mulated [28, 29]. Based on these plans, corporate level strategy is developed in a conference [30]. This strategy is implemented in a negotiation conference with the buying firm [31]. The results of the negotiation conference are: no sale [32]; contract awarded [33]; or further negotiations requested by the buyer involving plan modifications [34]. Marketing and engineering modifications are then performed [35, 36]. A corporate level meeting to evaluate and reconcile the modifications is then held [37]. Following this, a return is made to node 18 of the Q-GERT network where another negotiation conference with the buying firm is held [31]. At node 23, a successful sale has been made. Total costs when a successful scale is made are collected at node 23. Total costs of the negotiation when a sale has not been made are collected at node 22.

The GERT network model for the sales negotiation process allows an analyst to include probabilistic outcomes as modeled by the probabilistic branching at nodes throughout the network. In addition, recycling of activities is permissible as illustrated by the loop of nodes 18, 19, 20 and 21.

Nodes 22 and 23 are of special interest because they are terminal; node 12 is also singled out as a milestone because it denotes that all preliminary reports are favorable, and the final phase of the project is initiated. The GERT simulation program is instructed to collect statistics on these nodes, and for this reason, special notation is required. For example, node 12 contains the symbols 5/∞, L/F, 1, UF, 38, 12. It is read as follows. First, 5/∞ denotes that 5 activities (favorable reports) must be completed to realize this node the first time, while ∞ denotes that this node cannot be realized more than once. (Nodes 18, 19, 20 and 21 are nodes that can be realized more than once because they are contained in a network loop; in these cases, ∞ is replaced by the appropriate integer 1 or 2). The L/F symbol indicates that the time completion of the *L*ast (fifth) activity and the *F*irst activity are kept at this node. (Note, only the *F*irst activity completion times are collected at the other terminal nodes 22 and 23, because they are realized as soon as the *F*irst incoming activity is complete.) The last 4 entries indicate that on node 12, user function (UF) number 38 is used to compute attribute 1 which is cumulative project costs.

The main GERT program handles project time analysis and all computer outputs. It contains 145 lines (FORTRAN language), 111 of which deal with data describing the 37 network activities. The cumulative cost analysis is introduced by the *U*ser *F*unction, which in this case is two subroutines containing 38 lines of computer code, half of which are cost data. The *U*ser *F*unction in this case assumes that each activity has an associated fixed setup cost, and a variable cost that is linear in time. Other functions of time could be used by changing one FORTRAN statement. If node 22 (no sale) is realized, the *U*ser *F*unction administers company policy which requires that all work on the project is halted. Thus, the program stops all activities so that their remaining costs are not incurred and reflected in the computed total project cost.

**Table 10-2. Activity Descriptions for Sales Negotiations Network**

| Activity | | | Time in hours | | | Cost in dollars* | | Proba- |
|---|---|---|---|---|---|---|---|---|
| No. | Activity Description | Distribution | Mode | Min | Max | (a) | (b) | bility |
| 1 | Sales call by company salesman | Beta | 2.00 | 1.25 | 4.0 | 0 | 11. | 1.00 |
| 2 | Sales report to marketing vice president | Beta | 0.50 | 0.25 | 1.0 | 0 | 31. | 1.00 |
| 3 | Marketing vice president reports to president | Beta | 0.25 | 0.25 | 0.50 | 0 | 51. | 1.00 |
| 4 | Preliminary engineering report | Beta | 24.0 | 16.0 | 40.0 | 200. | 20. | 1.00 |
| 5 | Preliminary production report | Constant | 8.0 | N/A | N/A | 100. | 12. | 1.00 |
| 6 | Preliminary financial report | Constant | 16.0 | N/A | N/A | 50. | 10. | 1.00 |
| 7 | Preliminary marketing report | Constant | 8.0 | N/A | N/A | 50. | 10. | 1.00 |
| 8 | Preliminary purchasing report | Beta | 24.0 | 16.0 | 40.0 | 300. | 17. | 1.00 |
| 9 | Negative engineering report examined | Constant | 8.0 | N/A | N/A | 0 | 20. | 0.20 |
| 10 | Favorable engineering report examined | Constant | 8.0 | N/A | N/A | 0 | 20. | 0.80 |
| 11 | Production subcontracting investigated | Beta | 24.0 | 8.0 | 40.0 | 0 | 10. | 0.25 |
| 12 | Favorable production report examined | Constant | 8.0 | N/A | N/A | 0 | 20. | 0.75 |
| 13 | Favorable financial report examined | Constant | 10.0 | N/A | N/A | 60. | 20. | 0.85 |
| 14 | Negative financial report examined | Constant | 12.0 | N/A | N/A | 120. | 20. | 0.15 |
| 15 | Favorable marketing report examined | Constant | 12.0 | N/A | N/A | 61. | 20. | 0.80 |
| 16 | Negative marketing report examined | Constant | 16.0 | N/A | N/A | 248. | 20. | 0.20 |
| 17 | Favorable purchasing report examined | Constant | 2.0 | N/A | N/A | 61. | 17. | 0.85 |
| 18 | Negative purchasing report examined | Constant | 2.0 | N/A | N/A | 61. | 17. | 0.15 |
| 19 | Unfavorable production subcontract examined | Constant | 2.0 | N/A | N/A | 0 | 51. | 0.20 |
| 20 | Favorable production subcontract examined | Constant | 8.0 | N/A | N/A | 0 | 20. | 0.80 |
| 21 | Corporate level planning conference | Constant | 8.0 | N/A | N/A | 750. | 0 | 1.00 |

| | | | | | | | | |
|---|---|---|---|---|---|---|---|---|
| 22 | Sales call by marketing vice president and salesman | Constant | 16.0 | N/A | N/A | 400. | 31. | 1.00 |
| 23 | Engineering call by engineering vice president | Constant | 16.0 | N/A | N/A | 400. | 32. | 1.00 |
| 24 | Unfavorable sales call conference | Beta | 3.0 | 1.0 | 18.0 | 0 | 73. | 0.10 |
| 25 | Favorable sales call no conference | Constant | 0.0 | N/A | N/A | 0 | 0 | 0.90 |
| 26 | Unfavorable engineering call conference | Beta | 3.0 | 1.0 | 16.0 | 0 | 73. | 0.20 |
| 27 | Favorable engineering call no conference | Constant | 0.0 | N/A | N/A | 0 | 0 | 0.80 |
| 28 | Marketing negotiation plan formulation | Beta | 24.0 | 8.0 | 40.0 | 200. | 20. | 1.00 |
| 29 | Engineering design plan formulation | Beta | 80.0 | 40.0 | 160.0 | 800. | 26. | 1.00 |
| 30 | Corporate level strategy conference | Beta | 2.0 | 1.0 | 8.0 | 0 | 73. | 1.00 |
| 31 | Negotiation conference with buying firm | Beta | 6.0 | 2.0 | 16.0 | 400. | 73. | 1.00 |
| 32 | No sale | Constant | 0.0 | N/A | N/A | 0 | 0 | 0.40 |
| 33 | Contract awarded | Constant | 0.0 | N/A | N/A | 0 | 0 | 0.40 |
| 34 | Modifications requested by buyer | Constant | 0.0 | N/A | N/A | 0 | 0 | 0.20 |
| 35 | Modification of marketing negotiation plan | Beta | 12.0 | 4.0 | 20.0 | 200. | 20. | 1.00 |
| 36 | Modification of engineering design plan | Beta | 40.0 | 20.0 | 80.0 | 800. | 26. | 1.00 |
| 37 | Corporate meeting to reconcile modifications | Beta | 1.0 | 0.5 | 4.0 | 0 | 73. | 1.00 |

*Column (a) is the fixed cost for the activity while column (b) is the variable cost for the activity assumed to be linear over time.

## RESULTS OF COMPUTER SIMULATION

The Q-GERT model of the industrial sales negotiation process was simulated 500 times. Summaries of the time and cost results are presented in Figures 10-5 and 10-6. The estimates of the probability of losing a sale, node 22, is 0.844, that is, over 84 percent of the negotiations end in failure.

Looking at node 12, favorable reports received, it is seen that approximately 43 percent of the potential projects result in all five reports being favorable. Thus, 57 percent of the sales negotiations are turned down for internal reasons. The time estimates indicate that it takes over 38 days to decide that negotiations should be carried beyond the internal report phase. The cost data indicate that it costs over $3050 when all favorable reports are obtained. This information can be extremely useful for an analyst who is attempting to improve the sales negotiation process, as it provides trade-off data regarding the possibility of increasing the probability of favorable reports versus the time and costs required to obtain the favorable reports. By developing alternative networks up to node 12, such trade-offs can be made. The analyst should attempt to develop procedures for detecting when unfavorable reports will be issued. Possibly sequential reporting of the engineering, production, financial, marketing and purchasing reports should be made. Since engineering and marketing have the highest probability of issuing a negative report, possibly these two activities should be performed prior to the other reports. If this is done, the time required to reach the decision with regard to the preliminary reports will be extended.

The summary statistics for a lost sale indicate that it takes 62 days on the average to make this determination. Since node 22 can be reached from many points in the network, this time should have a wide variability which is the case as indicated by its standard deviation of 56.4 and its range of 23 to 288. The average cost associated with a lost sale is $4944. This indicates that when the project fails more money is expended in obtaining favorable reports than is put into the negotiation effort after the reports are obtained. When the project succeeds, which occurs only 15.6 percent of the time, the negotiation process takes a lengthy 182 days and costs on the average $11,947. Other statistical quantities concerning the time and cost when the negotiation is successful are shown in Figures 10-5 and 10-6.

The time and cost histograms associated with node 22, lost sale, are presented in Figure 10-7. These histograms illustrate that the distribution function associated with node 22 has discrete breaks due to the different paths that can be used to reach node 22. For example, failed reports occurred on 283 runs. $((1 - .434) \times 500 = 283)$. The histogram for node 22 shows 283 values in the range 20 to 50; hence, this cluster of values is associated with failed reports. The other values in the histogram for node 22 are for the times when a lost sale occurred after favorable reports were received.

```
GERT SIMULATION PROJECT SALES-14        BY PRITSKER
                   DATE  3/ 15/ 1977

          **FINAL RESJLTS FOR  500 SIMULATIONS**
```

**AVERAGE NODE STATISTICS**

| NODE | LABEL | PROBABILITY | AVE. | STD.DEV. | SD OF AVE | NO OF OBS. | MIN. | MAX. | STAT TYPE |
|---|---|---|---|---|---|---|---|---|---|
| 22 | LOST-SAL | .8440 | 62.1055 | 56.4425 | 2.7476 | 422. | 22.6712 | 287.7906 | F |
| 23 | SUCCESS | .1560 | 182.0164 | 49.1436 | 5.4512 | 78. | 125.7381 | 425.7386 | F |
| 12 | FAV-RPTS | .4340 | 38.1034 | 4.6936 | .3186 | 217. | 28.5247 | 50.9886 | F |

**Figure 10-5** Summary report for times associated with sales negotiation model.

**USER STATISTICS FOR VARIABLES BASED ON OBSERVATION AT TIME  30.095 IN RUN  500**

| | AVE | STD DEV | SD OF AVE | MINIMUM | MAXIMUM | OBS |
|---|---|---|---|---|---|---|
| FAV-RPTS | 3053.6797 | 141.4692 | 9.5036 | 2760.4418 | 3469.0602 | 217 |
| LOST-SAL | 4943.8782 | 3330.3886 | 162.1208 | 2475.4571 | 17774.4282 | 422 |
| SUCCESS | 11946.3841 | 2730.1616 | 309.1299 | 9524.0147 | 25238.0935 | 78 |

**Figure 10-6** Summary report for costs associated with sales negotiation model.

F STAT HISTOGRAM FOR NODE 22

LOST-SAL

| OBSV FREQ | RELA FREQ | CUML FREQ | UPPER BOUND OF CELL |
|---|---|---|---|
| 0 | 0 | 0 | 10.00 |
| 0 | 0 | 0 | 20.00 |
| 138 | .327 | .327 | 30.00 |
| 133 | .315 | .642 | 40.00 |
| 12 | .028 | .671 | 50.00 |
| 3 | .007 | .678 | 60.00 |
| 43 | .102 | .780 | 70.00 |
| 17 | .040 | .820 | 80.00 |
| 0 | 0 | .820 | 90.00 |
| 0 | 0 | .820 | 100.00 |
| 0 | 0 | .820 | 110.00 |
| 0 | 0 | .820 | 120.00 |
| 2 | .005 | .825 | 130.00 |
| 8 | .019 | .844 | 140.00 |
| 8 | .019 | .863 | 150.00 |
| 9 | .021 | .884 | 160.00 |
| 11 | .026 | .910 | 170.00 |
| 10 | .024 | .934 | 180.00 |
| 7 | .017 | .950 | 190.00 |
| 21 | .050 | 1.000 | +INF |
| TOTAL 422 | | | |

0 20 40 60 80 100

LOST-SAL

| OBSV FREQ | RELA FREQ | CUML FREQ | UPPER BOUND OF CELL |
|---|---|---|---|
| 0 | 0 | 0 | 2000.0000 |
| 203 | .481 | .481 | 3000.0000 |
| 80 | .190 | .671 | 4000.0000 |
| 0 | 0 | .671 | 5000.0000 |
| 38 | .090 | .761 | 6000.0000 |
| 25 | .059 | .820 | 7000.0000 |
| 0 | 0 | .820 | 8000.0000 |
| 0 | 0 | .820 | 9000.0000 |
| 10 | .024 | .844 | 10000.0000 |
| 33 | .078 | .922 | 11000.0000 |
| 13 | .031 | .953 | 12000.0000 |
| 3 | .007 | .960 | 13000.0000 |
| 8 | .019 | .979 | 14000.0000 |
| 6 | .014 | .993 | 15000.0000 |
| 1 | .002 | .995 | 16000.0000 |
| 1 | .002 | .998 | 17000.0000 |
| 1 | .002 | 1.000 | 18000.0000 |
| 0 | 0 | 1.000 | 19000.0000 |
| 0 | 0 | 1.000 | 20000.0000 |
| 0 | 0 | 1.000 | INF |
| 422 | | | |

0 20 40 60 80 100

**Figure 10-7** Histograms of times and costs associated with unsuccessful sales negotiation.

Histograms for the time and costs associated with a successful sales negotiation were also developed, but are not presented here. These histograms could be used to monitor company time and cost data associated with successful sales negotiations. They would indicate that over 20 percent of the time, the total sales negotiation cost will be greater than $14,000. Thus, we can expect 1 time out of 5 to have this high cost, given that the sales negotiation pattern follows the one described by the Q-GERT model. By making such comparisons, the company can maintain some control over the costs involved in their sales negotiations activities.

## SUMMARY DISCUSSION OF GERT

The above example is illustrative of the use of GERT in project management. In this case, the probability outputs in Figure 10-5 could have been computed analytically using the basic laws of probability. However, this is not true for the time and cost statistics in Figures 10-6 and 10-7. In a more complex hardware development project by Moder[10], it was found to be difficult to intuitively predict even the direction of the effect on time and costs of policy changes in a project plan. One change led to a reduction in both time and cost. GERT simulation is particularly useful in such cases.

Some 13 years have passed since the previous edition of this book in 1970. At that time it was expected that GERT would prove to be a useful tool in project management. The literature of the 70s attests to the validity of this forecast. For example, Vanston[19] used GERT to analyze the effects of various funding and administrative strategies on nuclear fusion power plant development. Halpin[7] used GERT to investigate the use of simulation networks for modeling construction operations. Another important application in this area (see reference 5) involved risk analysis of the construction schedule of the Alaskan pipeline. The effects of weather conditions on construction activities was modeled within the Q-GERT framework to estimate the distribution of both project time and costs performance. Many other applications could be cited.

It has been pointed out above that PERT network logic is a subset of GERT logic. Hence, the latter can be used to simulate PERT networks to obtain the probability distribution of project duration and activity criticality indices. At present, this application would require separate computer runs to produce a GERT simulation output and a conventional PERT output. A number of researchers (Wolfe[20] and Hebert[9]) are developing programs to produce both these outputs from a single computer run. When the objective is both project planning and control, this is the desired mode of operation and the development of such programs should provide a marked increase in the use of GERT in this area of application.

The most recent development in this area of stochastic networks by Moeller and Digman[11], has been called VERT, the acronym for Venture Evaluation and Review Technique. It is a computer-based mathematical simulation network technique designed to systematically assess the risks involved in undertaking a new venture and in resource planning, control monitoring and overall evaluation of ongoing projects, programs and systems. It involves even more *U*ser *F*unction types of interaction than GERT, and is perhaps a logical evolution to embrace the somewhat elusive third parameter of the time/cost/quality of performance triumvirate.

## REFERENCES

1. Arisawa, S. and S. E. Elmaghraby, "Optimal Time-Cost Trade-Offs in GERT Networks," *Management Science*, Vol. 18, No. 11, July 1972, pp. 589–599.
2. Eisner, H., "A Generalized Network Approach to the Planning and Scheduling of a Research Project," *Operations Research*, Vol. 10, No. 1, 1962, pp. 115–125.

3. Elmaghraby, S. E., "An Algebra for the Analysis of Generalized Activity Networks," *Management Science*, Vol. 10, No. 3, 1964, pp. 494–514.
4. Elmaghraby, S. E., *Activity Networks: Project Planning and Control by Network Methods*, John Wiley & Sons, New York, N.Y., 1977.
5. Federal Power Commission Exhibit EP-237, "Risks Analysis of the Arctic Gas Pipeline Project Construction Schedule," Vol. 167, Federal Power Commission, 1976.
6. Freeman, R. J., "A Generalized PERT," *Operations Research*, Vol. 8, No. 2, 1960, p. 281.
7. Halpin, D. W., "An Investigation of the Use of Simulation Networks for Modeling Construction Operations," Ph.D. dissertation, University of Illinois, 1973.
8. Hebert, J. E. III, "Critical Path Analysis and a Simulation Program for Resource-Constrained Activity Scheduling in GERT Project Networks," Ph.D. dissertation, Purdue University, 1975.
9. Hebert, J. E. III, "Project Management with PROJECT/GERT," ORSA/TIMS Joint National Meeting, November 1978, Los Angeles, Calif.
10. Moder, J. J., R. A. Clark, and R. S. Gomez, "Applications of a GERT Simulator to a Repetitive Hardward Development Type Project," *AIIE Transactions*, Vol. 3, No. 4, 1971, pp. 271–280.
11. Moeller, G. L. and L. A. Digman, "Operations Planning with VERT," *Operations Research*, Vol. 29, No. 4, July–August 1981, pp. 676–697.
12. Moore, L. J. and E. R. Clayton, *Introduction to Systems Analysis with GERT Modeling and Simulation*, Petrocelli Books, New York, N.Y., 1976.
13. Pritsker, A. A. B., "GERT: Graphical Evaluation and Review Technique," The Rand Corporation, RM-4973-NASA, Santa Monica, California, April 1966.
14. Pritsker, A. A. B., and W. W. Happ, "GERT: Graphical Evaluation and Review Technique, Part I. Fundamentals," *Journal of Industrial Engineering*, Vol. 17, No. 5, 1966, pp. 267–74.
15. Pritsker, A. A. B., and G. E. Whitehouse, "GERT: Graphical Evaluation and Review Technique, Part II. Applications," *Journal of Industrial Engineering*, Vol. 17, No. 5, 1966, pp. 293–301.
16. Pritsker, A. A. B., *The Precedence GERT User's Manual*, Pritsker & Associates, Inc., Lafayette, IN, 1974.
17. Pritsker, A. A. B., and C. Elliott Sigal, *Management Decision Making: A Network Simulation Approach*, Prentice-Hall, Inc., Englewood Cliffs, New Jersey, 1983.
18. Pritsker, A. A. B., *Modeling and Analysis Using Q-GERT Network*, Halsted Press Books (John Wiley & Sons), New York, N.Y., 1977.
19. Vanston, J. H., Jr., "Use of the Partitive Analytical Forecasting (PAF) Technique for Analyzing of the Effects of Various Funding and Administrative Strategies on Nuclear Fusion Power Plan Development," University of Texas, TR ESL-15, Energy Systems Laboratory, 1974.
20. Wolfe, P. M., "Using GERT as a Basis for a Project Management System," ORSA/TIMS Joint National Meeting, Los Angeles, California, November 1978.

## EXERCISES

1. Construct a table containing a list of criteria that you feel could be used to measure the utility of networking schemes. On the other axis list all the networking and project charting schemes you are familiar with, including hybrid forms of bar charts, milestone charts, or others. Then assign a rank or other value score to each scheme and criterion and summarize the score of each scheme. Which scheme received the highest score?

2. Compare the results of your scoring in the exercise above with those of other students and, if practical, with the opinions of persons experienced with a variety of networking schemes. As a result of these comparisons, comment on whether it is practical to evaluate networking schemes on a technical, objective basis.

3. The techniques of resource allocation (Chapter 7) are applicable to certain types of job-shop scheduling problems, especially where the sequence of jobs is fixed and the problems are reduced to questions of the loading of certain facilities or pools of resources and the avoidance of project (or job) delays. Discuss how generalized network concepts may provide solution methods to a wider class of job-scheduling problems.

4. Draw a GERT network for the following procedure used by a university patent committee.
   1. Inventor submits invention to patent committee.
   2. Above (1) initiates two concurrent activities:
      - 2.1. Technical or commercial reviews by experts; and
      - 2.2. Legal review of inventor's patent liability to contractors and university.
   3. Study of legal and technical reviews.
   4. The study in (3) leads to
      - 4.1. Favorable evaluation, or
      - 4.2. Unfavorable evaluation
   5. Favorable evaluation in 4.1 leads to
      - 5.1. Submission of invention to university patent attorney; or
      - 5.2. Submission of invention to outside patent corporation.
   6. Activity 5.1 leads to
      - 6.1. University marketing of patent; or
      - 6.2. University drops invention because of lack of patent protection.
   7. Activity 5.2 leads to
      - 7.1. Outside patent corporation rejects invention; or
      - 7.2. Outside patent corporation seeks patent.
   8. Activity 7.1 leads to
      - 8.1. Resubmission by university patent committee of patent to outside patent corporation for reconsideration; or
      - 8.2. University drops invention.
   9. Activity 7.2 leads to
      - 9.1. Patent denied so university drops invention; or
      - 9.2. Patent acquired so outside patent corporation markets patent.

**10.** Unfavorable evaluation in 4.2 leads to

**10.1.** Submission of invention to outside patent corporation; or

**10.2.** University drops invention.

**5.** Draw a GERT network for a general R&D process which includes the following activities or nodes: start project; problem definition; research activity; evaluate solutions; solution unacceptable more research; solution unacceptable redefine problem; project washout; develop prototype; redevelop prototype; solution implementation; project successful completion.

**6.** Consider drawing a GERT network for an embellishment of Problem 5. Suppose there are 4 projects and two research teams. Team 1 is to start on project 1, and team 2 on project 4. When a team completes a project (for any reason), it immediately moves on to a project that has not yet started. Team 1 is always to select the lowest project number available, while team 2 selects the highest project number available. The simulation is to continue until all 4 projects are complete. Note, at the end of the simulation period, one team will be idle, rather than helping the other team that is still working.

# 11
# COMPUTER PROCESSING

In Chapter 4, manual methods of making the basic critical path scheduling computations were presented. This chapter will describe how the same computations can be made by computers. The computer programs for CPM and PERT computations are available to anyone who has access to almost any general purpose business computer (including so-called minicomputers and microcomputers).

The treatment in this chapter begins with what the typical critical path software is and how one can make use of it. The actual software that one may find will vary somewhat in capacity, speed, features provided, and even sometimes in the computational results. Consequently, this text does not attempt to explain how each available program works. Rather, it is a treatment of how they all work in general, and what to look for in the way of comparative features.

The basic critical path scheduling computations are emphasized in this chapter. The roles of the computer in the more advanced topics are mentioned here but are covered more fully in the appropriate chapters.

## ADVANTAGES OF COMPUTERS

Computer processing of networks offers several advantages over manual calculations. Among these are:

1. The project activities can be easily sorted and listed in a variety of useful ways, such as by float (critical path listing), by early or late start dates

(showing what needs to be done next), by responsibility codes (providing lists for each supervisor or subcontractor), by activity numbers, and by other keys or combinations of keys. These sorted lists help simplify the management tasks.
2. Once the network is stored in the computer the project schedule can be easily updated, providing new listings based on actual progress on the project or changes in estimates. This is a key advantage in making critical path methods useful throughout a project's life cycle.
3. Certain analyses that go beyond the basic schedule calculations, such as time-cost trade-offs (Chapter 8), resource allocation (Chapter 7), and cost control (Chapter 5) are really practical only with the aid of computer programs.

In spite of these advantages, however, a computer is certainly not required or necessarily desired for all critical path applications. Computer processing does involve extra time and cost to transfer the network into a computer input format, and in many cases the extra cost is not justified. In other cases a suitable computer is simply not available.

## WHEN A COMPUTER IS NEEDED

In deciding whether to process a particular network by manual calculations or by computer, the following factors should be considered:

1. *Network size.* The larger the network, of course, the more likely that computer processing will be needed. However, there is no specific number of activities that would call for a computer, due to the influence of the other factors listed below.
2. *Computer availability.* Both a suitable computer and adequate critical path software must be available at a reasonable cost.
3. *Expected frequency of updating.* If the network is primarily a planning tool and is not likely to be revised and updated throughout the project, computer processing may not be worthwhile, even if the network is large.
4. *Desired output listings.* If printed listings of the network activities are desired, selected, and sorted by float, responsibility, start dates, activity numbers, etc., and if these lists are desired for each updating calculation on the project, then computer processing is the most practical answer.
5. *Advanced analyses.* In most cases computers and the appropriate software are required for time-cost trade-off, resource allocation, PERT statistics, and cost control analyses.
6. *Network format.* When the precedence network format is used, manual calculations of the basic schedule are complex, making the use of computer processing decidely advantageous.

Each potential user must consider these factors in relation to his project and the particular program available. A few hypothetical examples of such considerations are given below. These examples are not offered as firm guidelines for the types of situations described, but rather as indications of the considerations that should influence the decision of whether or not to use a computer.

CASE 1. *Heart Surgery.* A network of 75 activities has been prepared for the purpose of planning a complex surgical operation. The network will not be used after the plan has been worked out. *Recommendation:* In this case a manual computation and a time-scaled network would be the best means of working out a well-coordinated plan.

CASE 2. *Promotional Campaign.* A project to introduce a new consumer product involves about 80 advertising, manufacturing, distribution, and sales activities. The network is to be used for initial planning and coordination throughout the six-month project in order to assure proper timing of each phase. All the supervisors have had a short course in CPM. Management intends to update the network twice a month and distribute copies of the results to each of the key supervisors involved. *Recommendation:* In this case the frequency of updating indicates that computer processing would be faster and more economical than manual updating. Also, the computer can sort the output by responsibility, so that each supervisor can get an extracted report on only his activities.

CASE 3. *Construction of a Large Building in a Remote Location.* A five-story building is to be constructed in a small city where no critical path computer program is available. The local contractor is interested in applying CPM in planning the project, though, because this will be his first multi-story building project. The first draft of his network contained 270 activities. *Recommendation:* The manual computation of this network will be well worthwhile, perhaps saving the contractor a great deal of cost and confusion over the re-use of concrete forms, the phasing of carpentry, steel, electrical, and other trades, and by helping to solve other problems of coordinating multi-story construction that he has not yet experienced. (See also Exercise 7.)

CASE 4. *New Plant Start-Up.* This project includes construction of a plant, delivery and installation of machinery, recruiting and training a new labor force, and starting production. The key supervisors for the contractor, machinery suppliers, training and production departments have had no CPM training, and management considers it impractical to undertake such training at this time. The project manager's network contains 210 activities, and he intends to use it for schedule control during the project. A good CPM program is available. *Recommendation:* The pros and cons of using a computer in this case are not strong either way. The computer reports may be confusing and useless to supervisors not trained in how to read them. On the other hand, experience has shown that machinery suppliers and subcontractors tend to place greater effort on schedule control when the customer shows them periodic computer reports indicating that they will delay the project unless certain tasks are completed by certain dates. Here the psychological factors are more significant than the time or costs of processing. The decision could go either way, depending on the individuals involved.

## COMPUTER SOFTWARE

The term *computer software* refers to the recorded instructions or programs necessary to make a computer perform the functions desired, including the reading of input data, testing for errors, calculating, sorting and printing out reports, etc.

We shall use the term *critical path software* to describe the computer programs that perform at least the basic forward pass, backward pass and float computations associated with CPM or PERT networks. Usually the software is a series of programs (or *software package*) designed to operate on a particular model of computer but will accept any network data, provided the network data follow certain rules and do not exceed the capacity of the software. Thus new software does not have to be developed for each network.

The software is usually stored on a magnetic disk or tape.

Each software package has a booklet of instructions for the user, explaining how to prepare the network data for input, how to select various processing options, and how to read the output reports. By following these instructions carefully one can translate a network into the required input format and let the computer do all the critical path processing.

After the initial computation has been made, the computer may store the user's network on a magnetic tape or disk to await updating computations. A schematic diagram of these operations is given in Figure 11-1. At the time of updating, the user provides input data on the progress of the project and changes in the network. These data are then processed against the initial network the computer has stored. A schematic diagram of this updating function is shown in Figure 11-2.

## INPUT FOR INITIAL COMPUTATION

To illustrate how a typical critical path software package may be used, let us take the market survey network in Figure 11-3 and prepare it for computer processing. (In this case we will be employing the arrow scheme of networking. In-

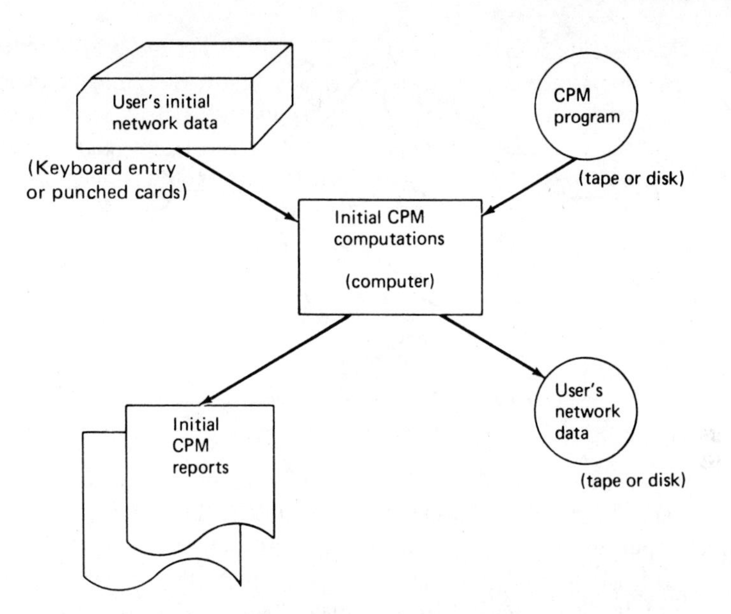


**Figure 11-1** Schematic flow diagram of initial CPM processing.

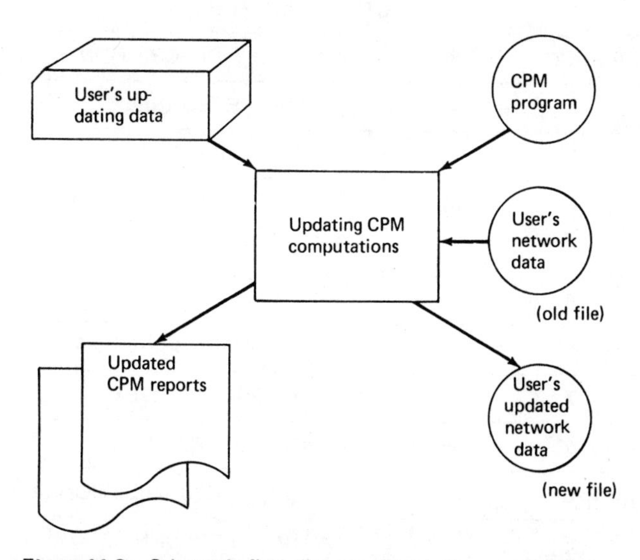


**Figure 11-2** Schematic flow diagram of updating computation.

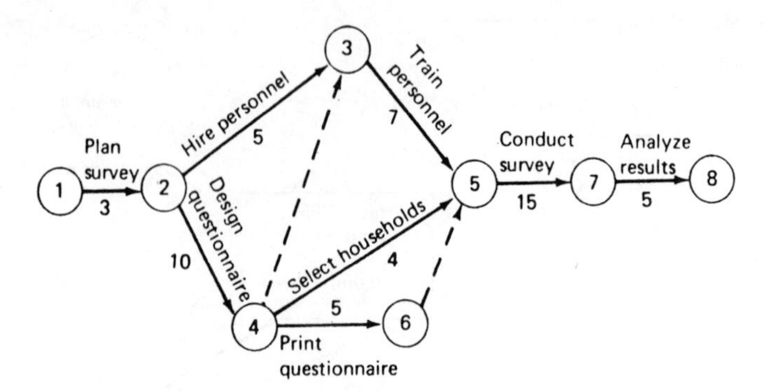


Figure 11-3 Network of survey project.

put procedures for networks in the node and precedence diagramming formats are different. Each software package contains instructions for input.)

Most input forms for arrow-scheme networks are similar to that shown in Figure 11-4. Data related to an activity is entered in one row of this form. The column labeled "pred" is for the predecessor event number of the activity, and "succ" is for the successor event number. The column labeled "$t$" is for the activity time estimate. (PERT input forms have columns for three time estimates.) The description column usually accepts a certain number of alphabetic or numeric characters, which are used to describe the activity.

COMPUTER INPUT FORM

| PRED | SUCC | t | DESCRIPTION |
|---|---|---|---|
| 2 | 3 | 5 | HIRE PERSONNEL |
| 1 | 2 | 3 | PLAN SURVEY |
| 2 | 4 | 10 | DESIGN QUESTIONNAIRE |
| 4 | 3 | 0 | DUMMY |
| 6 | 5 | 0 | DUMMY |
| 4 | 5 | 4 | SELECT HOUSEHOLDS |
| 5 | 7 | 15 | CONDUCT SURVEY |
| 7 | 8 | 5 | ANALYZE RESULTS |
| 4 | 6 | 5 | PRINT QUSS QUESTIONNAIRES |
| 3 | 5 | 7 | TRAIN PERSONNEL |

Figure 11-4 Basic input for critical path computer programs.

The way the form is filled out illustrates one of the rules that must be followed: all dummies must be entered into the form just as all other activities, except that dummies carry zero time estimates. The description "dummy" is optional. Note also that the activities are entered in the form in random order, which is permissible. (We have also assumed that the program to be used will accept events numbered at random; otherwise numbers 3 and 4 in the network would have to be reversed.) Other common rules of input are that each activity must be entered, and no activity may be entered twice. Programs that have scheduled date, calendar date, cost control, and other features have additional columns in the input forms for certain necessary data for these features.

After the form has been filled out and double-checked for errors or omissions, the form is given to a keyboard operator who transfers the information onto a computer input medium such as a magnetic disk, tape, or punched cards. The input medium is also checked for errors and then is fed into the computer system for processing.

In addition to the data on each activity, one must also provide certain general information and parameters for the network as a whole. These data would normally include the name of the network, whether the initial or updating computation is desired, and such other data as required by the options of the program. For example, programs having calendar dating capability normally would require that the network data include:

1. the network start date,
2. the network target completion date,
3. whether a 5-, 6-, or 7-day week is to be assumed, and
4. which holidays are to be included.

The requirements for such data and their input format vary with each program, therefore they are not illustrated here.

## OUTPUT FOR INITIAL COMPUTATION

The basic output report contains the same information as given by manual methods. The typical format is shown in Figure 11-5. Note that the first four fields contain the input data for each activity. The remaining fields provide the computed results for each activity. In this example the times are expressed in units, the unit being defined by the user. In most programs one may elect to obtain the output times in the form of calendar dates.

Note also that the output in Figure 11-5 is now in a particular order, although the activities were put in at random. The order here is by float or slack, from the lowest to the highest. This "slack sort" is one of the most commonly requested sorts of the output, because it lists the most critical activities first, and thus easily identifies the critical path, or paths.

**MARKET SURVEY PROJECT** **INITIAL SCHEDULE**

| *I* | *J* | DUR | DESCRIPTION | *ES* | *LS* | *EF* | *LF* | *S* |
|---|---|---|---|---|---|---|---|---|
| 1 | 2 | 3 | PLAN SURVEY | 0 | 0 | 3 | 3 | 0 |
| 2 | 4 | 10 | DESIGN QUESTIONNAIRE | 3 | 3 | 13 | 13 | 0 |
| 4 | 3 | 0 | DUMMY | 13 | 13 | 13 | 13 | 0 |
| 3 | 5 | 7 | TRAIN PERSONNEL | 13 | 13 | 20 | 20 | 0 |
| 5 | 7 | 15 | CONDUCT SURVEY | 20 | 20 | 35 | 35 | 0 |
| 7 | 8 | 5 | ANALYZE RESULTS | 35 | 35 | 40 | 40 | 0 |
| 4 | 6 | 5 | PRINT QUESTIONNAIRE | 13 | 15 | 18 | 20 | 2 |
| 6 | 5 | 0 | DUMMY | 18 | 20 | 18 | 20 | 2 |
| 4 | 5 | 4 | SELECT HOUSEHOLDS | 13 | 16 | 17 | 20 | 3 |
| 2 | 3 | 5 | HIRE PERSONNEL | 3 | 8 | 8 | 13 | 5 |

**Figure 11-5** Basic computer output report.

## INPUT FOR UPDATING

To update a critical path schedule after the project has begun and progress has been made, the computer programs may call for the following types of data:

1. Beginning date for activities that have begun.
2. Finish date for activities that have been completed.
3. New data for activities that have been changed.
4. Complete data for activities that have been added to the network.
5. Identification of activities that have been deleted from the network.

Most updating programs utilize only start and finish dates of activities, although a few will accept a "percent completion" figure.

For changed activities, normally only the changed field must be entered. This will usually be the time estimate or the description. The *I* and *J* numbers may not be changed, except by deleting the activity and adding a new one with the new numbers. The *I* and *J* numbers are the keys to identification of each activity in the computer files.

Care must be taken in deleting activities, to insure that errors or gaps are not created in the network. If activity 4-3 in the sample network were deleted the computer would accept the input and process it, although the result would be different than when 4-3 is included. However, if activity 3-5 were deleted (and not replaced by one or more other activities) the network would have more than one terminal event; some programs would halt on this condition.

In addition to the above data for the activities, updating input must include some general project information as in the input for the initial computation. An important fact for the updating run is the effective or "cut-off" date of the report. Some computations will involve the effective date. Also the effective date serves as the basis for certain error checks. For example, a reported actual date (for the start or finish of an activity) may not be later than the effective date.

| *Activity* | *Actual Start Date* | *Actual Finish Date* | *Change* |
|---|---|---|---|
| 1–2 | 0 | 2 | |
| 2–4 | 2 | 12 | |
| 4–6 | 12 | – | |
| 3–5 | | | Change time estimate to 10 days. |

**Figure 11-6** Progress on survey project as of the end of the twelfth day.

| *I* | *J* | DUR | DESCRIPTION | Update Code | Date |
|---|---|---|---|---|---|
| 1 | 2 | | | 1 | 0 |
| 1 | 2 | | | 2 | 2 |
| 2 | 4 | | | 1 | 2 |
| 2 | 4 | | | 2 | 12 |
| 4 | 6 | | | 1 | 12 |
| 3 | 5 | 10 | | 3 | |

**Figure 11-7** Input for updating computation.

To illustrate how typical updating data are entered, consider the survey project of Figure 11-3. Assume that at the end of day 12 progress has been made according to the information in Figure 11-6.

Assume that no other progress dates or changes have been recorded as of the end of day 12. These updating data may be entered on the program input form as shown in Figure 11-7. The key to the "Update Code" is as follows: 1 = start date; 2 = finish date; 3 = change. Note that only the activities that are actually involved in the updating are entered on the form.

It is emphasized that these procedures are presented only as representative of how some programs handle initial and updating information. The particular program that a user may obtain will probably differ from these examples in a variety of ways. This is especially true of the updating function. Consequently, the user must carefully study the instruction manual provided with the program to be used.

## OUTPUT FOR UPDATED SCHEDULE

To illustrate how updating input may be handled by a computer program, let us assume that day 40 is the established target date for completion of the market survey project, and that the computer program will make its backward pass from this target date in each updating run. (Some programs have this feature, others

**MARKET SURVEY PROJECT** **EFFECTIVE DATE: 12**

| I | J | Est Dur | Act Dur | Description | ES | LS | EF | LF | F |
|---|---|---|---|---|---|---|---|---|---|
| 1 | 2 | 3 | 2 | PLAN SURVEY | STARTED | 0 | FINISHED | 2 | |
| 2 | 4 | 10 | 10 | DESIGN QUESTIONNAIRE | STARTED | 2 | FINISHED | 12 | |
| 2 | 3 | 5 | | HIRE PERSONNEL | 13 | 5 | 18 | 10 | −8 |
| 3 | 5 | 10 | | TRAIN PERSONNEL | 18 | 10 | 28 | 20 | −8 |
| 5 | 7 | 15 | | CONDUCT SURVEY | 28 | 20 | 43 | 35 | −8 |
| 7 | 8 | 5 | | ANALYZE RESULTS | 43 | 35 | 48 | 40 | −8 |
| 4 | 5 | 4 | | SELECT HOUSEHOLDS | 13 | 16 | 17 | 20 | 3 |
| 4 | 6 | 5 | | PRINT QUESTIONNAIRE | STARTED | 12 | 17 | 20 | 3 |

**Figure 11-8** Update deport.

do not.) Under these conditions the computer output should be basically as shown in Figure 11-8.

The format of the updated report in Figure 11-8 is different from that of the initial report shown in Figure 11-5. The actual duration ("ACT DUR") now appears for the activities that have been completed. This figure is simply the difference between the reported start and finish dates, which are shown under the *LS* and *LF* columns, respectively. The preceding words "STARTED" and "FINISHED" indicate that the dates that follow are the actual reported dates rather than the computed *LS* and *LF* dates. Note also that the finished activities no longer show a float figure, and are listed first in the float sort.

The next four activities form the critical path, which has changed since the initial computation. The dummy activities have been automatically deleted from this report because they no longer play a role in the balance of the network. That is, the activities preceding the dummies have been completed, removing the precedence-constraint purpose of the dummies. (Some programs continue to list the dummies anyway. Other programs offer the user the option of omitting all dummies and completed activities from update reports.)

The update report, sorted by float, provides a straightforward statement of the project status. One sees at a glance which activities are completed and, of those that remain, which are most critical. The fact that the project is eight days behind schedule is also immediately apparent. The cause of the late condition can be discovered by looking at the records for activities preceding the current critical path and the initial activity on the critical path. The reader will see that activity 1-2 started on time and finished a day early. However, activity 2-3, which begins the new critical path, did not start as early as it could have. In fact, as of day 12 it still has not been reported started, meaning that it cannot start until day 13 at the earliest, which is 10 days after the *ES* and 5 days after the *LS* time. Furthermore, the estimated duration of activity 3-5 has been changed from 7 to 10 days. For these two reasons, then, the expected completion date

**Table 11-1. Optional Output Sequences ("Sorts")**

| *Sort Key* | *Application* |
|---|---|
| Activity Number (*I-J* or node numbers) | Enables quick location of information about individual activities. Useful in cross-referencing the graphic network with the computer output and for error-checking. |
| Float (or "Total Slack") | Organizes the activities into paths, with the most critical paths listed first. (Figure 11-8) |
| *ES* | Provides a chronological listing according to Early Start time of each activity, thus offering a form of "tickler file" by date. Late Start or any other computed date may be used instead of *ES*. |
| Responsibility Code | Each activity may be assigned a code representing the person, agency or subcontractor responsible for it. A sort on this code groups the activities by such responsibilities, providing a convenient report for subcontractors and others. |
| Designated Key Events or Activities | Provides a summary report for top management. |

of the project is now day 48 instead of the desired day 40. The total float has gone from the initial value of zero to -8, or 8 days behind schedule.

The float sort is not the only useful sequence of the output. For large networks an *I-J* sort is useful for quickly locating data on particular activities. A sort by one of the date columns can also be of value. For example, a sort by the *LF* column provides a list in chronological order, with the activities that should be finished first listed at the top. All of these sort options and others are usually available through simple *sort key codes* that can be specified by the user. Most users will specify two or three different sorts for each update report. A summary of some of the more useful sorts is given in Table 11-1.

## PRECEDENCE METHOD

As mentioned earlier, this chapter employs only the arrow scheme to illustrate computer input and output. For an example of computer reports under precedence diagramming, see Chapter 4, Figures 4-13, 4-14, 4-17, and 4-18.

## HOW TO FIND COMPUTER SERVICES

The availability of computer services is a twofold question; first, is a computer available, and second, is an adequate critical path program available for that computer? Organizations not possessing a computer may survey their city or nearby cities for computer installations that rent time on the equipment. The local representatives of the computer manufacturers are good sources for this information. Rental time is offered by many banks and industrial firms as well as computer service companies, and such services are often found in relatively small cities and towns. With the recent advances in mini- and microcomputers it has become possible to purchase one with critical path software for only two or three thousand dollars.

Having determined what equipment is available, one should determine next whether a suitable critical path program is available for that equipment. The organization operating the computer may have a suitable program in its library, the manufacturer may be able to supply one, or it may be necessary to seek a program from other sources. The appendix to this chapter lists a number of CPM and PERT programs available to the public. These programs differ significantly in a variety of ways, however, and the differences are worthy of close attention by the potential user.

Computer software companies are also good sources to investigate, since a number of them offer CPM software packages for sale. Purchasing a package is worthwhile if (a) it is expected that the package will be used often, (b) the package makes it possible to utilize the user's in-house computer, and (c) the package has valuable features not otherwise available.

It is also possible to obtain network computational service on a time-shared basis. The user provides the input through a rented or purchased terminal. The input is transmitted over telephone lines to a large computer at some other location, perhaps in another city, or via satellite to another country. The computer processes the input immediately, while it is simultaneously processing a number of other jobs for other users. The critical path output comes directly back to the user's terminal.

In any case, it is important to obtain and study the *user's manual* that is published for each available program. The manual describes the characteristics of the software and how to provide input for it. When attempting to use software for the first time, it is a good idea to test the published procedures with a small network such as the one in Figure 11-3. In this way one can verify the results by hand computation and also check the logic on updating, calendar dating, etc.

## COMPARATIVE FEATURES

In order to summarize the advantages and disadvantages of various critical path programs (the reader is reminded that the term *critical path programs* or *software* as used in this chapter refers to programs for CPM, PERT, and similar network methods), certain key features are listed below and commented upon.

The features discussed are by no means exhaustive of the points on which critical path programs differ. They are among the most significant, however, to the average potential user.

1. *Arrow vs. Node Format.* It is necessary that the network and the software be consistent in the use of either the arrow or the node network format. Thus, the available software might dictate which network format is to be used. An increasing number of software packages now will accommodate either the arrow or node format, or the more advanced precedence scheme.

If the user has already drawn a network in the arrow format and then finds that the only available software requires the node format, the network can be redrawn into the node format. Conversion from arrow to node is relatively easy. Conversion from node to arrow, however, may not be so easy unless the user is familiar with the pitfalls of arrow networking, such as the creation of false dependencies. (See "Common Pitfalls," Chapter 2.)

2. *Event Numbering.* A few of the oldest CPM programs require that events be numbered in ascending order, that is, each activity's successor event number must be larger than its predecessor event number. This is a severe restriction, for it inhibits the flexibility of the network and causes event-number bookkeeping problems. Most programs now permit random numbering of events (or activities in the node format), provided that no two nodes or activities receive identical numbers.

3. *Capacity.* The capacity of critical path programs is usually expressed in terms of the number of activities permitted. Capacities vary from a few hundred to at least 500,000 activities.

If a network exceeds the capacity of the program, it is not usually practical to divide the network into parts for separate computation. The interaction of forward and backward passes along all paths can make such subdivision highly complex. A better approach is to condense the network, as explained in Chapter 3. (See also Network Condensation, feature No. 12 below.)

4. *Calendar Dates.* Many programs have a calendar dating option, which will provide all output dates in the form of 02/03/85 or 03FEB85. To use this option the user needs only to input the base starting date for the first event in the project. A few other options are sometimes available, such as whether the calendar computation is to be based on 5-, 6-, or 7-day weeks, and whether there are holidays. The calendar dating option is highly desirable, although some industrial projects are scheduled by hours and would not use calendar dates. Some programs assume that start dates refer to the beginning of the day given, whereas finish dates are assumed to be at the end of the day given. This convention is logical and presents no problem as long as all the users are aware of it.

5. *Scheduled Dates.* Most programs will accept scheduled dates assigned to the terminal events in the network, and backward passes are made from these scheduled dates, rather than merely "turning around" on the terminal event's earliest expected date as computed in the forward pass. The acceptance of scheduled dates means that the slack figures will be related to the scheduled dates, and the critical path may have positive, zero, or negative float. (The critical path is defined as the path of least float.) Without this feature, one must per-

form hand or mental computations to determine the relationship of the expected completion date and the scheduled date.

In some programs scheduled dates are permitted at any intermediate event. In such programs the float computation is based on either the scheduled date or the computed latest allowed date, whichever is more constraining (earliest). This computation can result in a discontinuous critical path, i.e., one that may begin in the middle of a network.

Projects may indeed have more than one starting point and/or more than one objective, as discussed in Chapter 4. Where this is the case, the computations are inconvenient and sometimes incorrect if all the "dangling" events are artificially brought together to single initial and terminal events, which is a patch-up procedure sometimes used.

6. *Multiple Initial and Terminal Events.* Some critical path programs require that networks have only one initial event and only one terminal event. Other programs permit multiple initial and terminal events, which is advantageous for the situations described above. Also, if it is desired to merge two or more parallel projects for a single critical path computation, the multiple initial and terminal event capability greatly simplifies the mechanics of the procedure.

7. *Error Detection.* Critical path programs vary widely in their capability to detect and diagnose network errors and inconsistencies such as loops, nonunique activities, improper time estimates, and excessive terminal events. Most programs will not only detect loops and stop, but will print out the event numbers in the loop. This greatly aids in the manual search for the network or input error, which is usually two different events with the same number. Other types of error are less difficult to detect by programmed routines, and thus most programs will provide adequate detection and diagnosis.

8. *Output Sorts.* The way in which critical path output data are sequenced, or sorted, affects the utility of the report as a management tool. A list of the most useful sorts and their applications is given in Table 11-1. Most software offers all of these options and others.

Some programs also offer options to sort on major and minor keys, such as "ES within float" (or float major, ES minor key).

9. *Report Generator.* Although most programs provide their output in a certain fixed format, some programs permit a high degree of flexibility in format. By means of "report generator" routines, the user may select which information (from a list of 20 to 30 available fields of data) is desired and may specify the columnar sequence of the fields. Thus, the user can tailor the reports to particular needs, eliminating all undesired data. (See Figures 11-9 and 11-10.)

10. *Graphical Output.* In order to provide readable summaries for management, it is often desirable to convert tabular data into graphical formats. Graphical outputs of various kinds have become increasingly popular features of critical path software. These include bar charts, resource requirement distributions,

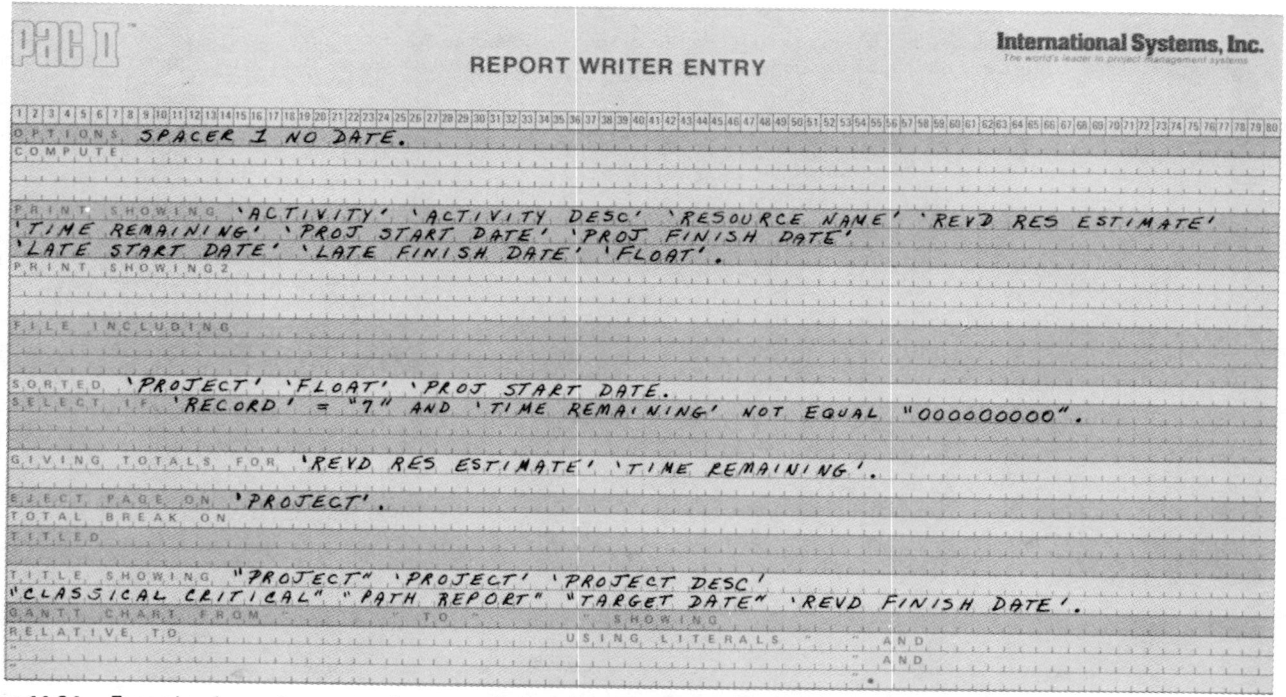

PAC II™

REPORT WRITER ENTRY

International Systems, Inc.
The world's leader in project management systems

OPTIONS SPACER 1 NO DATE.
COMPUTE
PRINT SHOWING 'ACTIVITY' 'ACTIVITY DESC' 'RESOURCE NAME' 'REVD RES ESTIMATE'
'TIME REMAINING' 'PROJ START DATE' 'PROJ FINISH DATE'
'LATE START DATE' 'LATE FINISH DATE' 'FLOAT'.
PRINT SHOWING 2
FILE INCLUDING
SORTED 'PROJECT' 'FLOAT' 'PROJ START DATE.
SELECT IF 'RECORD' = "7" AND 'TIME REMAINING' NOT EQUAL "000000000".
GIVING TOTALS FOR 'REVD RES ESTIMATE' 'TIME REMAINING'.
EJECT PAGE ON 'PROJECT'.
TOTAL BREAK ON
TITLED
TITLE SHOWING "PROJECT" 'PROJECT' 'PROJECT DESC'
"CLASSICAL CRITICAL" "PATH REPORT" "TARGET DATE" 'REVD FINISH DATE'.
GANTT CHART FROM " " TO " " SHOWING
RELATIVE TO USING LITERALS " " AND
" " AND
" ".

**Figure 11-9A** Example of report generator feature available with some software packages is illustrated here, first with the entry form. The user has specified a listing of all remaining uncompleted activities, sorted by float (primary key) and projected start date (secondary key).

PROJECT 072 TWO STORY STEEL/CONCRETE CLASSICAL CRITICAL PATH REPORT TARGET DATE 09/28/82 PAGE 1

| ACTI VITY | ACTIVITY DESCRIPTION | RESOURCE NAME | REVD RES ESTIMATE | TIME REMAINING | PROJECTED START | PROJECTED FINISH | LATE STAR DATE | LATE FINISH | FLOAT |
|---|---|---|---|---|---|---|---|---|---|
| 0010 | MOBILIZE ON SITE | LABOR FOREMEN | 2.0 | 2.0 | 04/05/82 | 04/06/82 | 04/05/82 | 04/06/82 | |
| 0090 | SITE EVACUATION | LABOR FOREMEN | 2.0 | 2.0 | 04/07/82 | 04/08/82 | 04/07/82 | 04/08/82 | |
| 0110 | FORM & POUR FOOTINGS | LABOR FOREMEN | 3.0 | 3.0 | 04/09/82 | 04/13/82 | 04/09/82 | 04/13/82 | |
| 0120 | ERECT STRUCTUAL STEEL | STEEL FOREMEN | 22.0 | 22.0 | 04/14/82 | 05/13/82 | 04/14/82 | 05/13/82 | |
| 0130 | FINE GRADE&GRAVEL FILL | LABOR FOREMEN | 2.0 | 2.0 | 05/14/82 | 05/17/82 | 05/14/82 | 05/17/82 | |
| 0150 | IN PIPE/COND/HEAT/SLAB | HEATING FOREMEN | 13.0 | 13.0 | 05/18/82 | 06/04/82 | 05/18/82 | 06/04/82 | |
| 0150 | IN PIPE/COND/HEAT/SLAB | PLUMBING FOREMEN | 13.0 | 13.0 | 05/18/82 | 06/04/82 | 05/18/82 | 06/04/82 | |
| 0160 | REIN/POUR/CURE GR SLAB | LABOR FOREMEN | 8.0 | 8.0 | 06/07/82 | 06/16/82 | 06/07/82 | 06/16/82 | |
| 0170 | FORM RFEINFORCE SLAB | LABOR FOREMEN | 5.0 | 5.0 | 06/17/82 | 06/23/82 | 06/17/82 | 06/23/82 | |
| 0190 | IN PIPE COND HT DUCTS | HEATING FOREMEN | 11.0 | 11.0 | 06/24/82 | 07/08/82 | 06/24/82 | 07/08/82 | |
| 0190 | IN PIPE COND HT DUCTS | PLUMBING FOREMEN | 11.0 | 11.0 | 06/24/82 | 07/08/82 | 06/24/82 | 07/08/82 | |
| 0200 | POUR/CURE/STRIP SLAB | LABOR FOREMEN | 5.0 | 5.0 | 07/09/82 | 07/15/82 | 07/09/82 | 07/15/82 | |
| 0570 | ERECT EXT MASON 1 FLR | MASONRY FOREMEN | 14.0 | 14.0 | 07/16/82 | 08/04/82 | 07/16/82 | 08/04/82 | |
| 0300 | ER EXT MASONRY TO ROOF | MASONRY FOREMEN | 14.0 | 14.0 | 08/05/82 | 08/24/82 | 08/05/82 | 08/24/82 | |
| 0280 | ERECT STUD PART 2FLR | LABOR FOREMEN | 5.0 | 5.0 | 08/25/82 | 08/31/82 | 08/25/82 | 08/31/82 | |
| 0380 | RN DKS/IN BXS/PIPE 2FL | ELECTICAL FOREMEN | 2.0 | 2.0 | 09/01/82 | 09/02/82 | 09/01/82 | 09/02/82 | |
| 0380 | RN DKS/IN BXS/PIPE 2FL | HEATING FOREMEN | 2.0 | 2.0 | 09/01/82 | 09/02/82 | 09/01/82 | 09/02/82 | |
| 0400 | IN DRY WALL 2ND FLOOR | LABOR FOREMEN | 5.0 | 5.0 | 09/03/82 | 09/10/82 | 09/03/82 | 09/10/82 | |
| 0410 | INTERIOR PAINTING 2FLR | LABOR FOREMEN | 7.0 | 7.0 | 09/13/82 | 09/21/82 | 09/13/82 | 09/21/82 | |
| 0500 | LAY ASPHALT TILE 2FLR | LABOR FOREMEN | 3.0 | 3.0 | 09/22/82 | 09/24/82 | 09/22/82 | 09/24/82 | |
| 0510 | CLEAN UP 2ND FLOOR | LABOR FOREMEN | 1.0 | 1.0 | 09/27/82 | 09/27/82 | 09/27/82 | 09/27/82 | |
| 0540 | FINAL INSPECTION | ADMINISTRATION | 1.0 | 1.0 | 09/28/82 | 09/28/82 | 09/28/82 | 09/28/82 | |
| 0100 | IN UGRD WTR & SWR LNS | PLUMBING FOREMEN | 2.0 | 2.0 | 04/09/82 | 04/12/82 | 04/12/82 | 04/13/82 | 1 |
| 0020 | ORDER/DEL STRUCT STEEL | ADMINISTRATION | 5.0 | 5.0 | 04/05/82 | 04/09/82 | 04/07/82 | 04/13/82 | 2 |
| 0590 | INSL & WOOD FURR 2FLR | LABOR FOREMEN | 2.0 | 2.0 | 08/25/82 | 08/26/82 | 08/30/82 | 08/31/82 | 3 |
| 0480 | IN HUNG CEILING 2FLOOR | LABOR FOREMEN | 4.0 | 4.0 | 09/13/82 | 09/16/82 | 09/16/82 | 09/21/82 | 3 |
| 0290 | IN GLAZE CAULK WND 2FL | LABOR FOREMEN | 3.0 | 3.0 | 08/25/82 | 08/27/82 | 08/31/82 | 09/02/82 | 4 |
| 0490 | HG DOORS/IN WD TRM 2FL | LABOR FOREMEN | 3.0 | 3.0 | 09/13/82 | 09/15/82 | 09/17/82 | 09/21/82 | 4 |
| 0210 | FORM REINF ROOF SLAB | LABOR FOREMEN | 5.0 | 5.0 | 07/16/82 | 07/22/82 | 07/23/82 | 07/29/82 | 5 |
| 0230 | POUR CURE STRP RF SLAB | LABOR FOREMEN | 4.0 | 4.0 | 07/23/82 | 07/28/82 | 07/30/82 | 08/04/82 | 5 |
| 0440 | PL WIRE/MT CIRCUIT BKR | ELECTICAL FOREMEN | 3.0 | 3.0 | 09/03/82 | 09/08/82 | 09/14/82 | 09/16/82 | 6 |
| 0450 | IN LT FIX/RECEPT/SWS | ELECTICAL FOREMEN | 5.0 | 5.0 | 09/09/82 | 09/15/82 | 09/17/82 | 09/23/82 | 6 |
| 0520 | CK OUT ELECT SYSTEM | ELECTICAL FOREMEN | 1.0 | 1.0 | 09/16/82 | 09/16/82 | 09/24/82 | 09/24/82 | 6 |
| 0530 | CK OUT HEATING SYSTEM | HEATING FOREMEN | 1.0 | 1.0 | 09/17/82 | 09/17/82 | [illegible]/27/82 | 09/27/82 | 6 |
| 0140 | PLUMB STRUC[illegible] | STEEL FOREMEN | | 3.0 | 05/14/82 | 05/1[illegible] | [illegible] | 06/04/82 | 12 |
| [illegible] | CERAMI[illegible] | [illegible]OR FOREMEN | | | 09/03/82 | [illegible] | [illegible] | [illegible]22/82 | [illegible] |

**Figure 11-9B** Result of input specification shown in Figure 11-9A is a report with critical path (zero float in this case) activities listed first. Illustrations of PAC II™ Report Writer courtesy of International Systems, Inc., King of Prussia, Pennsylvania.

PAC II — REPORT WRITER ENTRY — International Systems Inc.

```
OPTIONS
COMPUTE

PRINT SHOWING 'PHASE' 'ACTIVITY' 'RESOURCE' 'PROJ FINISH DATE'
'REVD FINISH DATE' 'CALENDAR'.
PRINT SHOWING 2

FILE INCLUDING

SORTED 'PROJECT'
SELECT IF 'RECORD' = "7"

GIVING TOTALS FOR

EJECT PAGE ON 'PROJECT'
TOTAL BREAK ON
TITLED

TITLE SHOWING "PROJECT DESCRIPTION" 'PROJECT DESC' "PROJECT" 'PROJECT'

GANTT CHART FROM "100182" TO "033183" SHOWING 'PROJ FINISH DATE'
RELATIVE TO 'REVD FINISH DATE'      USING LITERALS "+-" AND
"OCT    NOV    DEC    JAN    FEB    MAR    APR        " AND
"1      1      1      1      1      1      1          ".
GIVING COUNTS.
```

**Figure 11-10A** Specification of a Gantt chart by means of a report writer software package is illustrated in this example. Note columnar layout at end of form.

```
PRINT SHOWING 'PHASE' 'ACTIVITY' 'RESOURCE' 'PROJ FINISH DATE'
'REVD FINISH DATE' 'CALENDAR'.
SORTED 'PROJECT'.
SELECT IF 'RECORD' = "7".
EJECT PAGE ON 'PROJECT'.
TITLE SHOWING "PROJECT DESCRIPTION" 'PROJECT DESC' "PROJECT" 'PROJECT'.
GANTT CHART FROM "100182" TO "033183" SHOWING 'PROJ FINISH DATE'
RELATIVE TO 'REVD FINISH DATE' USING LITERALS "+-" AND
"OCT     NOV     DEC     JAN     FEB     MAR     APR         "  AND
"1       1       1       1       1       1       1           ".


PROJECT DESCRIPTION   VERY IMPORTANT PROJECT     PROJECT   PROJ 300                           DATE 04/22/81     PAGE    3

    ACTI  RES   PROJECTED  REVISED    OCT     NOV     DEC     JAN     FEB     MAR     APR
PH  VITY  ID    FINISH     FINISH     1       1       1       1       1       1       1

1   0010  JAN   10/08/82   10/08/82    -
1   0020  DOT   11/15/82   11/23/82                +++
1   0030  ART   10/15/82   10/15/82      -
1   0040  KEN   11/15/82   12/01/82                +++++
2   0050  BEN   12/07/82   12/06/82                       -
2   0060  DOT   11/09/82   12/08/82              +++++++++
2   0070  ART   02/10/83   12/30/82                              -------------
2   0080  DOT   12/14/82   12/16/82                        ++
2   0090  JAN   03/04/83   01/21/83                                    -------------
2   0090  KEN   02/17/83   01/07/83                                -------------
2   0100  PAT   03/18/83   02/04/83                                        -------------
```

**Figure 11-10B** Gantt chart resulting from entry in Figure 11-10A shows first the edited input instructions then the chart. Note that plus signs denote activities with positive float, minus signs for negative float. Illustrations of PAC II™ system courtesy International Systems, Inc., King of Prussia, Pennsylvania.

and even the network diagrams themselves. The charting devices used include the standard line printer on most computers, the CRT display screen, and continuous line plotters. For illustrations of a few of the graphic outputs available, see Figures 11-10 through 11-15.

11. *Updating.* A good software package will provide for updating by exception. The original network is stored on disk or tape, and progress reports are input to update and revise the network and schedule projections. The input transactions should consist only of changes in the network and actual progress, such as the activity start and completion dates since the last update as illustrated earlier in this chapter. The entire network need not be input again.

However, there are significant differences in how the updating function is handled among the available programs. For example, some programs permit only the input of finish dates of activities; the start dates are ignored. This means that if an activity has started late and is not yet finished, the late condition will not be recognized by the program, and the output will be incorrect to that extent. Also, if only finish dates are accepted, it is not possible to compute and record the actual duration of each activity for comparison with the estimated duration.

A similar but more serious error is to assume that, on a path having had some progress dates reported, the next unreported event date on the path will occur at its earliest time (*ES* or *EF*), regardless of the effective date of the input. To understand the effect of this assumption, review the market survey network and the sample progress report in Figures 11-6 and 11-8. Activity 2-3 is pertinent, for it has not started as of the reporting date, day 12. The correct handling of this condition is to compare the *ES* for the activity with the reporting date, and take the larger of the two as the basis for beginning the forward pass. If the report date is larger, one day may be added. Thus, since 2-3 had not started by the end of day 12, the earliest that it could now start is day 13. Therefore, the correct *ES* for 2-3 is 13, the *EF* becomes 18, and the project completion date is extended.

By the incorrect procedure, the *ES* for 2-3 would be assumed to be 2 (which is impossible since day 2 is now ten days ago), the *EF* would become 7, and this path of the project would apparently be a day ahead of schedule. Incorrect results of this type have caused misunderstanding and confusion for many users. Programs using this incorrect procedure should be avoided.

12. *Network Condensation.* Routines have been developed which in effect condense large networks into smaller ones. One such procedure involves three phases: (a) the condensation of large, detailed networks into smaller ones, (b) the integration of two or more condensed networks, and (c) the expansion of the condensed and integrated networks back into large, separate, detailed networks. The condensation phase involves the determination of the longest path between each sequential pair of preselected key events in the detailed network.

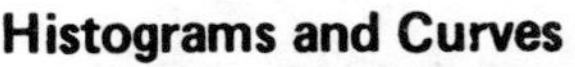

## Histograms and Curves

PROJECT BASIC1 TOTAL CONSUMPTION IS 36095. DOLLARS

This histogram shows the weekly cashflow.

DOLLARS 1750. 1500. 1250. 1000. 750. 500. 250. 0.0

5 52 102 151 201 250 297
N DEC JAN FEB MAR APR MAY JUN J AUG SEP OCT NOV DEC JAN FEB MAR
INTERVAL IS 1 WEEK(S)

Early (E) and late (L) schedule cumulative cashflow curves, indicating the total float in the network.

DOLLARS 35000. 30000. 25000. 20000. 15000. 10000. 5000. 0.0

5 52 102 151 201 250 297
N DEC JAN FEB MAR APR MAY JUN J AUG SEP OCT NOV DEC JAN FEB MAR

## Plotted Bar Charts

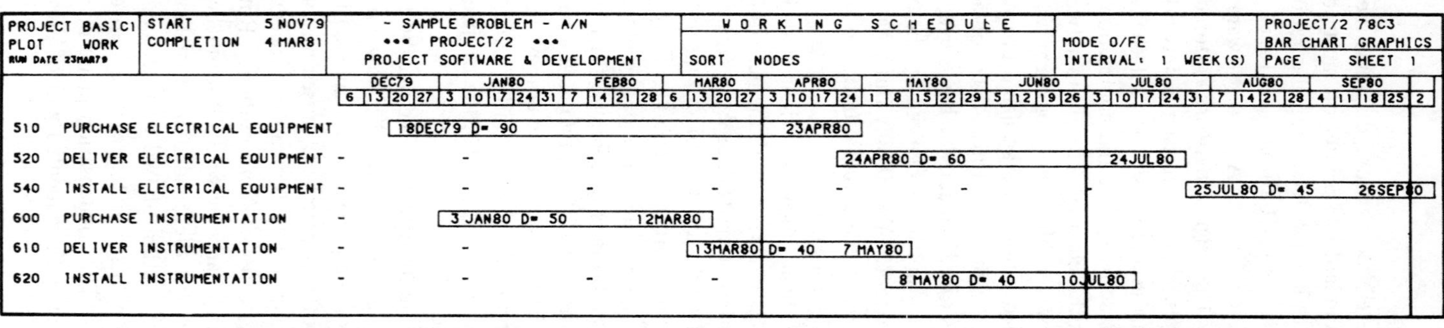


**Figure 11-11** Illustration of histograms, cumulative curves, and bar charts produced by the PROJECT/2 computer system. Courtesy Project Software & Development, Inc., Cambridge, Massachusetts.

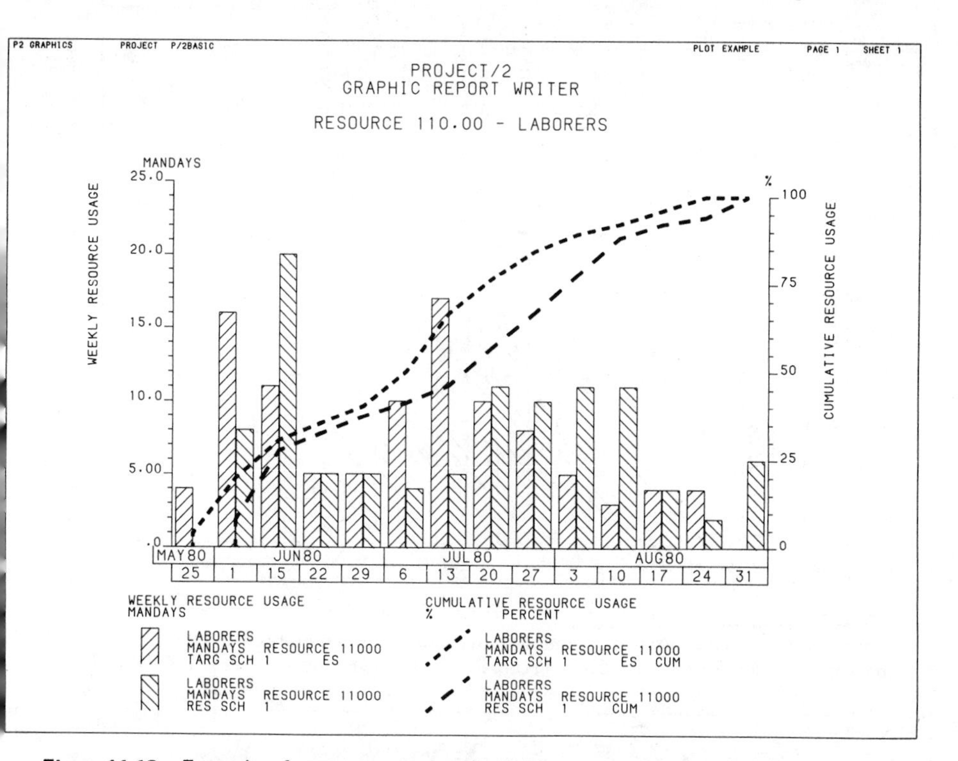


**Figure 11-12** Example of computer-produced graphic report of resource utilization, this one showing both weekly and cumulative figures for the resource under two different project schedules. Courtesy Project Software & Development, Inc., Cambridge, Massachusetts.

The result of this phase is a network consisting of the key events only, which are connected by single activities instead of groups of activities.

In the integration phase the computer processes two or more condensed networks as though they were a single network, each condensed network having one or more common events (interfaces) with another network. The results of this process are new earliest and latest times for each key event. The expansion phase, then, utilizes the new restraints on the interface events to compute new schedules for all activities in the detailed networks. The output of the expansion phase is in the same format as the output for the original detailed network, but the earliest and latest times and slack figures reflect the new restraints imposed on the interfaces.[1] Hence, this network condensation-integration-expansion procedure provides extraordinary capacity for processing the largest networks and groups of networks. It also aids in the preparation of summarized reports for management.

13. *Statistical Analysis.* Originally the controversy of the three-time estimate probabilistic approach versus the single-estimate deterministic approach to critical path computations provided a clear distinction between PERT and CPM.

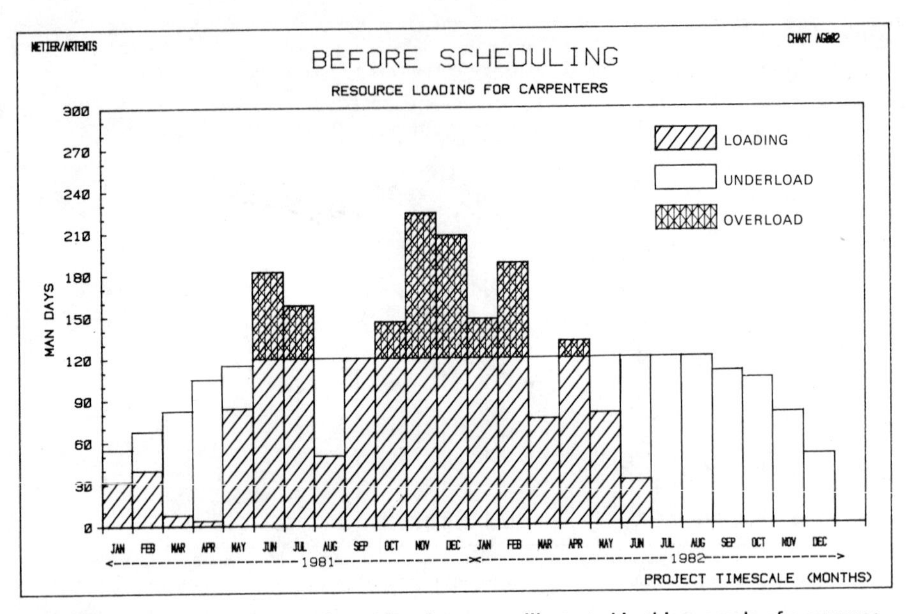


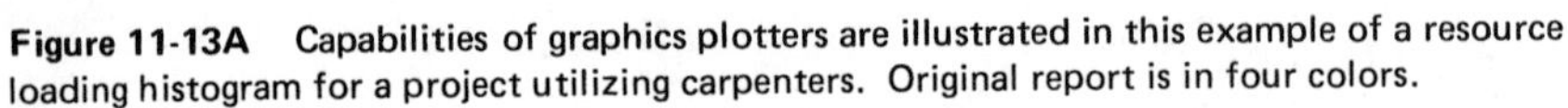

**Figure 11-13A** Capabilities of graphics plotters are illustrated in this example of a resource loading histogram for a project utilizing carpenters. Original report is in four colors.

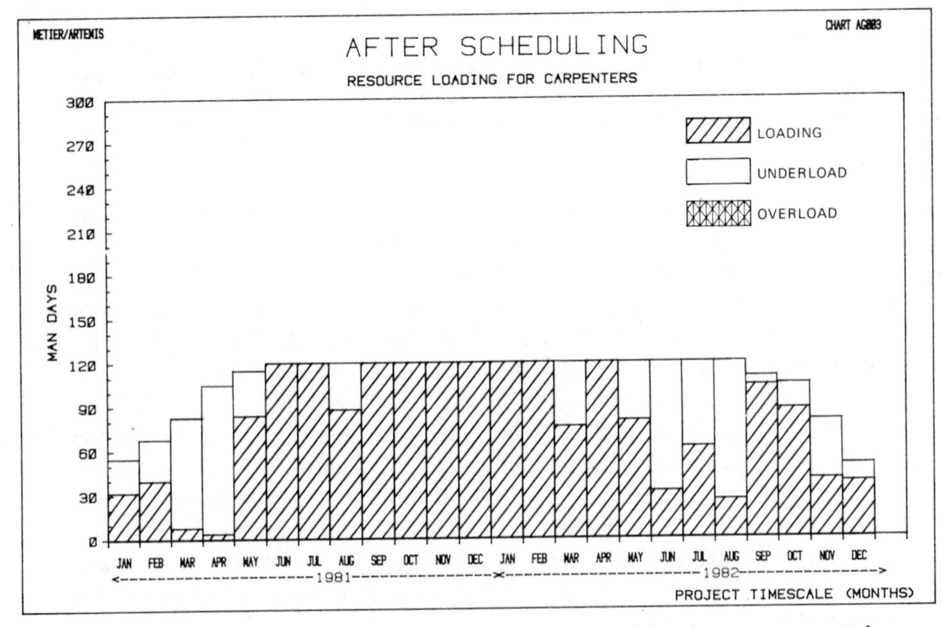


**Figure 11-13B** After the software has scheduled the project within the resource constraint, the resulting histogram displays the extended project duration and the new monthly requirements for carpenters. Illustrations of ARTEMIS system courtesy of Metier Management Systems, Inc., Houston, Texas.

While the distinction still exists among many critical path specialists and is maintained in this text, the two terms have been used loosely in practice and are no longer mutually exclusive. For example, some programs permit three time estimates in the input and obtain the usual weighted average of these figures, but do not make any summations of variance or computations of probability. Another system used by a government agency is labeled "PERT" but makes no use of three time estimates or statistics. Consequently, if one desires the probabilistic approach he must be careful not to rely on the title "PERT" or the three-estimate input, but must determine exactly what probabilities are computed by the programs in question. The most common probability computation is an approximation to the probability of meeting a scheduled date, usually the scheduled date for the completion of the project.

Those who are interested in the CPM (deterministic) approach but who find only PERT (probabilistic) programs available to them do not have a serious obstacle. Almost any probabilistic program may be used as a deterministic one by simply entering the single (expected) time estimate as each of the three estimates.

14. *Interactive Processing.* A trend in software is the facility for the user to interact with the program while processing is underway. Using keyboard data entry, for example, the program may make error checks on each set of activity data, and signal the user immediately when errors are found. After all the data are entered, initial computations may be performed and displayed so that the user can change the network in order to obtain a more acceptable schedule before printing out the reports. Frequent users of network methods find interactive processing advantageous.

## TRENDS IN SOFTWARE

In the first twenty years of critical path network technology the computer programs for network processing have undergone evolutionary change as well as proliferation. The growth in the market for commercial software and the features that have proven popular and unpopular provide a measure of the practical applications of network methods over the two decades. Trends for the future can also be discerned. (Material for this section is based on experience of the authors with several commercial programs, contacts with other users and software vendors, and several excellent surveys of software published between 1964 and 1980. (See References 2-6.)

The earliest software often had several inconvenient limitations. The programs were almost all designed solely for the arrow networking scheme. Many required $I < J$ node numbers for each activity, provided only one output tabulation, did not provide calendar dates, and were restricted to a single network with only one starting and one ending node. In the early sixties resource allocation algorithms were just being considered by graduate students and others and were not yet available in commercial programs.

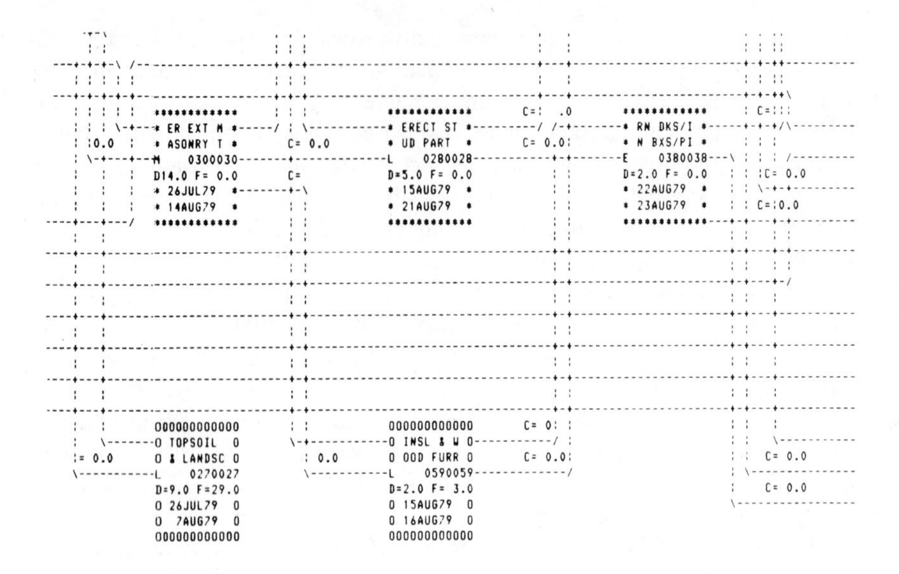


```
NETWORK FOR:  TWO STORY STEEL/CONCRETE STRUCTURE

SAMPLE ACTIVITY:   000000000000     S=START TO START LAG   (TIME IN   DAYS  )
                   0   NAME   0     C=FINISH TO START LAG
                   0   NAME   0     F=FINISH TO FINISH LAG
                   RES  ACT TAG
                   D=DUR F=TFLT
                   EARLY  START
                   EARLY FINISH
                   000000000000     *=CRITICAL ACTIVITY     X=COMPLETED ACTIVITY

                   C A L E N D A R   S P E C I F I C A T I O N S
                   ---------------------------------------------

START DATE IS  2APR79     WORKING   8. HRS/DAY ON MON  TUE  WED  THU  FRI
```

**Figure 11-14** The ability of computer systems to "draw" networks is displayed in this portion of a precedence network plotted on paper by a PAC graphics printer. Courtesy International Systems, Inc., King of Prussia, Pennsylvania.

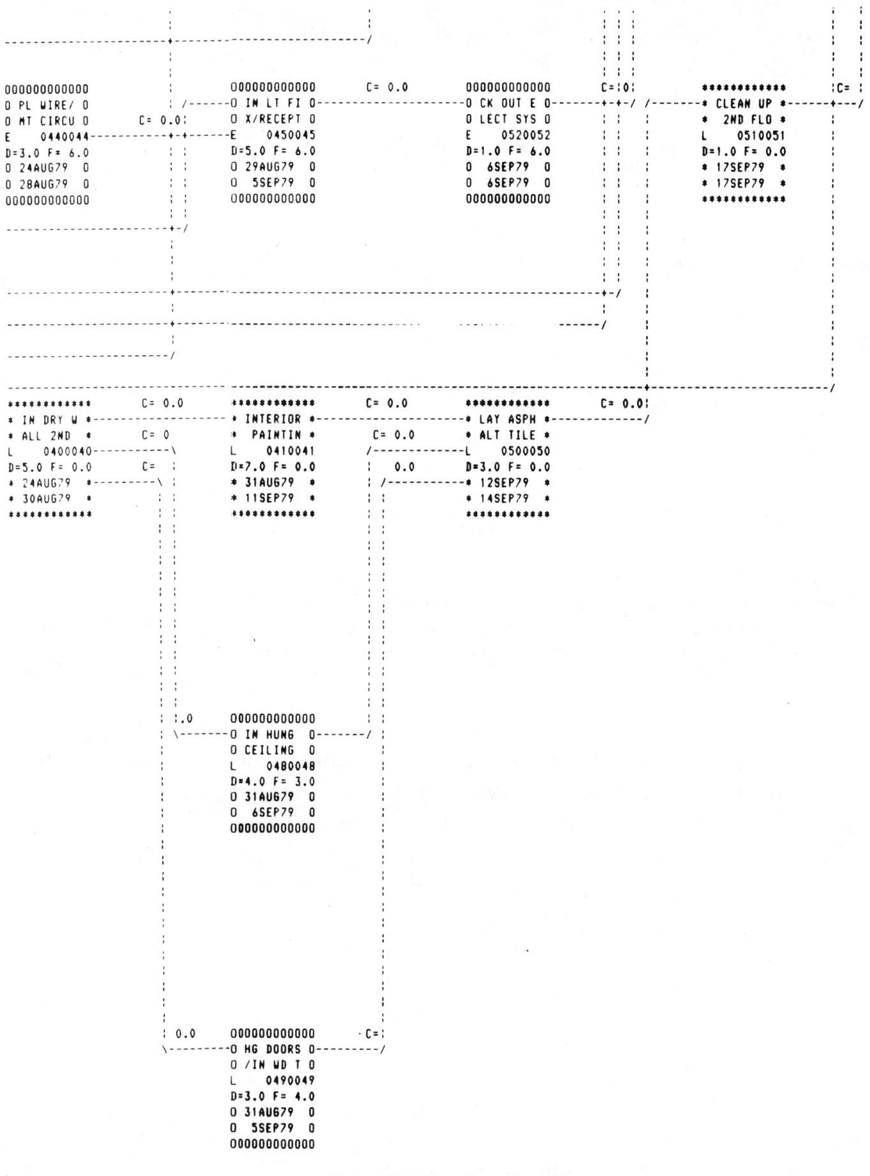


**Figure 11-14** Continued.

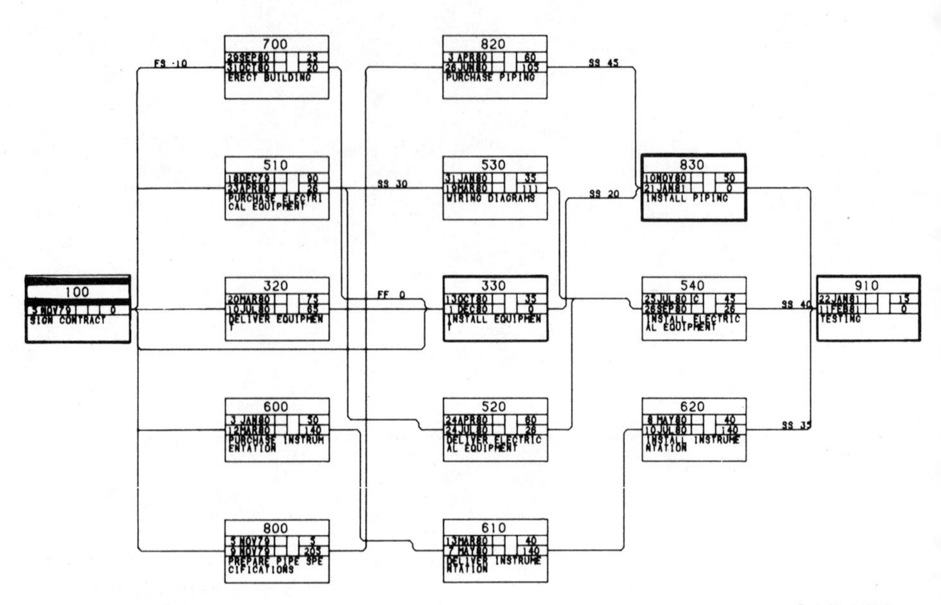


**Figure 11-15** Example of precedence network displayed on video terminal by QWIKNET software, part of PROJECT/2 system. With hand-held cursor the user may point to activities and indicate paths between them; the system then draws the precedence lines and prompts the user for lag relationships. Courtesy Project Software & Development, Inc., Cambridge, Massachusetts.

On the other hand, some of the earliest programs contained rather sophisticated mathematical features that have since almost disappeared from commercial software. The statistical and probability features originally associated with PERT have been the most notable casualties, due to the lack of user interest. Most programs now operate with a single time estimate per activity.

Similarly, users have not maintained interest in the powerful time-cost tradeoff procedure originated by Kelley and Walker in the first *Critical Path Method* computer program. While managers still acknowledge the validity and importance of the time/cost optimization concept (particularly with the dramatic inflation in construction and other costs in the late seventies), they tend to deal with the problem in other ways that do not require detailed time-cost analysis of each activity in a network. Another reason may be that resource allocation problems may be more constraining. Why expend a great effort to find an optimal time-cost trade-off if the resultant schedule is infeasible due to resource limitations? As yet, no software has been able to optimize resources as well as time-cost trade-offs.

Other interesting trends in the programs, indicating user needs and applications as well as advances in computer hardware and software, are categorized and commented upon below.

1. *Network Scheme.* The shift toward node or precedence diagramming has continued, judging from the software available.

2. *Graphic Reports.* Report formats have tended strongly toward such graphic displays as bar charts, resource histograms, and even network plots. These forms of output indicate emphasis upon extensive analysis of resource allocation problems, more use of network methods throughout the course of projects (using updating and progress reporting features such as revised network plots), and displays for clients or upper levels of management (who are informed and impressed more with graphs than with tables of data).

3. *Resource Allocation.* Although absent from the earliest programs, various kinds of resource allocation algorithms have now become common in the commercially available packages. These include time-constrained and resource-constrained procedures as well as a few optimizing routines. Popularity of these procedures indicates the value of network methods in analyzing an important practical problem in a wide range of project types–the utilization of limited resources.

4. *Hardware/Software Advances.* While the early programs required medium- to large-scale computers, the power of those machines is now available in mini- and microcomputers. Network software is now available for these very small configurations. Some programs also operate on an interactive basis with the user, through a keyboard or CRT terminal. Some programs are available through computer time-sharing services.

5. *Vendors.* Through the sixties most of the network software packages were available through the major computer manufacturers. The predominant trend in the seventies was toward the development of improved packages for sale or lease by software firms, some of which specialized in network services and software. Time-sharing services have also entered the business of critical path processing, which appears to have developed into a significant segment of the burgeoning computer software market in the U.S.

## REPRESENTATIVE SOFTWARE

To illustrate the variety and sources of software available in 1981, a small but representative list of programs is provided in Table 11-2. The list was compiled primarily from a comprehensive survey published by the Project Management Institute, plus other sources. The listing for each program was verified by the vendor. The vendors are listed at the end of this chapter.

All of the programs listed are available to the public in some manner through purchase, lease, or user charges. Not listed are any of the scores (or perhaps hundreds) of proprietary programs used internally by corporations, consultants, or universities.

The table does not attempt to list all of the special features offered, but is limited to a few of the more popular and significant features. All of the programs are presumed to incorporate the basic network scheduling calculations described in Chapter 4 and earlier in this chapter, including earliest and latest start and finish times for each activity, and total float. For more details on each program, including price information, contact the vendor.

The reader is cautioned not to infer any evaluations of *quality* or *cost-effectiveness* among the software packages listed in Table 11-2. It is possible that a program with comparatively few features may be among the most efficient, reliable, and cost-effective in certain applications. The authors disclaim any comparative evaluations of the programs listed, nor do we suggest that these are the best available at the time of this publication. For more up-to-date and qualitative information of this kind, review current literature and contact the Project Management Institute, P.O. Box 43, Drexel Hill, Pennsylvania 19026.

## SUMMARY

The question of when and how to use computer programs for network computation has been covered in this chapter, along with examples of input procedures and output reports. It is stressed that the software packages are readily available to the public and that they are not difficult to learn to use, even for persons totally unfamiliar with data processing. There are some significant differences in the features available in the programs, and the user should study these carefully before making a choice.

There is also a wide variety of software packages available for purchase or lease. The trends in the development of this software reveal the popularity of network methods in practical applications. A representative list of software commercially available is provided.

## SOFTWARE VENDOR ADDRESSES

Further details about features, prices, installation services and other information about the software packages in Table 11-2 should be addressed directly to the appropriate vendor as listed below. The list was prepared and verified in January, 1981. Addresses and contact names, of course, are subject to change after that date. The authors do not assume responsibility for further information on the packages or on vendor addresses.

| *Program Name* | *Vendor* | *Contact Person* |
|---|---|---|
| APECS | ADP Network Services, Inc.<br>180 Jackson Plaza<br>Ann Arbor, MI 48106 | Mr. Richard W. Rogers,<br>Manager, Project<br>Management Applications |

## Table 11-2. Representative Software

| PROGRAM NAME | SOURCE LANGUAGE AND HARDWARE REQUIRED | NETWORK SCHEME AND SIZE | Updating | Calendar Dates | Scheduled Dates | Bar Charts | Multi-networks | Cost Reporting | Resource Sched. | Time-Cost T/O | PERT Statistics | Network Plot | COMMENTS |
|---|---|---|---|---|---|---|---|---|---|---|---|---|---|
| APECS | MACRO 10, FORTRAN Available only through ADP Network Services | Arrow or Precedence 64,000 activities | X | X | X | X | X | X | X |  | X | X | Fully interactive. Integrated cost/schedule analysis. Many graphic output options. |
| ARTEMIS | Hewlett Packard 1000 mini with 20 Mbyte disc drive | Arrow or Precedence 32,000 activities | X | X | X | X | X | X | X |  | X | X | "User-friendly" hardware/software system with flexible printer or graphic reporting. |
| CPM | FORTRAN 50K, disk | Arrow 1,000 activities | X | X |  |  | X | X |  |  |  |  | Available on several timeshare services. |
| CPMIS | COBOL IBM 360/30 with 250K core, 4 tapes, 22314's | Arrow 30,000 activities | X | X | X | X | X | X | X |  |  |  | Network size a function of computer memory. Unlimited number of projects in a dataset. System is modifyable to user needs. |
| KRONOS | ANS COBOL 65K | Node/Precedence 1,000 activities | X | X | X | X |  | X |  |  |  |  | Capacity is expandable. Summary bar charting, other reporting features. |
| MicroPERT™ 2 | BASIC Tektronix 4051, 2 or 4, 32K memory and aux. disk or tape drive. | Arrow 440 activities | X | X |  |  |  | X |  |  | X | X | Interactive, menu-driven, many graphic and report output options. Resource and cost plots. |

**Table 11-2. Continued**

| PROGRAM NAME | SOURCE LANGUAGE AND HARDWARE REQUIRED | NETWORK SCHEME AND SIZE | Updating | Calendar Dates | Scheduled Dates | Bar Charts | Multi-networks | Cost Reporting | Resource Sched. | Time-Cost T/O | PERT Statistics | Network Plot | COMMENTS |
|---|---|---|---|---|---|---|---|---|---|---|---|---|---|
| MISTER | FORTRAN<br>From mini's to large scale computers | Arrow<br>200-10,000 activities | X | X | X | X | X | X | X | | | X | Available in 5 model sizes for small to large computers and work programs. |
| MPM (Multi-Project Management) | COBOL, FORTRAN<br>Available through General Electric Mark III System | Arrow<br>Unlimited number of 8,000 activity networks | X | X | X | X | X | X | X | | | | Cost curve reports. Resource forecasting. Unlimited number of resources per activity. |
| MSCS (Management Scheduling and Control System) | COBOL, FORTRAN, BAL<br>IBM 360/370<br>OS, MVT | Arrow or Precedence<br>42,600 activities | X | X | X | X | X | X | X | | | X | Resource constrained scheduling. Target reporting, selection criteria. Graphics interface. |
| OPTIMA 1100 | FORTRAN, COBOL, Assembler<br>1100 CPU | Arrow or Precedence<br>11,000 activities for time processing x 4095 networks | X | X | X | X | X | X | X | | | X | Data base available for user-defined reports. |
| OSCAR | FORTRAN and Assembler<br>Medium to large scale | Arrow or Precedence<br>Unlimited | X | X | X | X | X | X | X | | | X | An integrated database management system for schedule and cost control. |
| PAC II | ANS COBOL<br>Any computer with COBOL compiler and minimum core, tape or disk | Node or Arrow<br><br>Basically unlimited activities | X | X | X | X | X | X | X | X | | X | Skills scheduling by resource. Effectiveness scheduling.<br>In multi-project mode - develops critical path for each project including total float for each project. interactive facilities, and Report Writer. |
| PCM (Project Cost Model) | COBOL or FORTRAN<br>Most small or large machines | Arrow or Precedence<br>10,000 activities | X | X | X | X | | X | X | X | | | Resource optimization. Full cost and cash flow analysis and forecasting. |

| PROGRAM NAME | SOURCE LANGUAGE AND HARDWARE REQUIRED | NETWORK SCHEME AND SIZE | Updating | Calendar Dates | Scheduled Dates | Bar Charts | Multi-networks | Cost Reporting | Resource Sched. | Time-Cost T/O | PERT Statistics | Network Plot | COMMENTS |
|---|---|---|---|---|---|---|---|---|---|---|---|---|---|
| PMCS (Project Management and Control System) | FORTRAN<br>Large Scale | Arrow or Precedence<br>Unlimited | X | X | X | X | X | X | X | | | | Resource Leveling<br>Interactive Input and Reporting Capabilities. |
| PMS IV | Assembler<br>Any IBM 360/370, 303X, 43XX with supporting OS type operating system | Arrow or Precedence<br>8 million activities | X | X | X | X | X | X | X | | | X | Extensive cost-performance analysis. |
| PREMIS | Assembler<br>IBM 360 OS 256K | Arrow or Precedence<br>64,000 activities | X | X | X | X | X | X | X | | | X | Resource plots; report command language. |
| PROJACS | PL/1 and Assembler<br>IBM 360/370, 303X, 43XX | Arrow or Precedence<br>32,000 activities | X | X | X | X | X | X | X | | | X | Standard module (fragnet) library capability. |
| PROJECT CONTROL/70 | ANS COBOL<br>250K disk, 150K core | Precedence<br>Unlimited | X | X | X | X | X | X | X | | | | Report generator. |
| PROJECT/2 | ICETRAN<br>Any IBM OS/VS, any UNIVAC 1100 under EXEC 8; UNIVAC Series 90 under VMOS; any Digital Equipment VAX under VMS | Arrow or Precedence<br>32,000 activities | X | X | X | X | X | X | X | | | X | English language commands, report writer, resource plots, cost curves, bar chart plots, create networks on graphic terminal - QWIKNET. |
| PROMINI (Project Control on Mini Computers) | FORTRAN<br>64K words | Arrow or Precedence<br>32,000 activities | X | X | X | X | X | X | X | | | | Interactive input, report writer, resource plots. |

**Table 11-2. Continued**

| PROGRAM NAME | SOURCE LANGUAGE AND HARDWARE REQUIRED | NETWORK SCHEME AND SIZE | Updating | Calendar Dates | Scheduled Dates | Bar Charts | Multi-networks | Cost Reporting | Resource Sched. | Time-Cost T/O | PERT Statistics | Network Plot | COMMENTS |
|---|---|---|---|---|---|---|---|---|---|---|---|---|---|
| PROSE | FORTRAN<br>Any IBM OS/VS | Arrow<br>16,000 activities | X | X | X | X | X | X | X | | | X | Comprehensive cost control, resource histograms, cost and resource curves, on-line input and validation, report generator. |
| TOPS/SCHEDULE | FORTRAN IV<br>Any major processor | Arrow or Node<br>20,000 or more activities | X | X | X | X | | | | X | | | Input and output can be custom coordinated with existing data bases (accounting, payroll, sales, etc.) |

| *Program Name* | *Vendor* | *Contact Person* |
|---|---|---|
| ARTEMIS | Metier Management Systems, Inc.<br>10175 Harwin Drive, #100<br>Houston, TX 77036 | Mr. S. Wayne Wyatt |
| CPMIS | Glenn L. White Company<br>10560 Main Street, #406<br>Fairfax, VA 22030 | Mr. Glenn L. White |
| KRONOS | American Software, Inc.<br>343 East Paces Ferry Road<br>Atlanta, GA 30305 | Dr. Thomas L. Newberry |
| MicroPERT™ | Sheppard Software Company<br>1523 Coronach Avenue<br>Sunnyvale, CA 94087 | Mr. Leland C. Sheppard |
| MISTER | Shirley Software Systems<br>1936 Huntington Drive<br>South Pasadena, CA 91030 | Mr. Walter W. Shirley |
| MPM (Multi-Project Management) | Florida Power Corporation<br>P.O. Box 14042<br>St. Petersburg, FL 33733 | E. L. Schons |
| MSCS (Management Scheduling and Control System) | McDonnell Douglas Automation Company<br>P.O. Box 516<br>St. Louis, MO 63166 | Mr. Oscar H. Stepanek |
| OPTIMA 1100 | Sperry Univac<br>P.O. Box 500<br>Blue Bell, PA 19424 | J. D. Heitner |
| OSCAR | On-Line Systems, Inc.<br>115 Evergreen Heights Drive<br>Pittsburgh, PA 15229 | E. R. Artus |
| PAC II | International Systems, Inc.<br>890 Valley Forge Plaza<br>King of Prussia, PA 19406 | Mr. Joseph S. Herbets |
| PCM (Project Cost Model) | Project Software Ltd.<br>23 College Hill<br>London EC4R 2RT, England | Mr. Martin Barnes |
| PMCS (Project Management and Control System) | Honeywell Information Systems, Inc.<br>Large Information Systems Div.<br>P.O. Box 6000<br>Phoenix, AZ 85005 | Mr. G. E. Hanson |

| *Program Name* | *Vendor* | *Contact Person* |
|---|---|---|
| PMS IV | Scientific Marketing<br>IBM Corporation<br>1133 Westchester Avenue<br>White Plains, NY 10604 | Mr. Peter V. Norden |
| PROJACS | Scientific Marketing<br>IBM Corporation<br>1133 Westchester Avenue<br>White Plains, NY 10604 | Mr. Peter V. Norden |
| PROJECT/2 | Project Software and Development, Inc.<br>14 Story Street<br>Cambridge, MA 02138 | Mr. Ray Haarstick, Vice President, Sales |
| PROJECT CONTROL/70 | Atlantic Software, Inc.<br>320 Walnut Street<br>Philadelphia, PA 19106 | Mr. Ted Stein, Manager Network Sales |
| PROMINI (Project Control on Mini-Computers) | K&H Computer Systems, Inc.<br>48 Woodport Road<br>P.O. Box 4<br>Sparta, NJ 07871 | Mr. Sam Phelan |
| PROSE | Construction Industry Computer Consultants Ltd. (CINCOM)<br>620 Dorchester Blvd. West<br>Montreal, Quebec<br>Canada H3B-1N8 | U. Korngold or R. F. Gurr |
| TOPS/SCHEDULE | Hollander Associates<br>P.O. Box 2276<br>Fullerton, CA 92633 | Mr. G. L. Hollander |

## REFERENCES

1. Prostick, J. M., "Network Integration, a Tool for Better Management Planning," Presented at Meeting of Operations Research Society of America, Philadelphia, November 7–9, 1962.
2. Petersen, Perry, "Project Control Systems," *Datamation*, June 1979.
3. Phillips, Cecil R., "Fifteen Key Features of Computer Programs for CPM and PERT," *Journal of Industrial Engineering*, January–February 1964.
4. Smith, Larry A., and Peter Mahler, "Comparing Commercially Available CPM/PERT Computer Programs," *Industrial Engineering*, April 1978.
5. West, O. E. et al, *Survey of CPM Scheduling Software Packages and Related Project Control Programs*, Project Management Institute, Drexel Hill, Pennsylvania, January 1980.

6. Woolpert, Bruce, "Computer Software for Project Planning and Control," unpublished student paper, Stanford University, Graduate School of Business, May 1976.

## EXERCISES

1. Using the network of the survey project in Figure 11-3, update the project schedule with the following progress as of the end of day 8:

| *Activity* | *Actual Start Date* | *Actual Finish Date* | *Change* |
|---|---|---|---|
| 1-2 | 1 | 3 | |
| 2-4 | 3 | | Change time estimate to 8 days. |
| 2-3 | 5 | | No change in time estimate. |
| 4-6 | | | Change time estimate to 8 days. |

Assume that the project's scheduled completion day is day 37. Manually prepare an output report similar to the one in Figure 11-8 and answer the following questions:

a. Is the project on schedule?
b. What is the critical path?
c. List the features of a computer program that would be necessary to make the same updating calculations and produce the same report you just did by hand.

2. Using the network in Figure 11-3, assume that the same questionnaire is to be used in another survey project. The deadline for having the printed questionnaire on hand to conduct the other survey is day 15. Thus, day 15 is a scheduled completion date for activity 4-6. With this added constraint, manually compute the initial schedule for the project, producing a report similar to Figure 11-5. What features of a computer program would be needed to make the same calculations? Are these features available in all CPM software?

3. Redraw the network in Figure 11-3 using the activity-on-node scheme. Compute the initial schedule manually. What differences are there in the results shown in Figure 11-5? If you used a computer program, how would the input format differ from the one shown in Figure 11-4?

4. Repeat Exercise 1 using a CPM software package available to you. Discuss any differences in the output results.

5. Consider a research and development project for a new plastics product involving over 2000 activities, a budget of $1 to $1.3 million, and an estimated time frame of 20–26 months. Management wishes to use network analysis to plan and monitor progress on the project. An updated network schedule is desired the first of every month, including graphic presentations of (a) the

critical path activities and their schedule status, (b) summaries of costs to date compared with budget, and (c) listings of activities under the responsibility of each internal department and each outside vendor and contractor. If you are to prepare the project network and present the desired monthly reports: Would you want to use a computer program or make the analyses manually? If you decide to shop for computer software, what features would you seek? Would any of the programs listed in Table 11-2 not meet your needs?

6. Suppose that a particular software package offers the features of multiple initial and terminal events and the option of assigning different scheduled dates to the initial and terminal events. The software literature does not mention a "multi-network" feature. Could the package be used to process multiple networks that are interrelated? If so, construct an example of three projects that are interrelated but have different scheduled completion dates. Explain how the package could be used to properly compute the schedules of each project and all the activities involved.

7. Consider Case 3 on page 339. Suppose the contractor wanted to use a network for schedule updating and control throughout the project. What would be some of his options for manual or computer processing?

8. Redraw the network in Figure 11-3 using precedence diagram logic including the following special interdependencies:

| *Predecessor Activity* | *Successor Activity* | *Relationship* |
|---|---|---|
| 2-4 | 4-5 | Start-Start (Lag 5)<br>Finish-Finish (Lag 2) |
| 5-7 | 7-8 | Start-Start (Lag 5)<br>Finish-Finish (Lag 3) |

Compute the initial schedule manually. What differences are there in the results shown in Figure 11-5? Comment on the relative advantages of using a computer to process this network compared to problem 3.

# SOLUTIONS TO EXERCISES

## Chapter 2

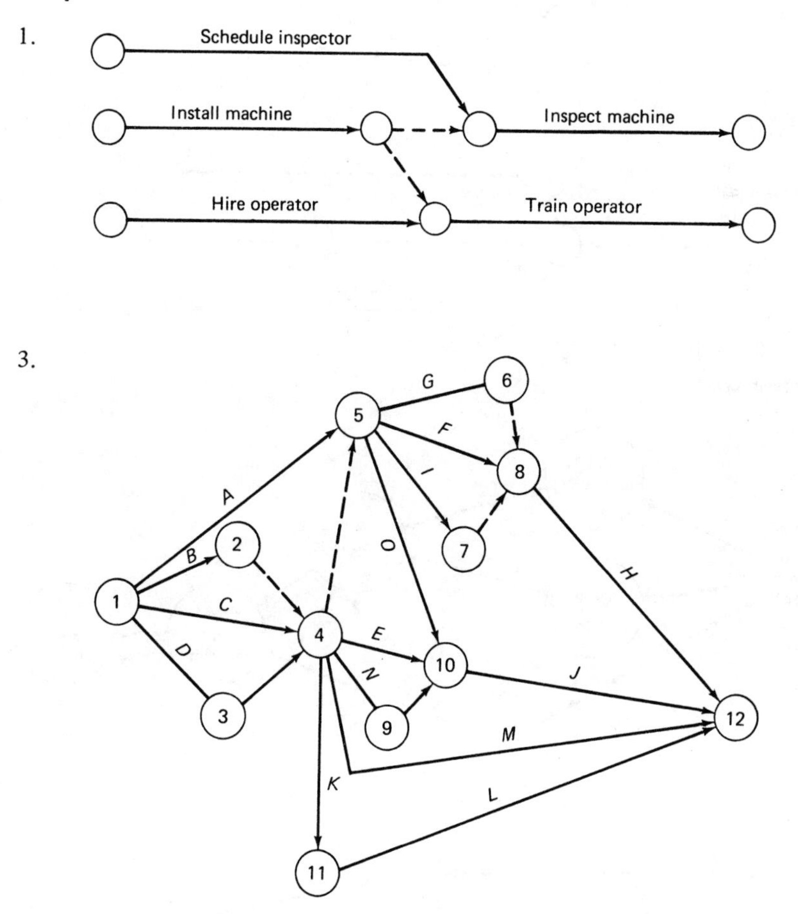


5. These are resource dependencies, indicating that there is only one crew of carpenters for the forming work. The dashed arrows from events 32, 37, and 42 also show that some subsequent activities are dependent on the use of forms that are removed in activities 31–32, 36–37, and 41–42. (The subsequent activities are not shown.) In this case the resource is the reusable forms.

7.

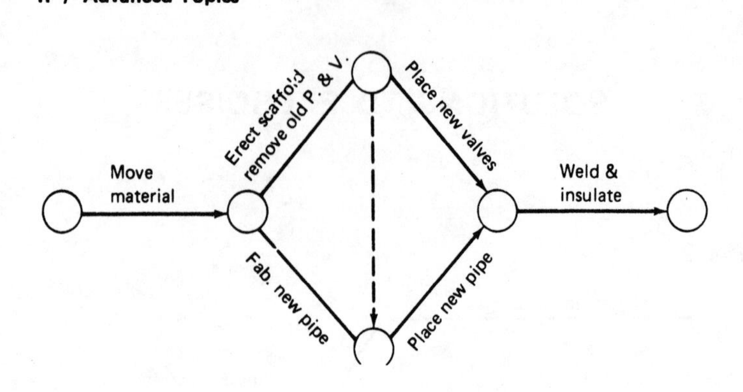


9.

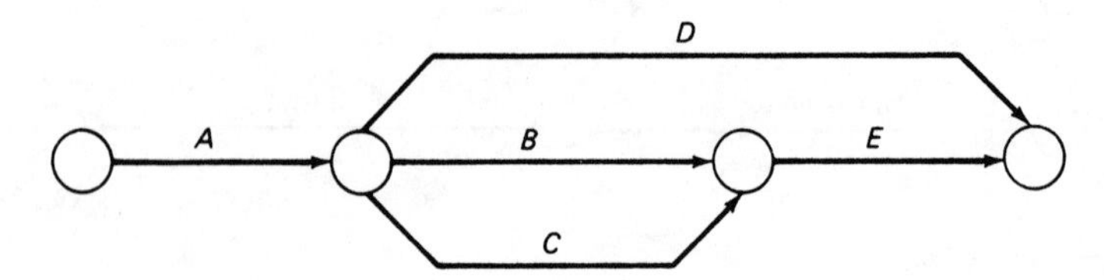


## Chapter 3

1.

3.

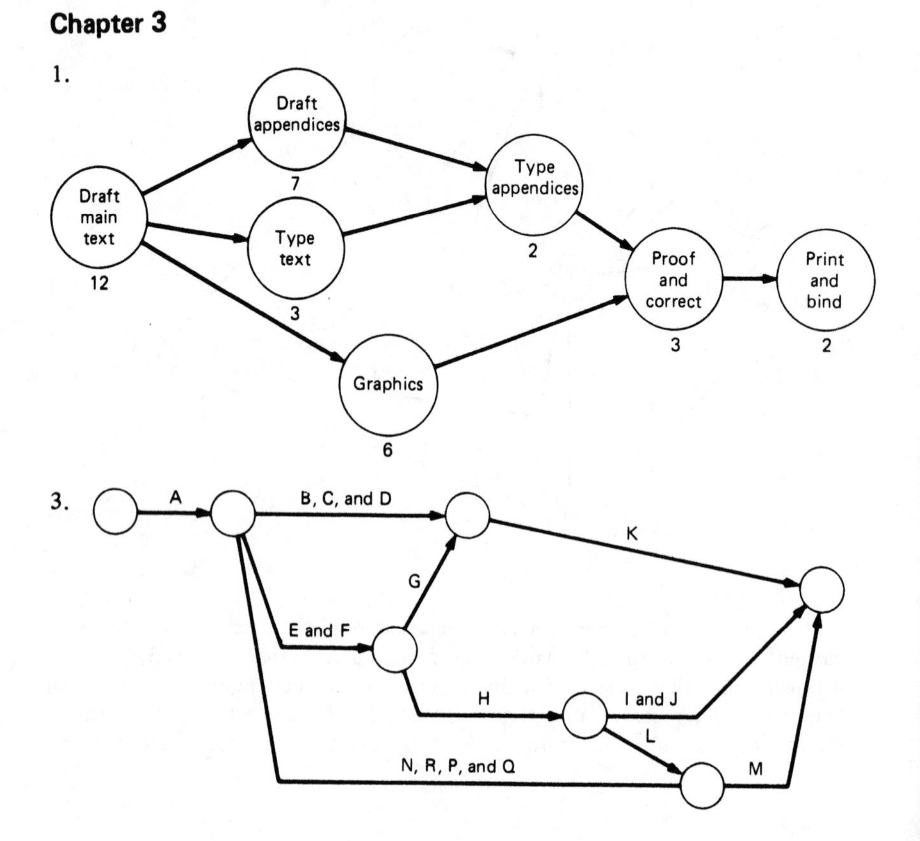

5. To the extent practical, plan activities in parallel rather than in series. (Often this means providing additional resources to avoid resource dependencies.) Anticipate long lead-times, such as fabrication of customized equipment, and plan for them to begin as early in the project as practical.
7. Inflation of time estimates tends to force those activities to become critical, thus increasing management attention and pressure.
9. The superintendent for a high-rise building project may want enough detail to work out the most efficient sequence for a floor, because the time saved would be accumulated over all floors. The president of a large firm is not likely to want much detail, but rather just the schedules for a few key events. A programming subcontractor would be especially interested in the true dependency relationships between programming activities and all other activities in the project.

## Chapter 4

1. a. No.
   b. 2.
   c. As late as possible, i.e., 6-7.
3. a. Critical Path 210-106-109  Slack = −2
   b. No effect for a time of 16; however, a scheduled time of 12 for event 106 would make $L_{106} = 12$, and thus the path 210-106 would have a new slack value of −4.
5. c. Latest allowable start time for the deactivation of the lines is 205 hours.
   d. Activities on the critical path are A-B-F-G-M-D-O.
7. a. Project is one day behind schedule. Path 3-4-5-8 has a slack of −1.
9. a. and b.

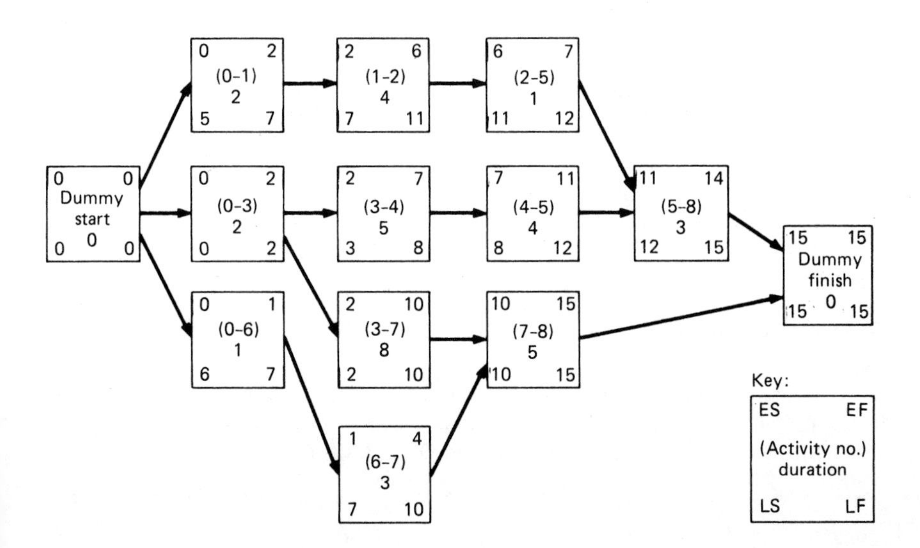

11. a., b.

| Activity | No Splitting Allowed | | | | | Splitting Allowed | | | | | Computation Values of $\alpha$ or $\beta$ |
|---|---|---|---|---|---|---|---|---|---|---|---|
| | ES | EF | LS | LF | F | ES | EF | LS | LF | F | |
| A | 0 | 10 | 0 | 10 | 0 | 0 | 10 | 0 | 10 | 0 | — |
| B | 3 | 11 | 8 | (16) | 5 | 3 | 11 | 8 | (17) | 5 | *$\beta = 8 - 4 = 4$ |
| C | 0 | 20 | 7 | 27 | 7 | 0 | 20 | 7 | 27 | 7 | — |
| D | (2) | 8 | 7 | 13 | 5 | (0) | 8 | 7 | 13 | 5 | $\alpha = 6 - 1 = 5$ |
| E | 10 | 22 | 10 | 22 | 0 | 10 | 22 | 10 | 22 | 0 | — |
| F | (13) | 27 | 13 | 27 | 0 | (12) | 27 | 13 | 27 | 0 | $\alpha = 14 - 5 = 9$ |
| G | (16) | (18) | 25 | 27 | (9) | (15) | (17) | 25 | 27 | (10) | — |

Differences in times noted by circled values.

c. Critical paths are the same in both cases, i.e.

A-10 (NC); FS0; E-12 (NC); FF5; F-14 (FC)

10 + 0 + 12 + 5 + 0 = 27

d. The allowance of splitting is not significant in this example. The critical path is unchanged.

## Chapter 5

1. Cumulative costs per period with the revised activity starts as indicated are as follows:

| Period | Cumulative Cost | Period | Cumulative Cost |
|---|---|---|---|
| 1 | $ 500 | 9 | $19800 |
| 2 | 1000 | 10 | 21000 |
| 3 | 2500 | 11 | 21900 |
| 4 | 4700 | 12 | 24000 |
| 5 | 6900 | 13 | 26100 |
| 6 | 10600 | 14 | 28200 |
| 7 | 13800 | 15 | 28900 |
| 8 | 17000 | | |

3. Overall to date, the results are as follows:

Heater House: Behind Schedule and Over Budgeted Cost
Tanks: Behind Schedule and Over Budgeted Cost
Site Work: Behind Schedule and Over Budgeted Cost

For most recent reporting period there are some bright spots:

| | |
|---|---|
| Heater House: | Behind Schedule and Over Cost |
| Tanks: | Behind Schedule but Under Cost |
| Site Work: | Ahead of Schedule and Under Cost |

Based on total performance to date, the project will experience a cost overrun of $22,700 unless some remedial actions are taken (assuming this is possible).

## Chapter 7

1. c. The internal tangent approach appears appropriate for Resource A because of the shape of the curve (relatively sharp increase in requirements between periods 3 and 13).

$$\text{Lower bound estimate} = \frac{99 - 15}{13 - 3} = 8.4 \text{ units/period}$$

The relatively gradual increase in requirements of Resource B over the entire project duration makes the internal tangent approach inappropriate.

$$\text{Average resource requirement} = \frac{98}{18} = 5.4 \text{ units/period}$$

d. *Case 1*

$$\text{Criticality index, resource A} = \frac{8.4}{8.0} = 1.05$$

$$\text{Criticality index, resource B} = \frac{5.4}{8.0} = 0.68$$

*Case 2*

$$\text{Criticality index, resource A} = \frac{8.4}{7.0} = 1.20$$

$$\text{Criticality index, resource B} = \frac{5.4}{7.0} = 0.77$$

e. Expected minimum project duration: determined by Resource A, i.e.,

$$\text{Resource A: } \frac{142}{8} = 17.75 \cong 18 \text{ days}$$

3. The final schedule obtained using the SIO rule to establish the OSS is shown below.

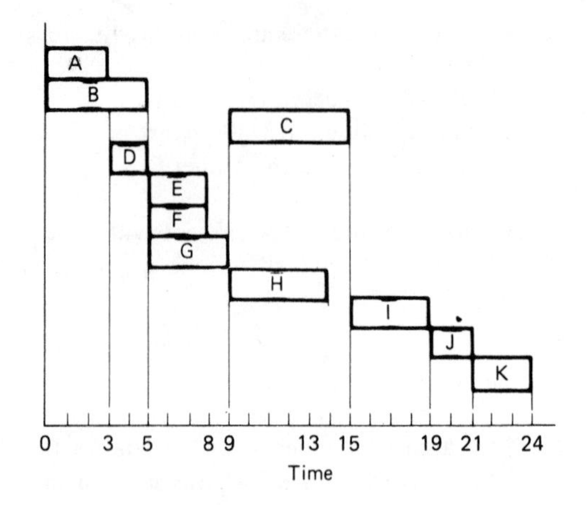


5. The final schedule is shown below.

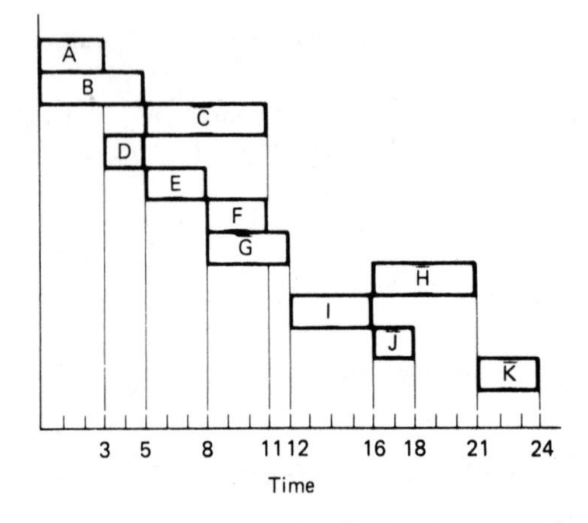


This sequence was obtained using the GRD rule to set the OSS and adding both resources together. If only resource A is used (the critical resource), the same solution is obtained.

## Chapter 8

1.

| *Project Duration* | *Direct Costs* | *Indirect Costs* | *Total Costs* | *Network Changes* |
|---|---|---|---|---|
| 12 | 610 | 900 | 1510 | – |
| 11 | 610 + 40 = 650 | 820 | 1470 | D ↓ 4 |
| 10 | 650 + 40 = 690 | 740 | 1430 | D ↓ 3 |
| 9 | 690 + 80 = 770 | 700 | 1470 | A ↓ 2, D ↑ 4, G ↓ 3 |
| 8 | 770 + 100 = 870 | 660 | 1530 | F ↓ 6, G ↓ 2 |
| 7 | 870 + 130 = 1000 | 620 | 1620 | B ↓ 5, D ↓ 3, F ↓ 5 |

3. Activity $i$–$j$.

Minimize $\quad (C_1 - C_2)\,\delta_1 - s y'_{ij}$

$$\text{Subject to: } d_1\delta_1 + d_2(1-\delta_1) + y'_{ij} + T_i - T_j \leqq 0$$

$$\delta_1 = 0 \text{ or } 1$$

$$0 \leqq y'_{ij} \leqq (d_3 - d_2)$$

$$(d_3 - d_2)(\delta_1 - 1) + y'_{ij} \leqq 0$$

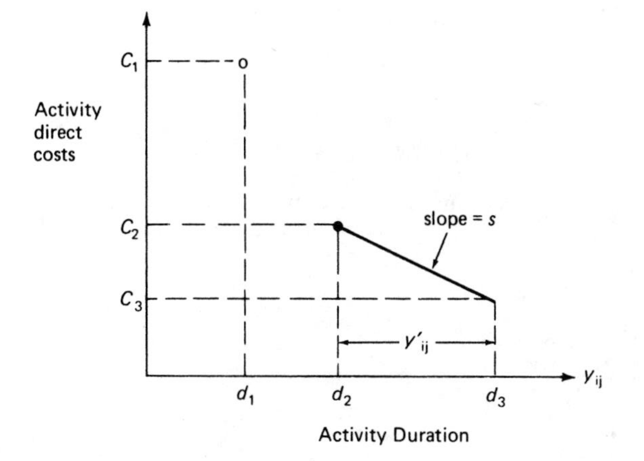


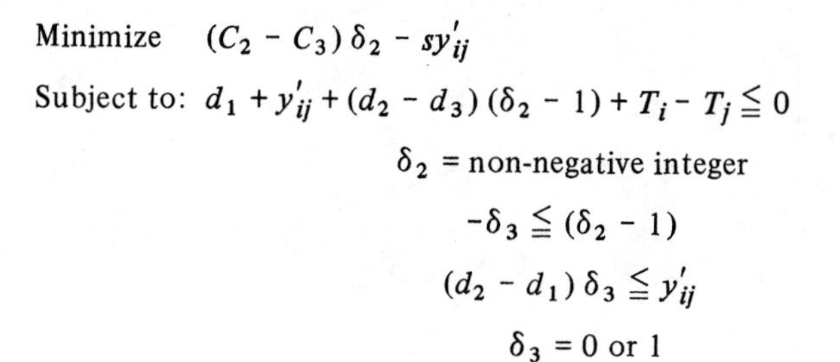

$$\text{Minimize} \quad (C_2 - C_3)\delta_2 - sy'_{ij}$$

$$\text{Subject to: } d_1 + y'_{ij} + (d_2 - d_3)(\delta_2 - 1) + T_i - T_j \leqq 0$$

$$\delta_2 = \text{non-negative integer}$$

$$-\delta_3 \leqq (\delta_2 - 1)$$

$$(d_2 - d_1)\delta_3 \leqq y'_{ij}$$

$$\delta_3 = 0 \text{ or } 1$$

$$0 \leqq y'_{ij} \leqq (d_2 - d_1)$$

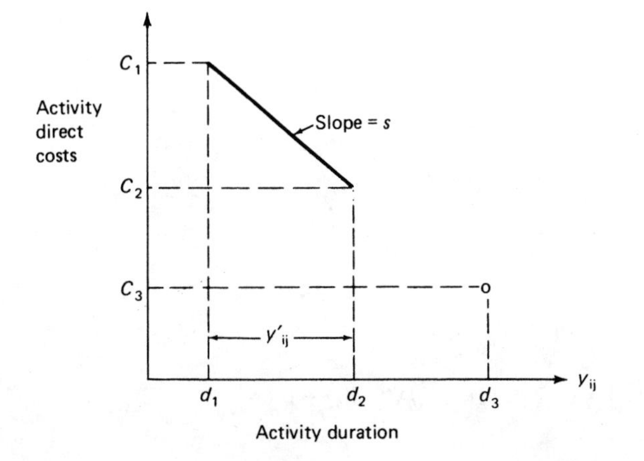

$$\text{Minimize} \quad (C_2 - C_3)\,\delta_2 - s_1 y'_{ij} - s_2 y''_{ij}$$

$$\text{Subject to:} \quad d_1 + y'_{ij} + (d_2 - d_3)(\delta_2 - 1) + y''_{ij} + T_i - T_j \leqq 0$$

$$(d_4 - d_3)(\delta_2 - 1) + y''_{ij} \leqq 0$$

$$-\delta_1 \leqq (\delta_2 - 1)$$

$$(d_2 - d_1)\,\delta_1 \leqq y'_{ij}$$

$$\delta_2 = 0 \text{ or } 1$$

$$0 \leqq y'_{ij} \leqq (d_2 - d_1)$$

$$0 \leqq y''_{ij} \leqq (d_4 - d_3)$$

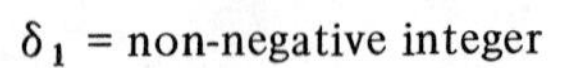

$$\delta_1 = \text{non-negative integer}$$

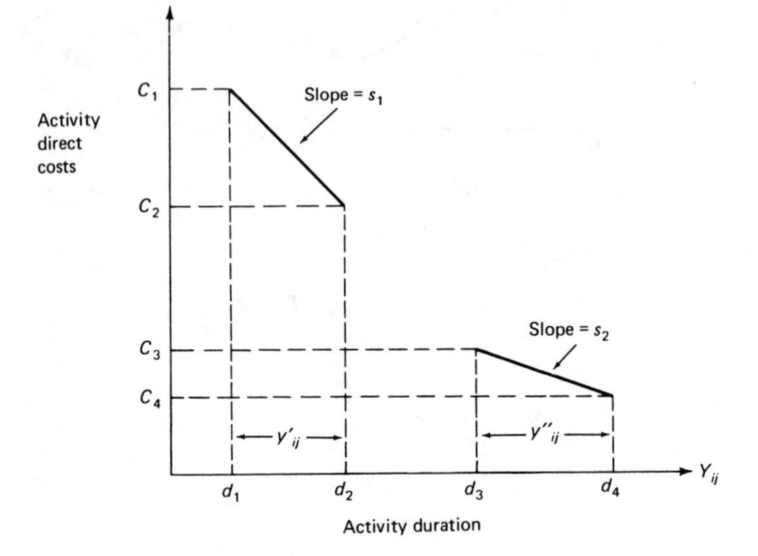


## Chapter 9

3. a.

| Activity | $a$ | $m$ | $b$ | $t_e$ | $V_t$ |
|---|---|---|---|---|---|
| 1-2 | 1 | 2 | 4 | $2\frac{1}{6}$ | 0.8789 |
| 2-3 | 3 | 5 | 7 | 5 | 1.5625 |
| 2-4 | 1 | 2 | 3 | 2 | 0.3906 |
| 4-3 | 1 | 3 | 5 | 3 | 1.5625 |

b. $7\frac{1}{6}$

c. 2.8320

d. $P\{T \leqslant 8\} = 0.69$; $P\{T \leqslant 9\} = 0.86$

e. $9.946 \cong 10$

5. $P\{T \leqslant 16 \mid T_7 = 10\} = P\{Z \leqslant (16\text{-}15)/\sqrt{0.88}\} = 0.86$

9. a. The activities that have two or more predecessor activities are 108, 113, 117, 119, 121, 125, 126, 127, 128, 129, 132, and 133. The predecessor events of these activities are the ones where a merge event bias may occur. Of these 12 cases, only activities 121, 125 and 126 will present merge event bias problems. For the others, the expected means of the merging paths are quite far apart, or else there is a great deal of correlation in the two paths, so that the bias will again be negligible, e.g., activity 132 and 129.

   It should also be noted that the merge event biases will occur on slack paths, and hence will not affect the critical path, and hence the conventional PERT analysis should be a good approximation in this example.

   b. The critical path is made up of activities 102, 105, 108, 112, 115, 118, 120, 123, 124, 128, 131, 132 and 133. The sum of the estimated mean times for this path is 382. The variance of this path can be approximated by considering only activities 105 and 118. The sum of their variances is 1221.3 and the approximate standard deviation is 35.

   c. $P\{T \leqslant 400\} = P\{Z \leqslant (400 - 382)/35\} = 0.70.$

## Chapter 10

1. Some of the alternative networking systems (or alternative computer packages) to consider include: 1. CPM, 2. PERT, 3. Precedence Diagramming, 4. Node Diagramming, 5. GERT, and 6. Bar Charts. Index these alternatives by $(i)$, so $i = 1, 2, \ldots, 6$, in this illustration.

   Some of the criteria to consider in choosing a networking system include: 1. Ease of drawing the initial network. 2. Ease of obtaining the initial data, . . . , 9. Ease of updating effectively. Index these criteria by $j = 1, 2, \ldots, 9$.

   Now assign weights, $w_j$, to each criteria to reflect their relative importance. Finally, for each network system $(i)$, assign a relative score, $s_{ij}$, to reflect how it achieves the $j$th criteria. The final score for each networking system is then computed from

$$\text{Score for System } (i) = S_i = \sum_{j=1}^{9} w_j s_{ij}$$

5.

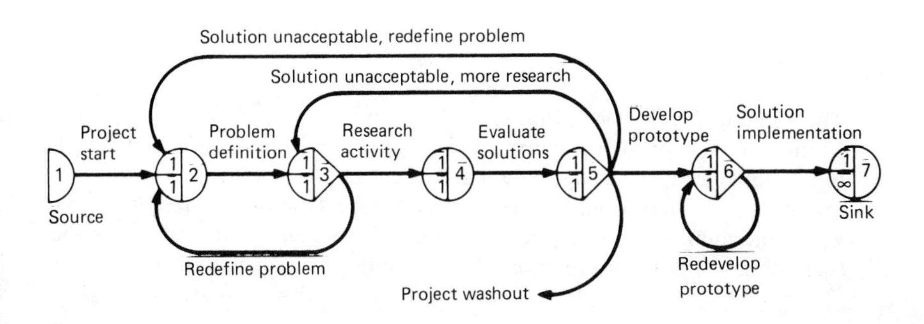

## Chapter 11

1. *Market Survey Project* *Effective Date: 8*

| *I* | *J* | *Est. Dur.* | *Act. Dur.* | *Description* | *ES* | *LS* | *EF* | *LF* | *F* |
|---|---|---|---|---|---|---|---|---|---|
| 1 | 2 | 3 | 2 | Plan | Started | 1 | Finished | 3 | |
| 2 | 4 | 8 | | Design | Started | 3 | 11 | 9 | −2 |
| 2 | 3 | 5 | | Hire | Started | 5 | 10 | 10 | 0 |
| 3 | 5 | 7 | | Train | 11 | 10 | 18 | 17 | −1 |
| 4 | 3 | 0 | | Dummy | 11 | 10 | 11 | 10 | −1 |
| 4 | 5 | 4 | | Select | 11 | 13 | 15 | 17 | 2 |
| 4 | 6 | 8 | | Print | 11 | 9 | 19 | 17 | −2 |
| 6 | 5 | 0 | | Dummy | 19 | 17 | 19 | 17 | −2 |
| 5 | 7 | 15 | | Conduct | 19 | 17 | 34 | 32 | −2 |
| 7 | 8 | 5 | | Analyze | 34 | 32 | 39 | 37 | −2 |

a. No. It is two days behind schedule.
b. 2-4-6-5-7-8
c. 1. Updating feature for actual start and finish dates.
   2. Activity revision feature.
   3. Use of actual start and finish times for schedule computations.

3.

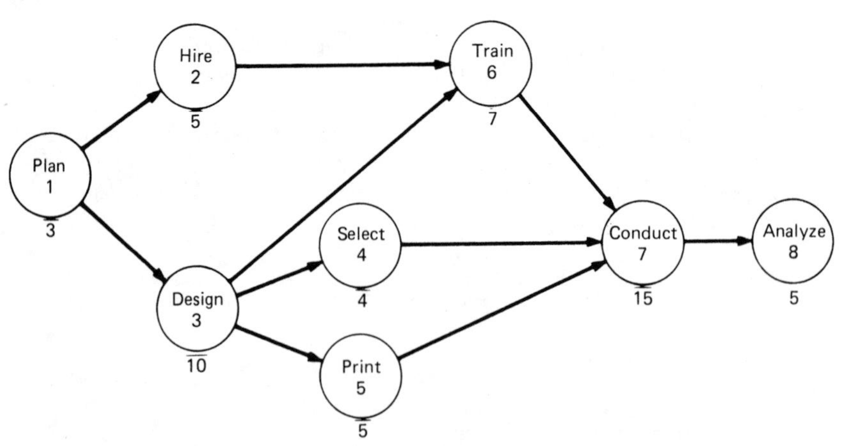


The computed schedule is the same. The input format would have one identifying number for each activity, rather than a pair of numbers (I and J). The input would also list the dependent activities for each activity.

5. Use of a computer would be recommended because of the size of the network, the number of updating calculations projected, and the various output sorts desired. Software features needed would include: updating, cost sum-

marization, graphic output, calendar dating, and sorting by responsibility code. Yes, several of the listed programs would be insufficient.

7. The contractor could use manual processing during the detailed planning phase of the project, in order to develop the most efficient sequence of activities for constructing each floor. He could then resort to a simpler network for monitoring the actual construction progress.

# INDEX

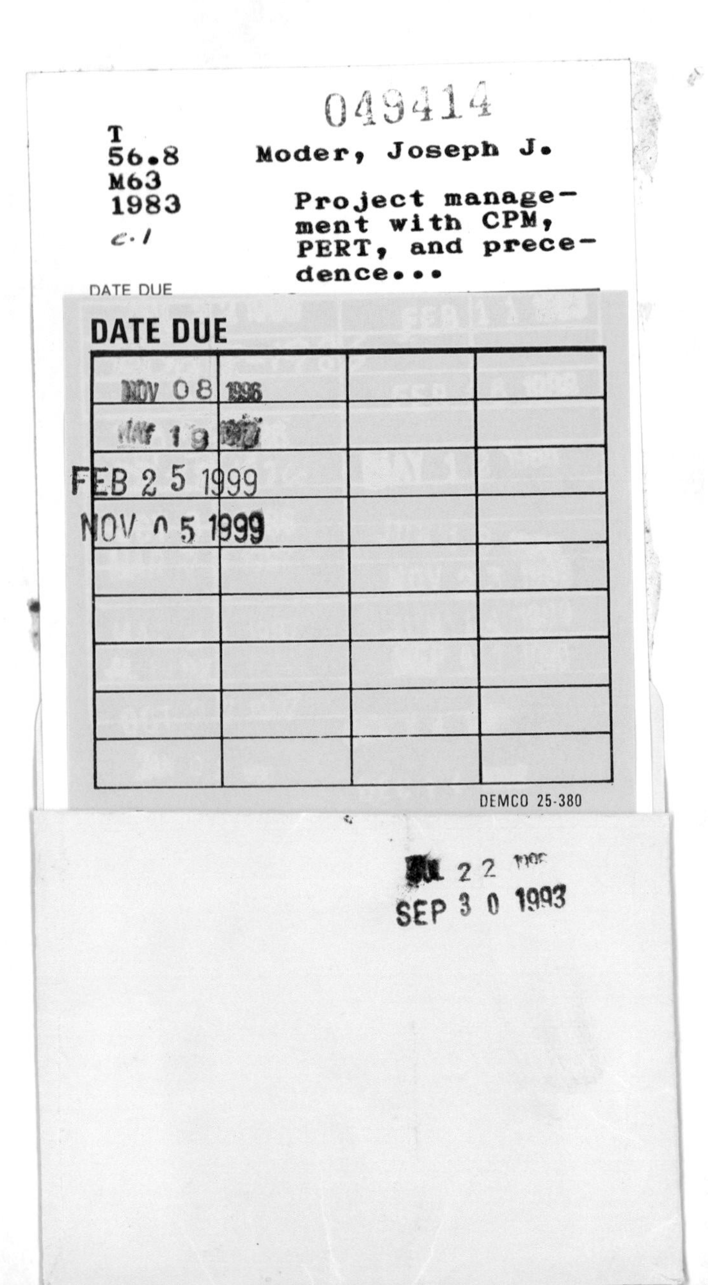